Global Climate Change and Coastal Tourism

Recognizing Problems, Managing Solutions and Future Expectations

Global Climate Change and Coastal Tourism

Recognizing Problems, Managing Solutions and Future Expectations

Edited by

Andrew Jones

The University of Malta, Malta
Small States Institute, The University of Malta, Malta

and

Michael Phillips

University of Wales Trinity Saint David, Wales, UK

CABI

CABI is a trading name of CAB International

CABI
Nosworthy Way
Wallingford
Oxfordshire OX10 8DE
UK

Tel: +44 (0)1491 832111
Fax: +44 (0)1491 833508
E-mail: info@cabi.org
Website: www.cabi.org

CABI
745 Atlantic Avenue
8th Floor
Boston, MA 02111
USA

Tel: +1 (617) 682 9015
E-mail: cabi-nao@cabi.org

A catalogue record for this book is available from the British Library, London, UK.

Library of Congress Cataloging-in-Publication Data

Names: Jones, Andrew L., 1958- editor. | Phillips, Michael R., 1962- editor.
Title: Global climate change and coastal tourism : recognizing problems, managing solutions, future expectations / edited by Andrew Jones, Michael Phillips.
Description: Boston : CAB International, 2017. | Includes bibliographical references and index.
Identifiers: LCCN 2017022422 (print) | LCCN 2017043258 (ebook) | ISBN 9781780648453 (ePDF) | ISBN 9781780648446 (ePub) | ISBN 9781780648439 (hbk : alk. paper)
Subjects: LCSH: Tourism--Environmental aspects--Case studies. | Coastal zone management--Case studies. | Sustainable tourism--Case studies. | Coastal ecology--Case studies. | Global climate change--Case studies.
Classification: LCC G156.5.E58 (ebook) | LCC G156.5.E58 G56 2017 (print) | DDC 338.4/79109146--dc23
LC record available at https://lccn.loc.gov/2017022422

ISBN-13: 9781780648439

Commissioning editor: Claire Parfitt
Editorial assistant: Emma McCann
Production editor: Shankari Wilford

Typeset by SPi, Pondicherry, India
Printed and bound in the UK by Bell & Bain Ltd, Glasgow, G46 7UQ

Contents

PART 2

Contributors

Abdullah Akbas, PhD Candidate, Research Assistant, Department of Geography, Ardahan University, Ardahan, Turkey. E-mail: abdullahakbas@ardahan.edu.tr

Nazan An, PhD Candidate, Researcher, Center for Climate Change and Policy Studies, Boğaziçi University, Istanbul, Turkey. E-mail: nazan.an@boun.edu.tr

Angela Anthonisz, Head of Subject, Events, Tourism and Hospitality, Faculty of Business and Law, The University of Northampton, UK. E-mail: angela.anthonisz@northampton.ac.uk

Salvador Anton Clavé, Professor of Regional Geographical Analysis, Research Group of Territorial Analysis and Tourism Studies (GRATET), Rovira i Virgili University, Spain. E-mail: salvador.anton@urv.cat

Paulo Borges, Department of Geosciences, University of the Azores, Ponta Delgada, Portugal. E-mail: paulo.js.borges@uac.pt

Stephen William Boyd, Professor of Tourism, Ulster University, Northern Ireland. E-mail: sw.boyd@ulster.ac.uk

Huong T. Bui, Associate Professor of Tourism and Hospitality, College of Asia Pacific Studies, Ritsumeikan Asia Pacific University (APU), Beppu, Oita, Japan. E-mail: huongbui@apu.ac.jp

Jeremy Buultjens, Associate Professor, School of Business and Tourism, Southern Cross University, Australia. E-mail: jeremy.buultjens@scu.edu.au

Helena Calado, MARE – Marine and Environmental Sciences Centre, and FCT, University of the Azores, Ponta Delgada, Portugal. E-mail: helena.mg.calado@uac.pt

Rita Cannas, Assistant Professor of Management and Business Communication, Department of Economics & Business, University of Cagliari, Sardinia, Italy. E-mail: rita.cannas@unica.it

James Andrew Graham Cooper, Professor of Coastal Studies, Ulster University, Northern Ireland. E-mail: jag.cooper@ulster.ac.uk

Janice Cumberbatch, Lecturer, Centre For Resource Management and Environmental Studies (CERMES), University of the West Indies (UWI), Cave Hill Campus, Barbados. E-mail: janice.cumberbatch@cavehill.uwi.edu

Osman Cenk Demiroglu, Postdoctoral Researcher, Department of Geography and Economic History, Umeå University, Umeå, Sweden. E-mail: cenk.demiroglu@umu.se

Denyse S. Dookie, School of International and Public Affairs (SIPA), Columbia University, New York. E-mail: dsd2123@columbia.edu

Danielle Edwards, The University of the West Indies Open Campus Dominica, Dominica. E-mail: daniellewards@gmail.com

Kristelle Francis, Centre for Resource Management and Environmental Studies (CERMES), University of the West Indies (UWI), Cave Hill Campus, Barbados. E-mail: kristelle.francis@gmail.com

Charles Galdies, Institute of Earth Systems, University of Malta. E-mail: charles.galdies@um.edu.mt

W.K. Athula Gnanapala, Senior Lecturer Gr. I, Department of Tourism Management, Sabaragamuwa University, Sri Lanka. E-mail: athulatmsusl@gmail.com/gnanapala@mgt.sab.ac.lk

C. Michael Hall, Professor, Department of Management, Marketing and Entrepreneurship, University of Canterbury, Christchurch, New Zealand. E-mail: michael.hall@canterbury.ac.nz

Azizul Hassan, PhD candidate, Cardiff Metropolitan University, Cardiff, UK. E-mail: m.hassan15@outlook.cardiffmet.ac.uk

Tim Heap, Principle Tutor and Research Manager, The University of Derby, Buxton, UK. E-mail: t.j.heap@derby.ac.uk

Julia Jabour, Senior Lecturer, Institute for Marine and Antarctic Studies, University of Tasmania, Private Bag 129, Hobart, 7001, Tasmania, Australia. E-mail: julia.jabour@utas.edu.au

Ian S. Jenkins, Associate Professor, University of Iceland/Háskóli Íslands. E-mail: jenkins@hi.is or dr.isjenkins@gmail.com

Andrew Jones, resident professor and Head of Department, Institute for Tourism Travel and Culture, The University of Malta and Islands & Small States Institute, The University of Malta. E-mail: andrew.jones@um.edu.mt

Mehmet Levent Kurnaz, Director, Center for Climate Change and Policy Studies, Boğaziçi University, Istanbul, Turkey. E-mail: levent.kurnaz@boun.edu.tr

Anh Le, Lecturer, Department of International Economics, National Economics University, Hanoi, Vietnam. E-mail: tuananhle7205@gmail.com

Tuan-Anh Le, Lecturer, School of Business and Tourism, Southern Cross University, Australia. E-mail: andrew.le@scu.edu.au

Hamzah Muzaini, Assistant Professor, Southeast Asian Studies Department, National University of Singapore. E-mail: hamzbm@hotmail.co.uk

Kiat Ng, Climate Change Impacts, Adaptation and Modelling Research Group, Faculty of Sciences, University of Lisbon, 1749-016 Lisboa, Portugal. E-mail: kenng@fc.ul.pt

Keith Nurse, Senior Fellow, Sir Arthur Lewis Institute for Social and Economic Studies and the World Trade Organization Chair, The University of the West Indies, Barbados. E-mail: keithnurse@me.com

Leonard Nurse, Senior Lecturer, Centre For Resource Management and Environmental Studies (CERMES), University of the West Indies (UWI), Cave Hill Campus, Barbados. E-mail: leonard.nurse@gmail.com

Tugba Ozturk, Assistant Professor, Department of Physics, Isik University, Istanbul, Turkey. E-mail: tugba.ozturk@isikun.edu.tr

Carlo Perelli, Associated Researcher, CRENoS, Centre for North South Economic Research, University of Cagliari and University of Sassari, Via S. Giorgio 12 - 09124 Cagliari, Italy. E-mail: carlo.perelli@crenos.unica.it

Michael R. Phillips, Pro Vice-Chancellor (Research, Innovation, Enterprise and Commercialisation), Coastal and Marine Research Group, University of Wales Trinity Saint David, Swansea, Wales, UK. E-mail: mike.phillips@uwtsd.ac.uk

Roya Rahimi, Lecturer in Tourism and Hospitality and Event Management, Department of Marketing, Innovation, Leisure and Enterprise, University of Wolverhampton, Wolverhampton, UK. E-mail: roya.rahimi@wlv.ac.uk

Yael Ram, Ashkelon Academic College, Ben Zvi 12, Ashkelon, 78211, Israel. E-mail: ramy@acad.ash-college.ac.il

Iraj Ratnayake, Senior Lecturer, Sabaragamuwa University, Sri Lanka. E-mail: iraj@mgt.sab.ac.lk

Thani Rashid Said, Cultural Geography Group, Wageningen University, the Netherlands. E-mail: thani.said@wur.nl

Òscar Saladié, Serra-Hunter Associate Professor, Department of Geography; and Research Group of Territorial Analysis and Tourism Studies (GRATET), Rovira i Virgili University, Spain. E-mail: oscar.saladie@urv.cat

Raquel Santos-Lacueva, Group of Territorial Analysis and Tourism Studies (GRATET), Department of Geography, Rovira i Virgili University, Spain. E-mail: raquel.santos@urv.cat

Daniel Scott, Professor and University Research Chair in Geography and Environmental Management and Executive Director of the Interdisciplinary Centre on Climate Change, University of Waterloo, Canada. E-mail: daniel.scott@uwaterloo.ca

V. Dao Truong, Research fellow, Tourism Research in Economic Environs & Society (TREES), North-West University, Potchefstroom, South Africa; Department of Tourism and Hospitality, National Economics University, Hanoi, Vietnam; Social Marketing Initiatives, Hanoi, Vietnam. E-mail: vandao83@yahoo.co.uk

M. Tufan Turp, Researcher, Center for Climate Change and Policy Studies, Boğaziçi University, Istanbul, Turkey. E-mail: tufan.turp@boun.edu.tr

René van der Duim, Personal Professor, Cultural Geography Group, Wageningen University, the Netherlands. E-mail: rene.vanderduim@wur.nl

Marta Vergílio, CIBIO – Research Center in Biodiversity and Genetic Resources/InBIO – Associate Laboratory, University of the Azores, Rua da Mãe de Deus, 13-A, 9501-801 Ponta Delgada, Portugal. E-mail: marta.hs.vergilio@uac.pt

Stephanie Verkoeyen, University of Waterloo, Waterloo, Ontario, Canada. E-mail: s.verkoeyen@gmail.com

About the Authors

Editors

Andrew Jones is currently a resident professor and Head of Department for the Institute for Tourism Travel and Culture at The University of Malta. He has also held positions at the University of Wales, International Faculty Cardiff, Swansea Business School and the University of Brunei. In his early career he qualified in Urban and Regional Planning and Economic Development and initially worked within these fields both in the public and private sectors before taking up a position at the University of Wales where he taught planning, economic regeneration, environmental management and tourism development. He has professional experience in planning, environmental management and tourism management at international, regional and local levels and has been an enthusiastic contributor to the academic tourism community within the UK and internationally for over 30 years. In this respect he has been an active practitioner, researcher and teacher since 1981. In his positions with both the University of Wales and University of Malta he has been involved in the universities' collaborative programmes and partnerships in Greece, China, India, Hong Kong, Singapore, Uzbekistan, Malaysia, Japan, Germany and the UK.

He completed his PhD on research investigating the relationship and tensions between cultural regeneration, the environment and the development of niche markets for tourism. Research associated with his academic experience over the last 20 years and the consequent research outputs have allowed him to develop a research profile based upon strategic/environmental planning, regeneration, economic and regional development, sustainable planning policy and the relationship between these dynamic processes and tourism development and in turn the development of sustainable/cultural (tourism) niche markets.

As such he has carried out numerous consultancy projects relating to economic regeneration, tourism planning, cultural regeneration and sustainability and has published a number of industry and academic articles, reports and conference papers. To this end he has travelled widely and has contributed to international conferences in Europe, North America, Asia and Australasia and spent 2 years as a visiting research fellow at the University of Brunei. His current research involves working jointly with the Islands and Small States Institute at the University of Malta in establishing a research platform for Island Tourism.

Michael Phillips has a BSc in Civil Engineering, an MSc in Environmental Management and a PhD in Coastal Processes and Geomorphology, which he has used in an interdisciplinary way to assess current challenges of living and working on the coast. He is Pro Vice-Chancellor (Research, Innovation, Enterprise and Commercialisation) at the University of Wales Trinity Saint David and also leads their Coastal and Marine Research Group. Professor Phillips' research expertise includes coastal processes, morphological change and adaptation to climate change and sea level rise, and this has informed his engagement in the policy arena. He has given many keynote speeches, presented at many major international conferences and evaluated various international and national coastal research projects. Consultancy contracts include beach monitoring for the development of the Tidal Lagoon Swansea Bay, assessing beach processes and evolution at Fairbourne (one of the case studies in this book), beach replenishment issues, and techniques to monitor underwater sediment movement to inform beach management. Funded interdisciplinary research projects have included adaptation strategies in response to climate change and underwater sensor networks. He has published more than a hundred academic articles and in 2010 organized a session on Coastal Tourism and Climate Change at UNESCO Headquarters in Paris in his role as a member of the Climate, Oceans and Security Working Group of the UNEP Global Forum on Oceans, Coasts and Islands. He has successfully supervised many PhD students, and as well as research students in his own university, advises PhD students for overseas universities. These currently include the University of KwaZuluNatal, Durban, University of Technology, Mauritius and University of Aveiro, Portugal. Professor Phillips has been a Trustee/Director of the US Coastal Education and Research Foundation (CERF) since 2011 and he is on the Editorial Board of the *Journal of Coastal Research*. He is also an Adjunct Professor in the Department of Geography, University of Victoria, British Columbia and Visiting Professor at the University Centre of the Westfjords. He was an expert advisor for the Portuguese FCT Adaptaria (coastal adaptation to climate change) and Smartparks (planning marine conservation areas) projects and his contributions to coastal and ocean policies included: the Rio +20 World Summit, Global Forum on Oceans, Coasts and Islands; UNESCO; EU Maritime Spatial Planning; and Welsh Government Policy on Marine Aggregate Dredging. Past contributions to research agendas include the German Cluster of Excellence in Marine Environmental Sciences (MARUM) and the Portuguese Department of Science and Technology.

Contributors

Abdullah Akbas is Research Assistant at the Geography Department of Ardahan University and also a PhD Candidate at the Geography Department of Istanbul University. His studies focus on hydrological, geomorphological and social aspects of climate change.

Nazan An is a PhD Candidate at the Institute of Environmental Science at Boğaziçi University. She graduated from the Department of Computational Science and Engineering at Boğaziçi University with an MSc in 2013. Her research fields are sustainability and vulnerability including socio-economic, demographic and environmental differences between developing and developed countries, measurement techniques for sustainable development, migration and climate change impacts, adaptation to climate change, carbon emissions related with the climate change impacts, and climate change impacts on agricultural production, extreme climate indices and crop yield variability.

Angela Anthonisz is Head of Subject, Events, Tourism and Hospitality at the University of Northampton where she teaches on the undergraduate and postgraduate programmes. Angela worked across a range of positions in the hotel and events industries before moving into academia in 1997. She has since taught in the UK, Hong Kong, Switzerland and Dubai and is currently in the final stages of her PhD. Angela's recent research has focused on the rapidly changing tourism and hospitality sectors in Dubai, with a particular focus on the impact of the EXPO and the implications this has for sustainability, strategy and human capital management.

Paulo Borges holds a PhD in Geology (Coastal Geology) from the Geosciences Department of the University of the Azores. Prior to his PhD, Paulo Borges had been working as a researcher since 1992, and is currently an Auxiliary Researcher at the University of the Azores and a collaborator at the Geology Centre of the University of Lisbon. He has carried out academic and applied research in coastal geology, mainly in the Azores archipelago, and published in several journals in these fields. His main scientific interests are sedimentology, geomorphology, coastal dynamics and coastal hazards, mainly on tsunamis, coastal storms, coastal flooding and coastal erosion. Based on his studies, particularly on the Azorean coastal hazards, he has recently started researching climate change impacts on the Azorean coastal zones. His scientific expertise and technical studies have been used and applied in the environmental planning and management of the Azorean coasts.

Stephen Boyd is Professor of Tourism at Ulster University in Northern Ireland, UK. A geographer by training, he has taught at Otago University in New Zealand, Staffordshire University in England and the University of Western Ontario in Canada. He has extensively researched and published in many areas of tourism but is more known for his work in heritage tourism, tourism and national parks, dark and political tourism, and the relationship between tourism and world heritage sites. He has had a long-term interest in Northern Ireland, and has published extensively on the changes this region has witnessed regarding tourism over the years. His most recent research has examined linear tourism space, co-authoring the first definitive text on tourism and trails with Professor Dallen J. Timothy (Arizona State University), and is now undertaking cross-cultural research on how visitors experience linear attractions, comparing the Wild Atlantic Way in Ireland with sections of the Silk Route in China. Other areas of research at present with his postgraduate students include community empowerment in islands, food tourism with a special focus on slow food, and urban sustainable tourism development. Stephen is visiting research professor at Luoyang Normal University in Henan Province, China.

Dr Huong T. Bui is Associate Professor of Tourism and Hospitality at Ritsumeikan Asia Pacific University (APU), Beppu, Oita, Japan. Her research interests are Asian tourism, heritage tourism and dark tourism in developing countries. She has conducted tourism research in South-east Asia, East Asia and Europe and published widely in international journals on topics including tourist behaviours, heritage conservation, destination marketing and tourism development policy.

Jeremy Buultjens is an Associate Professor in the School of Business and Tourism at Southern Cross University, Australia. He has taught in a number of units ranging from Economics and Indigenous Tourism through to Tourism Planning and the Environment. Jeremy's research interests include indigenous entrepreneurship, regional development, tourism in Sri Lanka and tourism in protected areas.

Helena Calado holds a degree in Geography and Regional Planning from the New University of Lisbon in Portugal and a PhD in Land Use Planning and Management from the University of the Azores in Portugal. She is currently a researcher at the Research Centre in Biodiversity and Genetic Resources and Professor at the Biology Department of the University of the Azores where she teaches Geography, Spatial Planning, Maritime Spatial Planning, Legislation and Environmental Management. Besides mentoring Master's students, she also manages a team of PhD students and researchers. With over 15 years of experience in land use planning and environmental impact assessments, she is currently focusing her research on coastal zone management plans, coastal hazards and mitigation measures, climate change impacts and implications, marine protected areas, and maritime spatial planning for small islands, particularly for the Azores. She has participated in numerous research projects and technical studies, and published numerous technical reports, in proceedings and in peer-reviewed academic journals and chapters.

Rita Cannas is an Assistant Professor of Management of the University of Cagliari, Sardinia, Italy in which she teaches the course of Business Communication. She holds a PhD in Economics (University of Bologna), a Master of Research Methodology in Business degree (University of Strathclyde) and a Marie Curie Fellowship (University of Malta) focused on sustainable tourism models. Before joining the academia she worked with some NGOs by coordinating national researches for sustainable tourism development in natural protected areas funded by the Italian Ministry of Environment. In 2003 she also coordinated the first Italian Master of Management for Sustainable and Responsible Tourism in Rome (Italy). Her research interests cover sustainable tourism management, particularly in the Mediterranean region, tourism seasonality, cultural and heritage tourism, and social entrepreneurship. Her latest publication is *Tourism Relationships between Sardinia and its Islands: Collaborative or Conflicting?* (2015, with E. Giudici).

Salvador Anton Clavé is a Full Professor of regional geographical analysis at Rovira i Virgili University where he serves as Director of the Doctoral Program in Tourism and Leisure. He has been a Visiting Research Scholar at the International Institute of Tourism Studies at George Washington University. He served as Director/Dean of the Tourism and Leisure School/Faculty of Tourism and Geography at the Rovira i Virgili University between 2002 and 2014. His research concentrates on the analysis of the evolution of tourism destinations, tourism destination design and planning, the impact of ICT on tourism and issues concerning tourism policies and local development.

Andrew Cooper is Professor of Coastal Studies at Ulster University in Northern Ireland, UK. He holds degrees in Geology from Queen's University Belfast and the University of Natal. He has been researching coastal processes and coastal management for over 30 years and has published widely on these subjects. His current interests include adaptation to climate change in coastal areas and the geological record of sea level change. He is co-author of *The World's Beaches* (University of California Press) and *The Last Beach* (Duke University Press), the latter being an urgent call for change in the relationship between people and beaches. Cooper is a Fellow of the Royal Society of South Africa and is a special visiting professor at the Federal University of Santa Catarina, Brazil and Honorary Research Professor at the University of KwaZulu-Natal, South Africa.

Dr Janice Cumberbatch teaches courses in Environmental Planning, Professional Skills for Natural Resource Management, Environmental Impact Assessment and Sustainable Tourism. Prior to joining CERMES, Janice Cumberbatch was the founder and Executive Director of Social and Environmental Management Services Inc. (SEMS), a private consulting firm that offered services in the areas of social and environmental impact assessments and social surveys; social and environmental policy research; social and environmental planning; community mobilization; and seminar and workshop design and facilitation. Dr Cumberbatch has published several articles in the areas of environmental management, social planning and sustainable development. Previously, Dr Cumberbatch worked for the United Nations Children's Fund (UNICEF), was the Deputy Director of a local heritage NGO, the Barbados National Trust and developed and coordinated the training portfolio for the Caribbean Natural Resources Institute (CANARI).

O. Cenk Demiroglu holds BA and MA degrees in Tourism Management from Boğaziçi University and Istanbul University, respectively, and two MSc degrees in Human Geography from Umeå University, with specializations in Tourism and Spatial Planning and Development. During 2011–2013, he managed research projects on the topic of climate change and ski tourism at the Western Norway Research Institute and Matej Bel University. He completed his PhD at the Geography Department of Istanbul University in December 2013 and went on to lead a post-doctoral research project on the climate change vulnerability of ski tourism in Turkey and Germany at the Istanbul Policy Center – Sabancı University – Stiftung Mercator Initiative. Having served as the Director of the School of Tourism and Hospitality at Istanbul Bilgi University and a lecturer at Boğaziçi University Tourism Administration Department from 2015 to 2017, Demiroglu currently pursues his career as a researcher at Umeå University Department of Geography and Economic History.

Denyse S. Dookie is a PhD candidate in the Sustainable Development programme at Columbia University. She holds an MA degree in Climate and Society from Columbia University, MA Development Studies/Economics of Development from the Institute of Social Studies in The Hague, as well as a PG Diploma International Relations and BSc Economics and Mathematics from the University of the West Indies (UWI) in Trinidad and Tobago. She has worked as a research fellow with the Inter-American Development Bank in Trinidad and Tobago, most notably with the Multilateral Investment Fund, in addition to several research assistantships/internships relating to sustainable, socio-economic and environmental development at UNDP, UNDESA, The Permanent Mission of Trinidad and Tobago to the UN, and with Caribbean development practitioners.

Danielle Edwards holds a Master of Laws (LL.M) in International Environmental Law from the American University Washington College of Law in Washington, DC, and a Bachelor of Laws (LL.B) from the University of the West Indies in Barbados. She is a practising Attorney-at-Law in the Republic of Trinidad and Tobago and The Commonwealth of Dominica. Danielle has served as legal consultant to The German Corporation for International Cooperation (GIZ), as law reviewer to the *Commonwealth Law Bulletin*, a publication of the Commonwealth Secretariat, and as legal contributor on Caribbean gender laws to the World Bank Group. Danielle is a Fellow of The Cropper Foundation, a not-for-profit organization based in Trinidad and Tobago, and a past Dean's Fellow of the International Legal Studies Program at American University. Her research interests include legal aspects of climate change adaptation for small island developing states, trade and energy law, and environmental policy.

Kristelle K. Francis is a recent MSc graduate from the Centre for Resource Management and Environmental Studies (CERMES), University of the West Indies, Cave Hill, Barbados where she obtained her degree in Natural Resource and Environmental Management with emphasis on Climate Change. Her primary interests include climate change, disaster risk management, tourism, costal/marine management and sustainable development. As a result of her education and enthusiasm surrounding these issues, she is eager to contribute to the formulation of policies on environmental sustainability, climate change (adaptation and resilience) and disaster risk management that helps to mobilise Caribbean countries through increased governance and the exchange of knowledge towards a much-needed sustainable natural resource and environmental era.

Charles Galdies holds a doctoral degree in Remote Sensing and Geographical Information Systems from the University of Durham (UK). Dr Galdies studied ways to improve small-scale weather and ocean forecasting in the central Mediterranean region using novel remote sensing observations of the ocean and the atmosphere. Between 2007 and 2011 Dr Galdies was Chief Meteorological Officer and Manager of the Malta MetOffice, which is the national weather service provider. During the same period, Dr Galdies was Malta's Permanent Representative (on behalf of the Government of Malta) with the World Meteorological Organisation. He currently advises the government on issues related to climate change. His main academic and research interests at the University of Malta are in aspects of weather and climate and on the tools to observe and analyse geophysical trends, with key specialization in Weather and Climate, Earth Observation Technology, Numerical Weather Prediction, Application of Geographic Information System for Earth Systems and Impacts of Climate Change.

Athula Gnanapala is a senior lecturer/HOD in the Department of Tourism Management, Sabaragamuwa University of Sri Lanka. His research interests include consumer behaviour in tourism including motivation and satisfaction, sustainable tourism planning and development, tourism and green marketing. Dr Athula Gnanapala has published scholarly papers and research reports on a variety of tourism issues. He completed his PhD in Tourism Management at the Xiamen University, Peoples Republic of China.

C. Michael Hall is a Professor at the University of Canterbury, New Zealand; Docent, Department of Geography, University of Oulu, Finland; Visiting Professor, Linneaus University, Kalmar, Sweden; and Senior Research Fellow, School of Tourism and Hospitality, University of Johannesburg, South Africa. He has published widely on tourism, environmental history and global environmental change.

Azizul Hassan is a member of the Tourism Consultants Network of the UK Tourism Society, and is currently working towards a PhD at Cardiff Metropolitan University. His main areas of research are: technology-supported marketing in tourism; innovative marketing dynamics; destination branding in tourism; cultural heritage tourism; heritage interpretation; and sustainable management/marketing alternatives for cultural heritage industries. He is a regular reviewer of *Tourism Management Perspectives*, *Tourism Analysis*, the *International Journal of Human Resource Management*, the *International Journal of Ecotourism*, the *eReview of Tourism Research (eRTR)* and the *International Interdisciplinary Business-Economics Advancement Journal*.

Tim Heap is a University Principal Tutor and Research Manager at the University of Derby, Buxton in the UK and has developed an extensive research profile in the field of tourism planning and destination management. He currently teaches at undergraduate, postgraduate and PhD level in the UK, and has international teaching experience in a number of countries including Switzerland, Zimbabwe and Malawi.

Dr Julia Jabour is a member of the Ocean & Antarctic Governance Research Program at IMAS. She has been researching, writing and lecturing on polar governance for more than 20 years. Most of her teaching and research is trans-disciplinary, involving examining current scientific and technical developments, determining their utility to the policy and law-making processes, and translating that information into user-friendly knowledge for uptake by non-specialist audiences. Julia has visited Antarctica six times, and been an advisor to the Australian Government at Antarctic Treaty Consultative Meetings on a number of occasions. Julia teaches into a variety of marine and Antarctic law and policy undergraduate and postgraduate units and has had teaching assignments in Iceland, Malaysia, Vietnam and New Zealand. She has strong links to the Arctic law community through hosting the 7th Polar Law Symposium in Hobart in 2014.

Dr Ian Jenkins is a geographer who has worked in the tourism and leisure industries for many years as a researcher, senior lecturer, consultant and director of several research units. This work has resulted in numerous publications including industry reports, conference papers, academic articles and book chapters; he has also been a peer reviewer for journal articles. Some of the research projects he has been involved with have resulted in legislative change and improved industry standards. In addition, he has undertaken work for prestigious organizations such as UNESCO, British Council, British Standards Institute, Health and Safety Executive, VisitWales and CEN. His research and consultancy expertise covers subjects such as: responsible/sustainable tourism, niche tourism development, risk and safety management and adventure tourism. Ian is currently an external examiner for the University of South Wales at both undergraduate and postgraduate level and also previously a PhD examiner at Cranfield University. Ian has been the co-editor of a special edition of the *Laureate Hospitality Journal*.

Levent Kurnaz received a BSc in Electrical and Electronic Engineering in 1988, an MSc in Electrical and Electronic Engineering and a BSc in Physics in 1990 from Boğaziçi University, an MSc in Physics in 1991 and a PhD in Physics in 1994 from University of Pittsburgh. Between 1995 and 1997, he worked as a research associate at Department of Chemistry in Tulane University. In 1997, he joined Boğaziçi University Physics Department as an Assistant Professor. Since 2005, he has been a Professor of Physics at Boğaziçi University. In 2014 he was appointed as the Director of the Center for Climate Change and Policy Studies at Boğaziçi University. Kurnaz has published one international book and 26 scientific papers in international journals. He has been a supervisor to 24 MSc and six PhD students for their theses.

Anh Le is a lecturer in the Department of International Economics at National Economics University, Hanoi, Vietnam, and is an official in the Department of Hotel Management at Vietnam National Administration of Tourism. He earned a MA in International Economics from the National Economics University in Vietnam and another MA in the same field from the International University of Japan.

Dr Tuan-Anh (Andrew) Le holds a PhD from Griffith University, Australia. Currently he works as a Lecturer in Tourism and Hospitality Management at The Hotel School Sydney (THSS), a partnership between the School of Business and Tourism, Southern Cross University, Australia, and Mulpha Australia. His area of research interests covers tourism, hospitality and sustainable development. Dr Le has more than 16 years of industry experience in various roles as master trainer, senior researcher, advisor and team leader. He has track records in provision of training and consulting services and proposal development which has been successfully funded by various international donors in Vietnam, for instance, the EuropeAid, Asian Development Bank (ADB), UNESCO and the Netherlands Development Organization (SNV).

Hamzah Muzaini is Assistant Professor (from August 2016) at the Southeast Asian Studies Department, National University of Singapore. He holds a PhD in Geography from Durham University, UK. His research interests include the material practices and cultural politics of war memory and commemoration, backpacking tourism, cultural heritage and theme parks, landscape approaches, and aspects of Singapore's transborder histories and geographies. He has published widely on these topics. He is co-author of *Contested Memoryscapes: The Politics of Second World War Commemoration in Singapore* (with Brenda Yeoh, Routledge, 2016).

Kiat Ng holds a BSc in Civil Engineering from McGill University in Canada, MSc in Environmental Engineering from Stanford University in USA, and PhD in Geography with emphasis on Spatial Planning from University of the Azores in Portugal. She has participated in Portuguese FCT-funded national project SMARTPARKS (Protected Areas' Planning and Management System for small islands) and is currently working on the cross-sectorial and cross-geographic European regions case study research task for FP7-funded EU project BASE (Bottom-Up Climate Adaptation Strategies towards a Sustainable Europe). Her experiences include water resources environmental consulting in California, and integrated coastal zone management, maritime spatial planning, climate variability and impacts, biodiversity conservation and protected areas for small islands. Her current research focuses on climate change adaptation, sustainable development and harmonious ways of living through an holistic approach and multidisciplinary studies.

Dr Keith Nurse is Senior Fellow at the Sir Arthur Lewis Institute for Social and Economic Studies and the World Trade Organization Chair at The University of the West Indies. He is the former Executive Director of UWI Consulting Inc. and Director of the Shridath Ramphal Centre for International Trade Law, Policy and Services. He is a member of the Economic Development Advisory Board, Government of the Republic of Trinidad and Tobago and serves on the executive bureau of the UN Committee for Development Policy, a subsidiary body of the United Nations Economic and Social Council. He has pioneered research on the creative industries, diasporic entrepreneurship, festival/heritage tourism, climate change and trade policy, and innovation governance and global value chains. Dr Nurse is the chair of CaribbeanTales Worldwide Distribution Inc. and the executive producer of the docudrama 'Forward Home: The Power of the Caribbean Diaspora'.

Dr Leonard Nurse's research focuses on human impact on coastal dynamics, integrated coastal management and the impact of climate change on small island environments. Dr Nurse is a member of the scientific team of the Intergovernmental Panel on Climate Change, which was awarded the 2007 Nobel Prize for its contribution to global climate change research. Prior to his appointment with CERMES, Dr Nurse held the posts of Director, Coastal Zone Management Unit, and Permanent Secretary, Ministry of Environment, Barbados. He has functioned in various professional capacities, including Vice-Chair, Intergovernmental Oceanographic Sub-Commission for the Wider Caribbean and Member, Scientific and Technical Advisory Panel of the GEF. Dr Nurse's continuous involvement in the scientific assessments of the IPCC dates back to 1990. He is currently Coordinating Lead Author for the chapter 'Small Island States' in the Fourth Assessment Report, already under preparation.

Tugba Ozturk is currently working as an Assistant Professor at the Department of Physics at Isik University. She received a BSc in Physics and Computer Engineering from Isik University and an MSc from the Department of Physics at Bogazici University. She completed a PhD in Physics at Boğaziçi University in 2014. Her research fields are regional climate modelling, climate extremes, and climate variability and change.

Dr Rolph Payet is an international policy expert, researcher and speaker on environment, climate and island issues, and was the first President and Vice-Chancellor of the University of Seychelles. He was Minister of Environment and Energy in the Cabinet of Seychelles from 2012 to 2014. He is currently United Nations Executive Secretary for the UNEP. Dr Payet has received numerous international awards and recognition for his work on islands, climate change and biodiversity. He is a lead author with the Intergovernmental Panel on Climate Change (IPCC) and was involved with the Third and Fourth Assessment Reports. In 2007, Dr Payet was elected Fellow of the Royal Geographical Society for his contributions to marine research. In January 2007, he was selected a Young Global Leader by the World Economic Forum. He received an honorary doctorate from the University of London in March 2016.

Carlo Perelli is Associated Researcher at CRENoS, University of Cagliari and University of Sassari, Italy. After an interdisciplinary MA in Mediterranean Studies between Venice, Montpellier and Madrid he received his PhD at The School for Advanced Studies in Venice Foundation, for a thesis on the implementation of sustainability policies in mass tourism coastal destinations. His interests are in tourism geography, planning for tourism, Integrated Coastal Zone Management (ICZM) and Tourism Carrying Capacity Assessment (TCCA) mainly in Sardinia (Italy), Morocco and Tunisia. Recently he has been working on tourism development in Morocco and Tunisia with a focus on heritage policies and halal tourism as product diversification strategies. In parallel to research activities he has been participating as a consultant in several EU-funded projects focusing on tourism in the MENA Region and notably on the following themes: ICZM and TCCA; water management in coastal destinations and heritage tourism.

Roya Rahimi completed her PhD in Tourism and Hospitality Management in 2013. She joined University of Wolverhampton, Business School as a lecturer in September 2014 where she teaches across tourism, hospitality, leisure and events subject areas. Her research interests are Customer Relationship Management (CRM), Organizational Culture, Human Resource Management, Gender Gaps and Equality and Diversity in Higher Education. Her work has been presented at various international conferences and has been published in a variety of journals and book chapters. She has experience of applying for national/international grants and H2020 funds. Her industry experience includes 7 years' experience working in the hotel industry in a number of international hotels in various countries. Roya is fluent in three languages (English, Turkish and Persian) and has familiarity with European and Asian cultures. Roya is a fan of using technology in her classes and uses simulating hospitality property management software for bringing the practice into theory and to make students familiar with real work experiences. She believes that high quality teaching, strong research and industry partnership go hand in hand given the application aspect of the tourism and hospitality field. She was nominated as inspirational lecturer of the year in 2015.

Yael Ram is Senior Lecturer in the Department of Tourism Studies, Ashkelon Academic College, Israel. Her research interests are sustainable mobility, cultural ecosystem services and sustainable tourism.

Iraj Ratnayake, PhD is currently attached to the Department of Tourism Management, Sabaragamuwa University of Sri Lanka, as a Senior Lecturer. His research and presentations mainly encompass development and promotion of special interest tourism and tourism planning in developing countries. Besides research and writing, he also participates actively in training, and consultation projects include tourism information systems, site planning, visitor management, product marketing and community tourism.

T. Rashid Said joined the Cultural Geography Group at Wageningen University as a PhD candidate in June 2014. His research mainly focuses on climate change along with other multiple stressors with a particular interest to the communities' livelihoods in small island developing states. Before he joined Wageningen University, he was teaching Geography and Environmental Studies for undergraduate students from 2009 at the State University of Zanzibar (SUZA) in Tanzania.

Dr Òscar Saladié is a Serra-Hunter Associate Professor in the Department of Geography (Universitat Rovira i Virgili). He is involved in the BA in Geography, the Master's degree in Territorial Management and the Master's degree in Environmental Law. He is a member of the Research Group on Territorial Analysis and Tourism Studies and, since 2012, Director of the Dow/URV Chair of Sustainable Development. His research is focused on environmental issues (climate change, waste management and mobility), especially as they relate to tourism activities. In his position as Director of the Dow/URV Chair of Sustainable Development, he is leading line 5 of the URV Environmental Plan.

Raquel Santos-Lacueva is a PhD fellow in the research Group of Territorial Analysis and Tourism Studies (GRATET) in the Department of Geography (Rovira i Virgili University). She holds a BA in Political Science and Public Administration (University of Barcelona) and an MSc in Regional and Environment Planning (University of Zaragoza). Her research focuses on climate change, tourism and public policy.

Daniel Scott is a Professor and University Research Chair as well as the Executive Director of the Interdisciplinary Centre on Climate Change at the University of Waterloo (Canada). He has worked extensively in the areas of climate change and the global tourism sector, including collaborations with the United Nations World Tourism Organization, United Nations Environment Programme, World Meteorological Organization, and World Travel and

Tourism Council. Dr Scott has been a contributing author and expert reviewer for the Intergovernmental Panel on Climate Change Third, Fourth and Fifth Assessment Reports.

Dao Truong is a lecturer in the Department of Tourism and Hospitality at National Economics University, Hanoi, Vietnam. He is also a research fellow in Tourism Research in Economic Environs & Society (TREES) at North-West University, Potchefstroom, South Africa, and is co-founder of Social Marketing Initiatives. Dao earned his PhD in Management from the University of Canterbury in New Zealand. His research has appeared in refereed journals, including the *Journal of Sustainable Tourism*, *Consumption Markets & Culture* and *Journal of Macromarketing*, among others.

M. Tufan Turp is a PhD Candidate at the Institute of Environmental Sciences at Boğaziçi University. He earned a BSc from the Physics Department at Yildiz Technical University in 2006, and an MSc from the Computational Science and Engineering Department at Boğaziçi University in 2010. As a member of the Boğaziçi University Center for Climate Change and Policy Studies, he conducts research in climate science. He has expertise in regional climate modelling and the impacts of climate change. He has also worked as a part-time lecturer at the Isik University Health Services Vocational School. He is currently studying on the projections of climate change over the Australasia region for his PhD thesis.

René van der Duim has worked since 1991 at the Department of Cultural Geography of Wageningen University where he now is Personal Professor. His research focuses on the relation between actor–network theory and tourism and on the tourism–conservation–development nexus, with a particular interest in eastern Africa. He is a co-editor of *Actor-Network Theory and Tourism: Ordering, materiality and multiplicity*, published in 2012 with Routledge, *Tourism Encounters and Controversies: Ontological Politics of Tourism Development*, published with Ashgate in 2015 and *Institutional Arrangements for Conservation, Development and Tourism in Eastern and Southern Africa*, published with Springer in 2015. He has also published his work in journals such as *Annals of Tourism Research*, *Journal of Sustainable Tourism*, *Journal of Ecotourism*, *Tourist Studies*, *Tourism Geographies*, *Society & Natural Resources*, *Environmental Conservation* and *Conservation and Society*. He is also chair of the European Association for Leisure and Tourism Education and Research, ATLAS.

Marta Vergílio holds a Degree in Environmental Engineering from Coimbra College of Agriculture and a Master's degree in Environmental Engineering from the University of Aveiro. Since graduation, she has collaborated in different projects related to environmental management and spatial planning. She has also collaborated on the research Project 'SmartParks – Planning and Management System for Small Islands Protected Areas'. Currently, Marta Vergílio is developing a PhD project entitled 'Ecological Structure in Spatial Plans: Adaptation and Implementation in the Azores Autonomous Region'. Her main scientific interests are environmental planning and management, protected areas, green infrastructures and ecosystem functions and services. She has also co-authored in technical reports and peer-reviewed academic journals.

Stephanie Verkoeyen is a PhD candidate at the University of Waterloo, Canada, in the Faculty of Environment. Her doctoral research involves applying a cognitive development framework to the study of climate change adaptation to understand why scuba divers intend to change (or not change) their behaviour in response to coral bleaching.

Foreword

Coastlines are synonymous with tourism and everywhere they are now under significant threat from climate change. Coastal and marine tourism is an economic driver and a significant part of the economy for most Small Island Developing States (SIDS) and developing countries. However, it is also significant in Western economies such as the United States and Australia, where it is the largest industry, employer and earner of foreign currency. Tourism provides livelihoods for millions of people in coastal and island destinations, but these livelihoods have become highly vulnerable to direct and indirect impacts of climate change such as extreme storms and sea level rise. Coastal erosion, physical damage to infrastructure, flooding, water shortages and contamination are common occurrences as assets are often located within short distances of the shoreline. High vulnerability is often coupled with low adaptive capacity, and this is especially the case for SIDS and coastal destinations in developing countries. Therefore, it is important that we innovate to find solutions to these problems and my own research into sustainable tourism and climate change has focused on these issues for the Seychelles and elsewhere. Through its case study approach, this book provides the basis and in some case the business case for policy-makers and other stakeholders to address climate change impacts on tourism's underpinning resources. For example, Antarctic tourism operators are now preparing for the polar shipping code implementation, which will enhance protection of this unspoiled region already facing the impacts of climate change

The strong seasonality and market competitivity of beach tourism needs to be considered, as this in itself can be exacerbated by climate change. For example, in many beach destinations the main tourism season coincides with dry seasons and low water availability, for example Morocco, which has aggravated water management and environmental issues. However, impacts of climate change and global warming will vary quite a lot in different coastal regions, so it is very good timing that this book is now being published. It contains an important global spread of issues from the continents to the poles and details climate change impacts for different tourism resources and case studies. These include: interventions in Zanzibar; vulnerability in Mexico, Morocco and Vietnam; policy and planning in Barbados, Israel and New Zealand; and comparisons of climate change effects on island tourism including the Azores, Bangladesh, Iceland, Ireland, Malta, Sardinia and Sicily. Sustainability and improving resilience are themes developed with examples from the Turquoise Coast and Sri Lanka. The case study from Dubai has the tantalizing title of a

'Theme Park Approach to Climate Change', and threatened developments considered in other book chapters include a harbour and UNESCO World Heritage Site.

Nature-based tourism relies on a diversity of resources which vary greatly in time and space. They will be affected by climate change in various ways and it is often difficult to assess the magnitude of impacts. Although ecosystems can be highly vulnerable there are good opportunities to design effective adaptation strategies for ecotourism and nature-based destinations. There are many and varying activities that can be developed in natural areas, where ecosystems also provide good adaptation options. For example, the chapter on Riviera Maya gives an overview of non-climatic stressors that determine vulnerability of this Caribbean coastline to climate change, a region including the second longest barrier reef system in the world.

It is well known that an evaluation and understanding of the present day status of a coastline is fundamental in deciding whether to actively manage or to refrain from intervention. However, climate change effects will increasingly focus management decisions on addressing future costs and consequences. This will mean that politicians and communities will have to make difficult future decisions as to how their coastal areas will be managed. The case study example of Fairbourne, where a decision taken as part of a shoreline management strategy, means that approximately 400 homes are going to be lost to the sea. These homes have lost economic value and the community will gradually relocate, although how this will be done is yet to be decided. The consequences of implementing this strategy give an insight into difficulties that are going to be more widely faced by global coastal communities in the coming years.

In conclusion, this book relates climate change, the major issue of our time, to tourism and economic consequences. The book has specific case studies which deal with future competitiveness, political and cultural dynamics, and lack of management. It shows that both non-climatic and climatic drivers have a direct impact on the social, economic and environmental aspects that make coastal zones vulnerable. Useful suggestions are provided for integrating human activities with environmental change and in my opinion it will become a reference for addressing future expectations. Therefore, this book is ideal for decision makers, practitioners, coastal managers, academics, students and the general public. This book comes at an opportune time since the UN has named 2017 as the Year of Sustainable Tourism.

Dr Rolph Payet
Executive Secretary
UNEP

Preface

It gives me great pleasure to present this follow-up study of Climate Change and Coastal Tourism which follows in the footsteps of our previous volume titled *Disappearing Destinations*, which was published by CAB International in 2011 (Jones and Phillips, 2011). This new volume presents the latest discussions and cases on the topic and in turn provides a framework for understanding the complex relationships between assessing such problems, meeting expectations from all stakeholder groups and finding solutions to the issues and challenges presented. Such concepts were key conclusions drawn from *Disappearing Destinations* (2011).

In this respect the growth of coastal tourism destinations and their relationship with coastal environments and climate change have become ever more topical. However, the growing uncertainty and complexity of issues has over more recent times often raised more questions than answers. Indeed, the first publication, *Disappearing Destinations*, did provide a platform for discourse and suggested parameters from which problems and issues highlighted could be tackled. Since then this has increasingly been set against the bounds for tourism growth, particularly for coastal destinations, island destinations, beaches and beach resorts, which have become ever reliant on sustained growth for economic success. With more recent and more alarming predictions of climate change and sea level rise in the last year or two, such destinations are now becoming further threatened by climate-induced damage and economic uncertainty. In this context forecasts still continue to predict that some of the world's most famous tourist destinations such as Australia's Great Barrier Reef, archipelagos of the Indian Ocean, Caribbean and Pacific or low-lying coastal fringes of many tourism destinations will suffer unprecedented damage because of changing climate patterns. *Disappearing Destinations* (2011) suggested that some destinations could be permanently closed to tourists by mid-century or at least face severe restrictions to visitor numbers and perhaps even sharp increases in access costs. Conclusions from the case studies presented illustrated a complex relationship of concurrent processes or 'continuums' that integrate at differing levels and stages of the tourism destination management cycle. Conclusions proposed that policy and management responses should address coastal tourism destination management, and try to balance the often complex, dynamic and symbiotic processes between problem recognition, meeting stakeholder expectations and delivering concrete solutions in order to address and ameliorate problems and threats at hand.

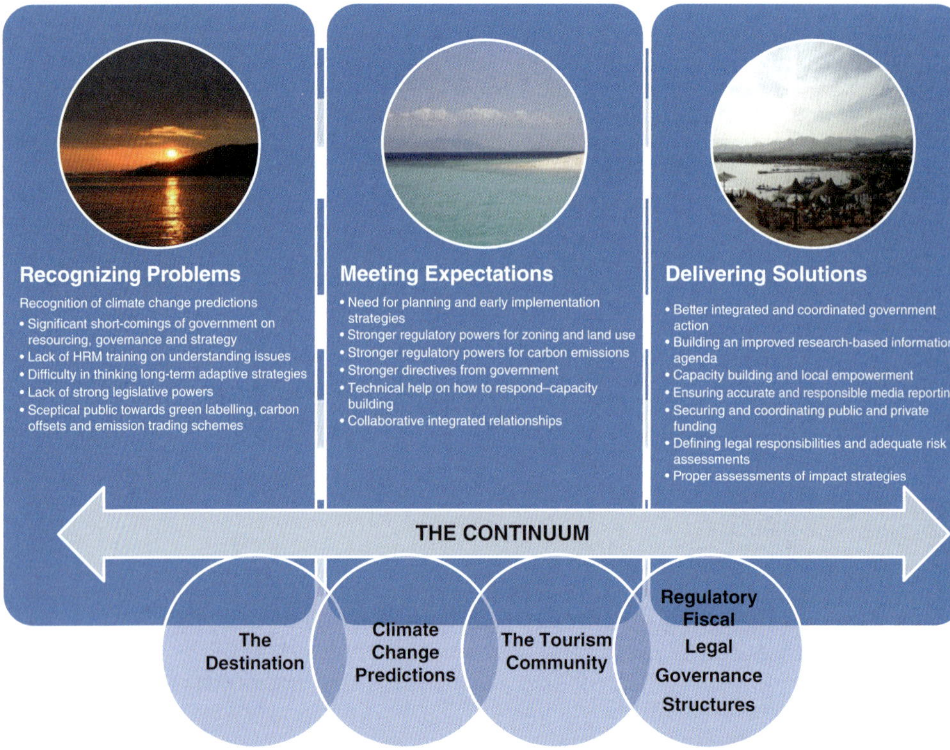

Fig. 1. Disappearing destinations – the coastal tourism destination management continuum: recognizing problems, meeting expectations and delivering solutions (Jones and Phillips (2011)).

The conclusions forecasted clear warnings, that within a short- to medium-term time period destinations across the world including cases from North America, Australasia, Asia, Antarctica and Europe could experience severe damage, in turn forcing the closure or economic demise of many coastal tourism destinations.

Despite such predictions, tourism growth trends and consequent spatial demands are showing continued signs of exponential growth, which in turn may continue ultimately to intensify current concerns regarding predicted climate threats and raise further questions concerning the ultimate sustainability of such destinations. It is the consequences of such phenomena and dynamics that continue to threaten the long-term future of coastal tourism environments and their ultimate survival. It is now recognized that assertive and innovative actions identifying management strategies that protect coastal tourism infrastructure and resources are now more urgently required. This includes destinations within both the developed and developing world, where the destruction of both natural and tourist environments through changing climate phenomena has no discrimination. That said, all is not doom and gloom. Despite many examples of unsustainable coastal development from around the world, there are now political moves across continents – North America, Australasia, Asia and Europe – where tourism coastal management strategies to tackle climate change are firmly on national and international agendas.

This book, and the chapters within, thus aims to update the last volume, *Disappearing Destinations* (2011), and present new cases that discuss current threats and consequences of climate predictions on coastal tourism destinations. In this context predicted changes and implications for management and policy at such destinations are globally assessed.

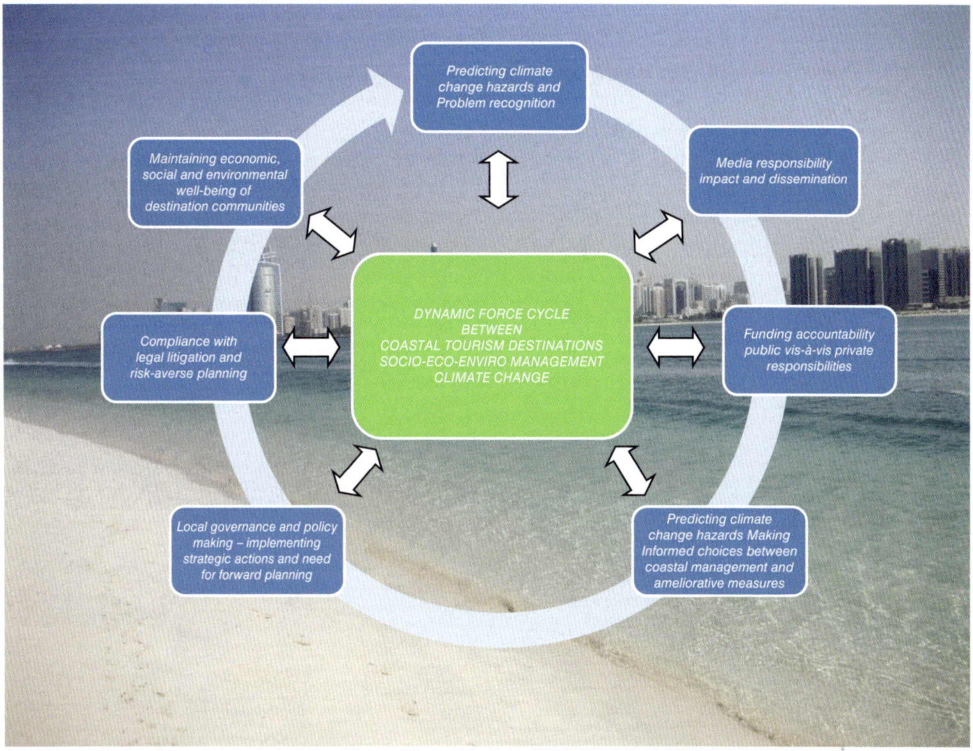

Fig. 2. Dynamic force cycle between coastal tourism destinations, socio-economic enviro management and climate change (Jones and Phillips (2011)).

From the analysis of specific tourism case studies, local impacts of climate change on coastal tourism infrastructures are evaluated and consequences for tourism development gauged. The validity and practicality of management options to tackle the complex nature and juxtaposition between tourism, climate change and coastal zone management are considered, including an evaluation of management responses and consequent policy choices. Whilst conclusions from the new cases demonstrate a renewed vigour in efforts to combat climate change through numerous mitigation and adaptation techniques, progress nevertheless remains patchy. Whilst there is now a growing recognition that coastal protection measures should be linked to integrated management processes, the cases often highlight that stakeholder perceptions and policy implementation measures sometimes ignore such imperatives, often resulting in inappropriate or ill-informed management responses. Contemporary integrated management strategies are thus considered and advocated for managing coastal tourism destinations in order to offset pressures from predicted climate change. From an assessment of socio-economic, environmental and political standpoints recommendations are made for innovative mitigation and adaptation measures to ameliorate projected impacts on coastal tourism infrastructure.

The book chapters are divided into two key parts. In the first part, Chapters 1 to 7 explore the theoretical and contextual frameworks of climate change processes and their general relationship with coastal tourism destinations. Such concepts, for example, evaluate the relationship and juxtaposition between climate change and tourism, coastal zone management, climate predictions and explore emerging issues relating to tourism destinations, sustainability and economic well-being.

In Part 2, Chapters 8 to 25 provide a more applied practical discussion, which draws upon a range of international case studies from leading academics, professionals and practitioners. These case studies illustrate contemporary issues in the development of coastal tourism destinations, perceived impacts of climate change, and review suggested or actioned ameliorative measures and evaluations of these. The case studies are based on specific coastal tourism typologies, including established beach resorts, ecotourism destinations, island destinations and adventure/alternative coastal destinations, and are drawn from key regions of the world including the Americas, Europe, Australasia and Asia.

The final chapter draws together the key themes and lessons that can be synthesized from such case examples and suggests possible frameworks for the most effective and feasible means of tackling the threats to tourism destinations from climate change today.

Andrew Jones
2017

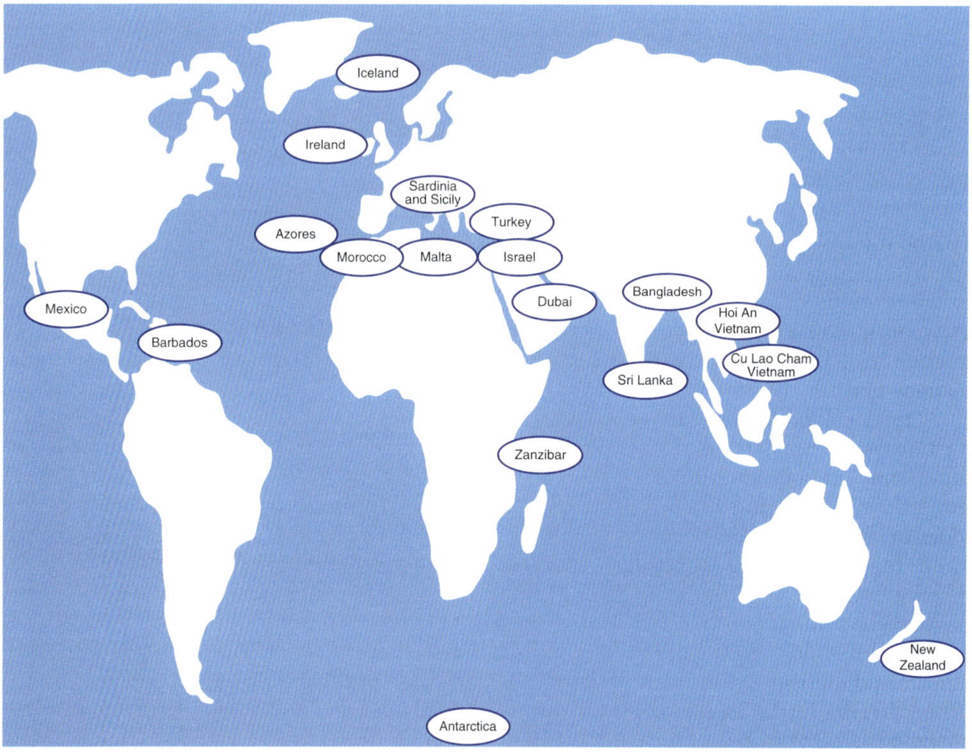

Fig. 3. Global case studies in this volume.

1 Introduction – Coastal Tourism and Climate Change: Current Narratives and Discourse

A. Jones*

Institute for Tourism Travel and Culture, The University of Malta; and Islands & Small States Institute, The University of Malta

Background: From 'Sun, Sea and Sand' to the IPPC – Problem Recognition and Identification

This book focuses on the contemporary current strategic management issues that are critical to the growing complexity of relationships between global tourism, predicted climate change and policies for tourism coastal management.

The development of coastal tourism destinations forms a major part of our understanding of the notions and concepts regarding the growth and definitions of modern-day tourism. Indeed, within the context of symbiotic relationships the concept of tourism development along coastlines is one that has become synonymous with the three 'S's – 'sun, sea and sand'. This relationship between climate and tourism growth has thus been one of the driving forces in the phenomenon of emerging global tourism markets. Such destinations now form a well-documented account of the historical development of modern-day tourism and our basic understanding of it. However, one of the key issues confronting coastal environments is the continued growth of tourism development and the impact of such on coastal zones. In the mid-1990s organizations such as the United Nations began to highlight such issues, particularly in developing tourist regions such as the Caribbean (UNEP, 1997). Indeed, the European Commission in 1999 launched its own policy statements on integrated coastal zone management (ICZM) – *Integrated Coastal Zone Management: A Strategy for Europe* (European Commission, 1999), whilst in the USA there had been a longer policy framework for regional coastal zone management, for example, the 1972 Coastal Zone Management Act (Office of Ocean and Coastal Resource Management/OCRM, 2004).

This was followed by other assessments at the time, which framed both the narrative and discourse on predicted climate change and impacts of such on coastal zones and tourism destinations, for example, Environmental Scientist (1999, 2000), Lohmann (2002), Nature (2002) and Phillips and Jones (2006).

Indeed, the symbiotic relationship between coastal tourism destinations and amenable climates is one that in some respects has now become a paradox with climate now threatening to destroy the very nature of tourism that, in the past, it has so successfully encouraged. Predictions from Povh (2000) also

*E-mail: andrew.jones@um.edu.mt

provide some sobering thoughts. His assertions have predicted that three-quarters of the world's population will be living within 60 km of the shoreline by 2020, and as a consequence suggested that there will be increased tensions between coastal leisure and tourism developments and the unpredictable but growing threats from climate change.

Much debate has been coalesced over the last 15 years or so. In this respect it has been increasingly argued that some predictions, especially with respect to sea level rise, storm surges and rising temperatures, will have significant consequences for the future management of the coastal zone and coastal tourism destinations. Authors such as Granja and Carvalho (2000), Vilibic *et al.* (2000), Jensen *et al.* (2001), Hall and Higham (2005), Phillips and Jones (2006) and Jones (2009) have already highlighted such issues.

The Stern Review (HM Treasury – Cabinet Office, 2005) also concluded during this time that there was clear scientific evidence to show that emissions from economic activity, particularly the burning of fossil fuels for energy, were causing changes to the Earth's climate. The Intergovernmental Panel on Climate Change (IPCC, 2007) also provided stark warnings. The 2007 study predicted that global warming would happen faster and be more devastating than previously thought and concluded that climate change would be far more destructive and have earlier impact than was first estimated. Predictions suggested devastating storms will increase dramatically; sea levels will rise over the century by around half a metre; snow will disappear from all but the highest mountains; deserts will spread; oceans will become acidic, leading to the destruction of coral reefs and atolls; and deadly heat waves would become more prevalent. It is now widely accepted that the findings from both the IPCC Report (2007) and the Stern Review (HM Treasury – Cabinet Office, 2005) have provided the framework and initial evidence to confirm predicted changes to climate and consequences for the environment and economic well-being. These are now firmly recognized as affecting the planet in potentially adverse ways. It was seen, at the time, that such predictions would also ultimately adversely impact on many coastal environments, particularly island destinations, and that impacts for coastal tourism destinations would, in turn, be far reaching. The most recent IPCC (2014) Fifth Assessment Report has also (with stark warnings) reconfirmed that the process of climate change is accelerating with profound impacts from rising temperatures and extreme weather (IPCC, 2014). In this context the report also highlighted that tourism economies across the world will not escape from such events and will in turn be severely affected.

The relationship between climate change and tourism is now generally well documented. Over the last few years, the relationship between climate change and tourism has generated much debate and discourse which has stemmed from initial research by, for example, Agnew and Viner (2001), Lohmann (2002), Smithers (2006) and Viner (2006). In this context, Smithers highlighted the fact that some of the world's most famous tourist destinations could be closed to visitors by 2020 and beyond because of worries about climate change. Areas particularly highlighted in, for example, the Mediterranean included tourist areas such as Pueta de Marrozon and the Murcian coastline of Spain, the island of Crete, the Amalfi coast of Italy and Athens including the Attica region of Greece. Again, such sentiments have more recently been supported and evaluated by authors such as Becken and Hay (2007), Gössling (2011), Hall (2011), Jones and Phillips (2011), Ranade (2012), Scott *et al.* (2012), Singh (2012), Prideaux and McKercher (2014), Flannery (2015), Hall and Gössling (2015) and Romm (2015) and organizations such as UNEP (2009), UNEP/OECD (2010) in their current assessments.

From such evidence, it seems that there is still a growing concern that pressures from climate change set against continued tourism growth is placing unprecedented socio-economic strains on coastal tourism destinations. Authors such as Booker (2009), Hulme (2009), Dessler and Parson (2010), Giddens (2011), Henson (2011) and Romm (2015) continue to claim that the future predictions of climate change are not an exact science. However, current assessments of extreme impacts, together with implications for socio-economic and physical disruption continue, none the less, to provide evidence of growing threats. In this context the well-being of many tourism destinations will remain at best uncertain and in the longer term may be severely compromised with perhaps critical consequences for future sustainability.

To date, it is probably true to claim that climate and environmental change is increasingly seen as one of the major long-term threats facing global economies both in the developed and developing world. As such, tourism does not escape, especially those regions that are reliant on tourism-based economies. In this respect, many low-lying coastal regions are specifically at risk from adverse climate change. Evidence of this process is already underway with many examples of coastal tourism destinations experiencing at least early signs of stress or significant signs of negative impact. Ridderstaat *et al.*'s (2014) work on the island of Aruba in the Caribbean, Meyer-Arendt's (2011) work on the Gulf coastline of Louisiana, USA, Wilson and Turton's (2011) work on the Queensland coast and Great Barrier Reef in Australia or Jones's review of coastal destination issues in the UK (Jones, 2011) together with Jones and Phillips' broader strategic assessments on coastal tourism issues (2011) are good illustrations in this respect. Authors such as Prats (2011), Scott *et al.* (2012) and Singh (2012) also provide other contemporary assessments.

Earlier examples have also provided both lessons and evidence. The Townsend and Harris (2004) review of the potential fatal outcomes from inaction are pertinent in this respect. In a similar light, research by Epaedia (2005) also suggested that the biggest driver of development in the European coastal zone in recent years has been the demand for tourism and the growing concern on the need for more sustainable management strategies to offset continued growth demands and adverse climate threats.

Evidence from Epaedia's research in 2005 showed that the Mediterranean coasts of France, Spain and Italy were receiving in excess of 75 million, 59 million and 40 million visitors, respectively. Obvious concerns are raised regarding such growth and further concerns are raised on future growth predictions along the Mediterranean coastlines (Epaedia, 2005). Using another Mediterranean example, the agency's research suggested that in French coastal regions alone tourism provided approximately 50% of jobs, generating more revenue than fishing or shipping and that peak population densities on the Mediterranean coasts of France and Spain reached 2300 people per square kilometre, more than double the winter populations. It estimated a further 40% increase in peak populations for the future, thus emphasizing the critical economic, social and environment interrelationships at play within such destinations. Benoit and Comeau (2005) have also highlighted some of the key pressures on the coastal fringe. From their research for 'Plan Bleu', a UNEP/ EU initiative, key pressures are identified with tourism being highlighted as one of the key coastal environment protagonists.

In the USA, Houston (2002) reported that travel and tourism had become the largest US industry, employer and earner of foreign exchange and that beaches were the major factor in this tourism market. He further identified beach erosion as the number one concern of Americans who visit beaches. Research by the US Army Corps of Engineers (1994) illustrated that 33,000 km of shoreline within the USA were experiencing some kind of erosion and that 4300 km were critical. Their findings considered this a serious threat to tourism and therefore a major threat to the national economy. Work by authors such as Dharmaratne and Braithwaite (1998) whose research in the Caribbean, and Williams and Micallef's (2009) work on beach tourism, has also stressed the importance of beaches to national economies.

In this context climate models suggest a future warming of 0.2–0.3°C per decade with sea levels expected to rise at a rate of 4–10 cm per decade. An increase in extreme weather events such as floods and storms is also expected. A rise of 4–10 cm per decade does not seem like a rise that will adversely affect destinations but, as the IPCC predict this will potentially cause major problems, particularly for coastal areas (IPCC, 2007, 2014).

Similarly, Greenpeace (2007) issued controversial warnings by predicting a hypothetical future 'post climate change' Spanish coastline at La Manga: using computer modelling a 'virtual' visual analysis illustrated the consequences of severe flooding if steps were not taken to stop the effects of severe environmental damage caused by climate change. In this respect, Greenpeace advocated a much more strategic approach to offset such threats by promoting a much more vigorous approach to problem recognition and stakeholder engagement and in turn encouraging wider impact adaptation and amelioration measures. However, such predictions, as well as proposed actions, still remain controversial. Others, such as The World Wildlife Fund report (WWF, 2007), suggested that the tourism industry's heavy reliance on the local environment and climate to sell holidays means that it could face serious challenges as a result of climate change. UNESCO's (2007) assessment of impacts on world heritage sites has already illustrated and predicted that many of the world's tourist sites may be under threat from climate change particularly through rising sea levels, increased flooding risks and depleted marine and land biodiversity. Such predictions claimed that this could have disastrous effects on over 830 designated UNESCO world heritage sites. The UNWTO (2007) Climate Change and Tourism: Responding to Global Challenges report – a response to the UN Davos climate change conference – highlighted a number of regional threats across the globe where significant impact of climate on tourism destinations was predicted. This included North America, Northern Europe, the Mediterranean, Caribbean, Indian Ocean, Africa, Asia and Australasia, with no region escaping predicted threats (see Chapter 5 for further details).

More recent reports from both industry and governmental organizations such as the 'Tourism 2023' report by Forum for the Future (2009) or UNEP (2009) and UNEP/OECD (2010) have reinforced such concerns. Together with these global assessments more local evidence assessing coastal destinations and national tourist economies has been reviewed by Williams and Micallef (2009) and Mushi (2011) who highlight the key economic and social impacts from climate change on coastal tourist communities. The European Union Environmental Agency (2012) has also explicitly predicted adverse climate impact upon European economies, particularly those that are heavily dependent on tourism in Southern Europe (EEA, 2012).

Such predictions are still not an exact science and there still remains a gap in measurable empirical research on the subject. None the less, a report by Quiret (2011) has suggested that decaying ecosystems can account significantly for a decline in tourism GDP. Despite, however, the lack of empirical data, there has been much other discourse. For example, in 2009, the consulting firm KPMG claimed that tourism is one of the global industries least prepared and one of the most vulnerable to environmental and climate change. It suggested that the tourism industry has yet to come to terms with the associated risks and costs it is facing as threats from heatwaves, droughts and rising sea levels are just some of the factors that will continue to adversely impact upon the industry, especially in terms of social conflict and continued economic viability (KPMG, 2009). In the same year a review by the United Nations Environment Programme (UNEP) (2009) in association with the French Government and World Tourism Organisation (UNWTO) highlighted growing concerns between the need for better integrated coastal management and the need to adapt tourism destinations for climate change (UNEP, 2009). In this context, Jones and Phillips' (2011) review of *Disappearing Destinations* also highlighted the need for a much more coordinated and strategic approach; an approach that promotes a three-'pronged' management push to ensure: (i) problem recognition; (ii) meeting stakeholder expectations; and (iii) delivering sustainable solutions. Their review of specific global cases illustrated current practices and challenges and provided a platform from which to determine new ideas and concepts for future policy directions.

Problem Responses and Contemporary Policy Frameworks

However, despite much rhetoric on proposed and potential solutions the discussions tend to raise more questions than provide answers.

Despite such negatives and political inertia in some quarters, there have been some positives as well, with some concerted effort from, for example, the travel and tourism industry to address the challenges from climate change. The World Travel and Tourism Council (WTTC) (2009, 2014) and Responsible Travel (2014) have been at the forefront of this push by promoting accountability and responsibility by endorsing travel and tourism development awards that recognize good practice in sustainable tourism and carbon management. Other travel conglomerates such as TUI (2014) have followed suit, advocating environmental and social responsibility.

Such developments and predictions should, however, also be considered within the context of the continued growth of broader global tourism markets. Despite the current economic gloom, forecasts for global tourism remain buoyant and predictions, however conservative, show that world tourism statistics are set for further growth over the next decade (UNWTO, 2014). These figures present quite a conundrum. The demand across the regions for such growth raises the ever-mounting question of how growing demand for tourism can be sustained, balanced or for that matter strategically managed in the light of the ongoing predictions for climate change and its consequences.

Paradoxically, such issues have become quite complex, with adverse climate events and associated assessments for environmental damage now threatening to destroy coastal tourism destinations. Predictions also suggest that this will also be exacerbated by ever-increasing concerns and debates over the continued need and merits for remedial actions such as 'hard' and 'soft' mitigation measures (e.g. hard engineering options, smart technology and smart design options, skills and training through capacity building) to offset such problems. Who takes responsibility for the implementation and funding of such actions also remain key issues that remain (Agarwal and Shaw, 2007; Kunreuther and Erwann, 2007; Gössling, 2011; Jones and Phillips, 2011; Prats, 2011). As such, the evidence or science, although not exact, has predicted unambiguous consequences for coastal destinations where both predictions for adverse climate change and unprecedented levels of tourism growth appear to be on an escalating collision trajectory.

Although there has been much media attention since 2005, concerns and possible solutions are not just a contemporary phenomenon. There have been some earlier responses to predicted threats, particularly relating to carbon emissions. As early as 1999 Viner and Agnew (p. 2) recommended that:

> the tourism industry itself must take action to reduce its contribution to global greenhouse gas emissions. For example, in destinations, changes to energy supply should be introduced, creating a shift from fossil fuel to renewable sources of energy such as wind, biomass and solar power. This needs to be coupled with changes to planning procedures and laws, so that more opportunities for renewable energy sources can be developed. More stringent efficiency standards and a compulsory energy rating scheme could also be employed in buildings, such as hotels. Transport to, from, and around resorts and within destinations, is another key area where changes could be made.

The United Nations has also made some headway in the 15-plus years since it began addressing climate change. The UN's 1997 Kyoto Protocol to address carbon emissions set binding targets for carbon emissions, but the absence of support from the USA made the protocol weaker than many had hoped (Henson, 2006). The second International Conference on Climate Change in Davos, Switzerland (UNWTO, 2007) established new agreements that the tourism sector must rapidly respond to climate change and progressively reduce its greenhouse gas (GHG) contribution. This, it suggested, would require action to mitigate its GHG emissions, derived especially from transport and accommodation activities; adapt tourism businesses and destinations to changing climate conditions; apply existing and new technology to improve energy efficiency; and secure financial resources to help poor regions and developing countries (UNWTO, 2007). The consequent United Nations Copenhagen Climate Change Conference in December 2009, although hailed as a further step, was disappointing in this respect and failed to reach agreement (Vidal *et al.*, 2009). The Doha Climate Change Agreement in 2012 went some way to address the balance between climate and tourism and future ways forward. However, more recent initiatives, such as the United Nations Framework Convention on Climate Change (UNFCCC) held in Paris during December 2015, have been seen as a watershed in establishing binding agreements and as such bode improved prospects for tackling climate change threats in the short- to medium-term futures (Harvey, 2015).

Summary

From such evidence it seems increasingly apparent that coastal tourism and its relationship with climate change are now established topics of research, increasingly discussed within international policy contexts. In this respect, coastal destinations, beaches and beach resorts have become synonymous with socio-economic growth but in turn are becoming increasingly threatened by climate change and associated environmental and economic disruption and damage. It is the consequences of such phenomena that will ultimately impact upon the long-term future of coastal tourism environments and, of course, their continued survival. With respect to such, it would seem increasingly propitious to identify management strategies that on the one hand recognize climatic threats and on the other protect tourism infrastructure and coastal resources, especially in areas significantly reliant on the tourism industry for their economy. This includes destinations small and large, both in the developed and developing world, as coastal regions are places where the impacts from climate change have no discrimination. As a result, across the world, in the USA, Australasia, Asia and Europe, coastal zone strategies to tackle such challenges are now firmly on the political agenda.

Clearly, in the second decade of this new millennium, two factors are clear, one suggests that tourism is having a major environmental impact on many coastal areas and the second suggests that potential threats from climate change are likely to create considerable adverse impacts unless managed effectively. Thus we find an increasingly clear juxtaposition and paradox emerging between, on the one hand, tourism itself, creating many undesirable impacts on the coastal zone and on the other, climate change threatening to adversely impact on coastal tourism destinations, ultimately threatening the very nature, character and socio-economic well-being of many tourist coastal environments.

In summary this edited volume will explore such issues and discuss the consequences of current and future tourism growth within coastal destinations and the threats and implications that are predicted from climate change. The book is thus divided into two parts. Chapters 1–7 focus on the current theoretical and conceptual issues that are currently in the policy and enviro-socio-economic arenas. For example, Chapters 1, 2, 3 and 4 assess and evaluate contemporary relationships between climate change, coastal zone management and tourism and the current policy and management concepts pertaining to these three dynamic and increasingly interrelated topics. Chapters 5, 6 and 7 discuss emerging strategic interrelated themes on regional assessments and knowledge gaps for coastal and island destinations and, in turn, their implications for economic well-being.

In Part 2 of this book, chapters 8–25 will present more specific global case examples illustrating the practical relationship between climate change and coastal tourism at local destinations. The case examples from Asia, Europe, North and Central America, the Caribbean, Australasia and Antarctica assess implications of current and predicted impact of climate change on coastal tourism destinations and discuss options for ameliorative management and policy measures that can be adopted to help offset predicted effects. Conclusions from such case examples are evaluated and consequences for tourism development outlined. The validity and practicality of management options to tackle the complex nature and juxtaposition between tourism, climate change and coastal zone management are explored and considered, including an evaluation of management responses and consequent policy choices. Recommendations are made to ameliorate projected impacts on coastal tourism infrastructure by making choices that include a range of strategic policy measures and applied management options for coastal destinations and to explore new opportunities for alternative sustainable tourism development.

References

Agarwal, S. and Shaw, G. (2007) *Managing Coastal Tourism Resorts*. Channel View Publications, Clevedon, UK.

Agnew, M.D. and Viner, D. (2001) Potential impacts of climate change on international tourism. *International Journal of Tourism and Hospitality Research* 3(1), 37–60.

Becken, S. and Hay, J.E. (2007) *Tourism and Climate Change: Risks and Opportunities*. Channel View Publications, Clevedon, UK.

Benoit, G.E. and Comeau, A. (2005) *Sustainable Future for the Mediterranean: The Blue Plan's Environment and Development Outlook*. UNEP, Earthscan, London.

Booker, C. (2009) *The Real Global Warming Disaster: Is the Obsession with 'Climate Change' Turning Out to be the Most Costly Scientific Blunder in History?* Continuum International Publishing Group, London.

Dessler, A. and Parson, E. (2010) *The Science and Politics of Global Climate Change: A Guide to the Debate*, 2nd edn. Cambridge University Press, Cambridge.

Dharmaratne, G.S. and Braithwaite, A.E. (1998) Economic valuation of the coastline for tourism in Barbados. *Journal of Travel Research* 37(2), 138–144.

Environmental Scientist (1999) Living in the greenhouse. *Environmental Scientist* 8(1), 1–3.

Environmental Scientist (2000) The cost of climate change – UK report is a world first. *Environmental Scientist* 9(3), 1–2.

Epaedia (2005) Available at: www.eea.europa.eu (accessed 1 June 2017).

European Commission (1999) *Towards a European Coastal Zone Management: General Principles and Policy Options*. European Commission, Brussels.

European Environmental Agency (EEA) (2012) *Climate Change, Impacts and Vulnerability in Europe 2012*. EEA 12/2012. European Environmental Agency, Copenhagen.

Flannery, T. (2015) *Atmosphere of Hope: Searching for Solutions to the Climate Crisis*. Atlantic Monthly Press, New York.

Forum for the Future (2009) Tourism 2023, Forum for The Future. Available at: www.forumforthefuture.org/project/tourism-2023/overview (accessed 30 May 2016).

Giddens, A. (2011) *The Politics of Climate Change*, 2nd edn. Polity Press, Cambridge, UK.

Gössling, S. (2011) *Carbon Management in Tourism: Mitigating the Impacts on Climate Change*. Routledge International Series in Tourism, Routledge, Abingdon, UK.

Granja, H.M. and Carvalho, G.S. (2000) Inland beach migration (beach erosion) and the coastal zone management (the experience of the northwest coastal zone of Portugal). *Responsible Coastal Zone Management. Periodicum Biologorum* 102(1), 413–424.

Greenpeace (2007) *Photoclima-Photoclimate-Spain*. Greenpeace, Madrid, Spain.

Hall, M.C. (2011) Climate change and its impacts on tourism: regional assessments, knowledge gaps and issues. In: Jones, A. and Phillips, M. (eds) *Disappearing Destinations: Climate Change and Future Challenges for Coastal Tourism*. CAB International, Wallingford, UK.

Hall, M.C. and Gössling, S. (2015) *The Routledge Handbook of Tourism and Sustainability*. Routledge, London.

Hall, M. and Higham, J. (2005) *Aspects of Tourism: Tourism Recreation and Climate Change*. Channel View Publications, London.

Harvey, F. (2015) Paris Climate Change Agreement: The world's greatest diplomatic success, *The Guardian*, 14 December 2015.

Henson, R. (2006) *The Rough Guide to Climate Change*. Rough Guides Limited, London.

Henson, R. (2011) *The Rough Guide to Climate Change*, 3rd edn. Rough Guides Limited, London.

HM Treasury – Cabinet Office (2005) *The Economics of Climate Change: Stern Review*. HM Treasury, HMSO, London.

Houston, J.R. (2002) The economic value of beaches – A 2002 update. *Shore and Beach* 70(1), 9–12.

Hulme, M. (2009) *Why We Disagree about Climate Change: Understanding Controversy, Inaction and Opportunity*. Cambridge University Press, Cambridge.

Intergovernmental Panel on Climate Change (IPCC) (2007) *Climate Change 2007*, the Fourth IPCC Assessment Report. UNEP, Cambridge University Press, Cambridge.

Intergovernmental Panel on Climate Change (IPCC) (2014) *Fifth Assessment Report on Climate Change 2014*. UNEP, Cambridge University Press, Cambridge.

Jensen, J., Bender, F. and Blasi, C. (2001) Analysis of the water levels along the German North Sea coastline. In: Ozhan, E. (ed.) *Medcoast 01: Proceedings of the fifth International Conference on the Mediterranean Coastal Environment*. Medcoast, Ankara, Vol. 3, pp. 1129–1140.

Jones, A.L. (2009) Climate change impacts on UK coastal tourism destinations. *Proceedings of Council for Australia Universities CAUTHE Annual Conference 2009 on Sea Change: Tourism and Hospitality in a Dynamic World*, Fremantle, Perth, Australia.

Jones, A.L. (2011) UK coastal tourism destinations: Assessment of perceived climate impacts: issues for destination management, local governance and public policy making. In: Jones, A. and Phillips, M. (eds) *Disappearing Destinations: Climate Change and Future Challenges for Coastal Tourism*. CAB International, Wallingford, UK.

Jones, A.L. and Phillips, M. (eds) (2011) *Disappearing Destinations: Climate change and the Future Challenges for Coastal Tourism*. CAB International, Wallingford, UK.

KPMG (2009) *Climate Changes Your Business*. KPMG, UK.

Kunreuther, H.C. and Erwann, O. (2007) Climate change, insurability of large-scale disasters and the emerging liability challenge. *National Bureau of Economic Research*, issue 12821. NBER, Cambridge, Massachusetts.

Lohmann, M. (2002) Coastal resorts and climate change. In: Lockwood, A. and Medlick, S. (eds) *Tourism and Hospitality in the 21st Century*. CAB International, Wallingford, UK, pp. 285–287.

Meyer-Arendt, K. (2011) Grand Isle, Louisiana: A historic U.S. Gulf Coast resort adapts to hurricanes, subsidence, and sea level rise. In: Jones, A. and Phillips, M. (eds) *Disappearing Destinations: Climate Change and Future Challenges for Coastal Tourism*. CAB International, Wallingford, UK, pp. 203–217.

Mushi, R. (2011) *Climate Change and the Coastal Environment: Implications on Coastal Tourism in Bagamoyo District, Tanzania*. LAP LAMBERT Academic Publishing, Saarbrücken, Germany.

Nature (2002) *When Doubt is a Sure Thing*. News Feature. *Nature* 418, 476–478.

Office of Ocean and Coastal Resource Management/OCRM (2004) Coastal Zone Management Act. Available at: https://coast.noaa.gov/about (accessed 1 June 2016).

Phillips, M. and Jones, A. (2006) Erosion and tourism infrastructure in the coastal zone: problems, consequences and management. *Tourism Management* 27(3), 517–524.

Povh, D. (2000) Economic instruments for sustainable development in the Mediterranean region. *Responsible Coastal Zone Management. Periodicum Biologorum* 102(1), 407–412.

Prats, L. (2011) *Researching Coastal and Resort Destination Management: Cultures and Histories of Tourism*. Palibrio, Bloomington, Indiana.

Prideaux, B. and McKercher, B. (2014) *Climate Change and Tourism in the Asia Pacific*, 1st edn. Routledge, London.

Quiret, M. (2011) Maldives in a troubled paradise: time runs out on an environmental deadline. *Time Magazine*, 17 October.

Ranade, P.S. (2012) *Climate Change and Tourism*. SBS, London.

Responsible Travel (2014) Responsible Travel Awards. Available at: www.responsibletravel.com/awards (accessed 1 June 2017).

Ridderstaat, J., Oduber, M., Croes, R., Nijkamp, P. and Martens, P. (2014) Impact of seasonal patterns of climate on recurrent fluctuations in tourism demand: evidence from Aruba. *Tourism Management* 41, 245–256.

Romm, J. (2015) *Climate Change – What Everyone Needs to Know*, 1st edn. Oxford University Press, Oxford.

Scott, D., Hall, M.C. and Gössling, S. (2012) *Tourism and Climate Change: Impacts. Adaptation and Mitigation*. Routledge, Abingdon, UK.

Singh, T.V. (2012) *Critical Debates in Tourism*. Channel View Publications, Clevedon, UK.

Smithers, R. (2006) Tourist hotspots at risk of closure. *The Guardian*, 22 September 2006.

Townsend, M. and Harris, P. (2004) Now the Pentagon tells Bush: Climate Change will destroy us, secret report warns of rioting and nuclear war; threat to the world is greater than terrorism. *The Observer*, London, 22 November 2004.

TUI (2014) *Better Holidays Better World: Sustainability Strategy 2015–2020*. TUI Group, Luton, UK.

UNEP (1997) Coastal Tourism in the Wider Caribbean Region: Impacts and Best Management Practices. *CEP Technical Report* No. 38.

UNEP (2009) *ICZM for Coastal Tourism Destinations Adapting to Climate*. UNEP.

UNEP/OECD (2010) *Climate Change and Tourism Policy in OECD Countries*. UNEP/OECD.

UNESCO (2007) *Case Studies on Climate Change and World Heritage*. UNESCO, Venice, Italy.

United Nations World Tourism Organisation (UNWTO) (2007) *Proceedings of 2nd International Conference on Climate Change and Tourism*, Davos, Switzerland.

United Nations World Tourism Organisation (UNWTO) (2014) *Tourism Highlights, 2014 edn*. UNWTO, Madrid.

US Army Corps of Engineers (1994) Shoreline protection and beach erosion control study. Phase 1: Cost comparison of shoreline protection projects of the US Corps of Engineers. Water Resources Support Center, Washington, DC.

Vidal, J., Stratton, A. and Goldenburg, S. (2009) Low targets, goals dropped: Copenhagen ends in failure. *The Guardian*, London, 19 December 2009.

Vilibic, I., Leder, N. and Smircic, A. (2000) Storm surges in the Adriatic sea: an impact on the coastal infrastructure. *Responsible Coastal Zone Management. Periodicum Biologorum* 102(1), 483–488.

Viner, D. (ed.) (2006) Tourism and its interactions with climate change. *Journal of Sustainable Tourism* 14(4), 317–322

Viner, D., Agnew, M. and World Wide Fund for Nature (WWF) (1999) *Climate Change and Its Impacts on Tourism*. WWF, Norwich, UK, p. 2.

Williams, A. and Micallef, A. (2009) *Beach Management Principles and Practice*. Earthscan, London.

Wilson, R. and Turton, S. (2011) The impact of climate change on reef-based tourism in Cairns, Australia – Adaptation and response strategies for a highly vulnerable destination. In: Jones, A. and Phillips, M. (eds) *Disappearing Destinations: Climate Change and Future Challenges for Coastal Tourism*, CAB International, Wallingford, UK, pp. 233–253.

World Wide Fund For Nature (WWF) (2007) *Environmental Report 2007*. WWF, London.

World Travel and Tourism Council (WTTC) (2009) *Leading the Challenge on Climate Change*. WTTC, London.

World Travel and Tourism Council (WTTC) (2014) Tourism for Tomorrow Awards. Available at: www.wttc.org/tourism-for-tomorrow-awards (accessed 6 May 2016).

2 A Rapidly Changing Climate in an Era of Increasing Global Carbon Emissions

C. Galdies*
Institute of Earth Systems, University of Malta, Malta

Introduction

Humanity continues to blaze the path towards the increased extraction and burning of fossil fuels without a full understanding of its consequences. With seemingly no end to this finite resource, new drilling and increased extraction opportunities have brought the price of this commodity down. Meanwhile, current assessments on the impact of increased levels of CO_2, which are primarily generated from the burning of such fuels, point towards the consequential effects of extreme natural events in light of heatwaves and droughts, heavy rainfall, floods and sea level rise on communities (IPCC, 2014). Climatologists highlight the urgent need to cut down drastically CO_2 emissions in view of the longevity of airborne carbon present in the atmosphere (Archer, 2005) and have subsequently placed a red flag on the resulting persistence of the induced warming (Solomon *et al.*, 2010) which can entrench inevitable and highly undesirable consequences.

Vigilance of the Climate System

The climate system is defined as an interactive and multi-component system consisting of the atmosphere, the hydrosphere, the cryosphere, the land surface and the biosphere, all of which are influenced by various external forcing factors, such as the sun. The direct effect of human activities on the dynamics of the climate system is termed as internal forcing (IPCC, 2016).

The vigilant monitoring of the climate system is continuously garnering new empirical evidence of its rapidly changing nature. The future prediction of a positive trend in global warming from increasing atmospheric levels of greenhouse gases (GHGs)[1] is now robust (IPCC AR5, 2014). This change is quickly moving outside the boundaries of human experience, with the occurrence of unpredictable, geographically distant and disastrous events, which are already stressing societies around the world (NOAA, 2016). This does not exclude the fact that cyclic fluctuations of our climate during the past million years did not disrupt salient land and ocean processes. There is now ample scientific evidence to prove that the climate with time widely fluctuated between Ice Ages and warm interim periods, as a result of which there have already been major biological extinctions.

So why is so much concern given to our climate system in view of its perennially

*E-mail: charles.galdies@um.edu.mt

changing nature? Put succinctly, the difference between today's global temperature and the last Ice Age[2] lies only in an increase of 5°C. As a result the answer to this legitimate question is provocatively straightforward; one need only focus on this temperature range and compare it with the present-day rate of temperature increase. The similarity in the variation between the current post-glacial temperature (i.e. pre-industrial global air temperature) and the predicted increase in global temperature up till 2100[3] under moderately high anthropogenic forcing conditions is striking (IPCC, 2013). One may instantly realize that the rate of increase in temperature will occur in just over 100 years as opposed to 7000 years. The airborne CO_2 concentration that has already accumulated in the atmosphere will continue changing the Earth's radiation budget in centuries to come, let alone its expected increased levels.[4]

So far the *global* temperature anomaly of 0.87°C since 1880 is considered to be quite significant. In certain parts of the world, such as the central Mediterranean (Fig. 2.1), this rate of increase has been shown to be higher during recent years (Galdies, 2012).

The latest 'long-term' simulations of future climate were produced by the Coupled Model Inter-comparison Project Phase 5 (CMIP5).[5] Simulations are made of both the 20th century (based on past, natural and anthropogenic forcing) and of the 21st-century climate (based on four assessments called Representative Concentration Pathways – RCPs; van Vuuren *et al.*, 2011). CMIP5 simulations of the global annual mean temperature show that it is expected to increase over time (Fig. 2.2), especially over particular geographical regions. For example, under the RCP 2.6 scenario the annual mean temperature values produced by the HadGEM2-ES

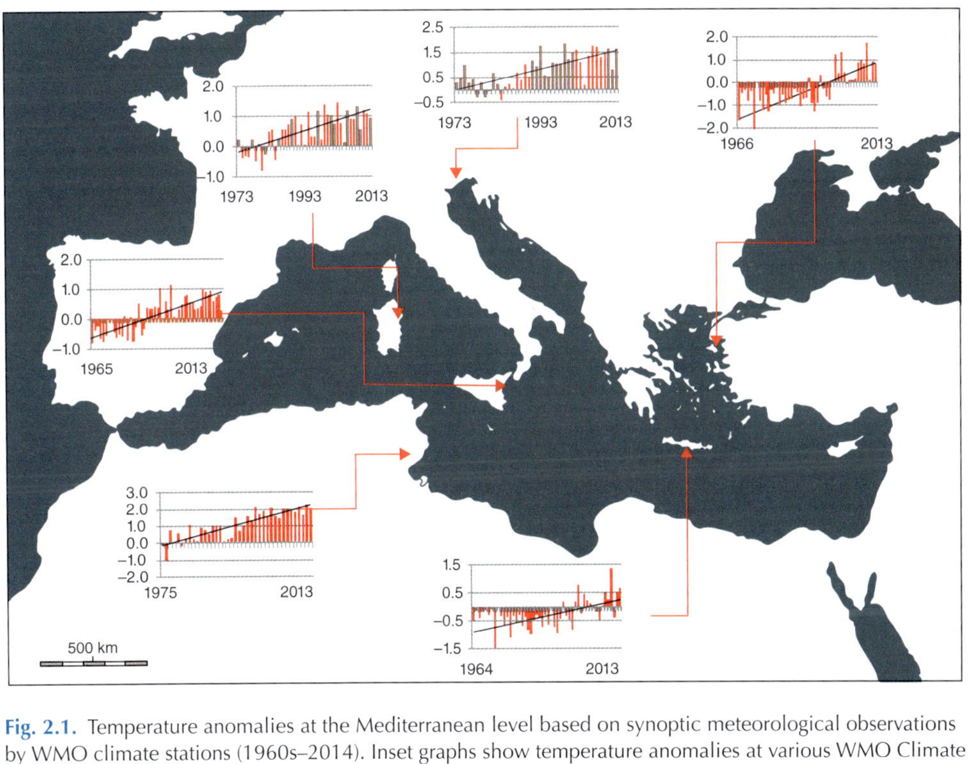

Fig. 2.1. Temperature anomalies at the Mediterranean level based on synoptic meteorological observations by WMO climate stations (1960s–2014). Inset graphs show temperature anomalies at various WMO Climate Centres in the region and thus climatic variability of the region (Galdies, 2015). WMO recommends 30 years as a standard period for the analysis of climate anomalies (Folland *et al.*, 1990), and the climate period between 1961 and 1990 at the individual locations was used as a typical baseline to calculate site-specific temperature anomalies, in conformity with the IPCC and other official Climate Centres.

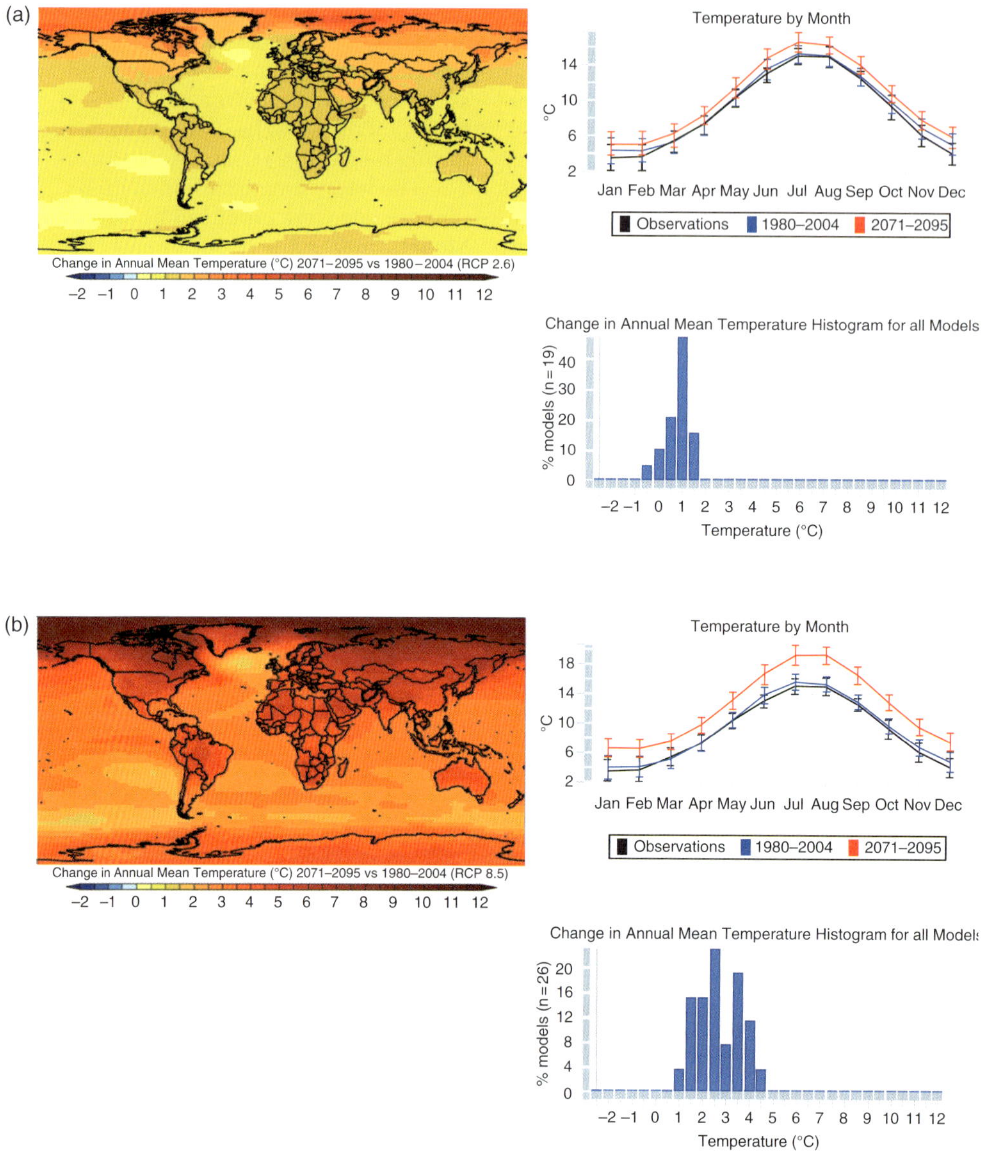

Fig. 2.2. (a) projected change (°C) of the mean temperature surface air temperature at the end of this century (2075–2095) relative to the recent past (1986–2005) for the lower RCP 2.6 (top left) and (b) for the higher RCP 8.5 scenarios (bottom left). Individual figures were generated by the author and adapted using USGS CMIP5 Global Climate Change Viewer. Both projections shown have been produced by UK's HadGEM2-ES climate model. The sets of graphs on the right show the resultant mean monthly temperature produced from all CMIP5 models for the UK (Alder and Hostetler, 2013; Alder *et al.*, 2013).

model are projected to increase especially over the northern hemisphere, unlike that simulated under RCP 8.5, which shows an overall increase of temperature worldwide. Taking the UK as an example, the projected mean CMIP5 model temperature increase is expected to be between 2°C and 8°C (Fig. 2.2) depending on the RCP used by the model.

The crux of the matter lies in the speed at which the current climate is changing in that it is greater than any previous changes detected in what are known as *proxy* indicators of the ancient climate (such as ice- and sediment-cores). Suffice to say that at the time of writing the current global concentration of CO_2 to 405.75 ppm[6] was last seen around 15 million years ago[7] (Tripati *et al.*, 2009). This rapid perturbation to the entire climate system is making it more difficult for both humankind and nature itself to adapt quickly to altered states of its various components. While climate sceptics invoke the role of natural forcing behind such a rapid change, an overwhelming majority of scientists rule out any natural changes in external forcing by the sun, volcanic activity, or variations such as the El Niño-Southern Oscillation (ENSO; EPA, 2016) as being the main culprit of the present change in the climate.

The long-term, irreversible shift of the climate system as a response to continued carbon emissions is now well documented in the scientific literature (IPCC, 2013). A total of five IPCC assessments since 1990 have mainly assessed the projections of climate change and their impact for the 21st century. The group of experts responsible for these two tasks continues to uphold its views on matters relating to the time lag invoked by an increasing carbonized atmosphere, in other words, even if carbon emissions are kept constant or reduced by 2100, atmospheric CO_2 concentrations and surface temperatures would remain high, and that sea level would continue to increase for thousands of years to come. This situation will remain so unless an efficient and large-scale carbon sequestration and storage mechanism is used to immediately capture all of the surplus airborne carbon as well as that produced at source. Carbon sequestration occurs very slowly by a variety of natural processes, which stands in stark contrast with the hundreds of millions of years in order to produce the fossil fuels reserves we have at our disposal today (National Research Council, 2015).

Ironically, the advance in climatology is sometimes impacted by perplexing discoveries that challenge our current understanding of the magnitude of anthropogenic forcing on our climate system. The scientific process is obliged to unravel such conflicting results and review them in the light of further experimentation and understanding. Xie (2016) took up this important task by attempting to explain how the observed recent slowdown in global warming (better known as the 'climate hiatus') came about during periods of increased carbon emissions. This occurred at a time when the rate of global mean temperature decreased by a factor of two during the period 1998–2012 when compared to the previous 15 years. This has of course become embroiled in scientific and political debate (Fyfe *et al.*, 2016).

Xie (2016) documented the required theoretical and practical studies and observations made on Earth's energy balance of the ocean in order to explain this discovery. This thorough analysis has placed the tropical Pacific decadal variability[8] in an important position as a modulator of the climate system and as being the main cause of this climate hiatus; however, its mechanisms still need to be elucidated. For many researchers this unexpected discovery was seen as an opportunity for further research, highlighting the need to continue understanding the climate system as a precondition for its accurate prediction.

Climate Change Impacts and their Implications on Global Security

Anthropogenic climate change has been linked to an increase in the frequency and intensity of a range of disruptive environmental events (IPCC, 2014). Such events are already discernible and causing stress to communities (Fig. 2.3) in the form of damage to local and globally integrated systems that support human well-being, such as public health (Mellor *et al.*, 2016), ecosystem health and services, food supply (Tripathia *et al.*, 2016) and infrastructure. NASA's Global Climate Change Portal (2016) cites the lengthening of growing season, changes in precipitation patterns, increased incidences of droughts and heatwaves, intensification of hurricanes, sea level rise, and de-icing of the Arctic Ocean as distinctive current and future impacts. Another equally significant impact

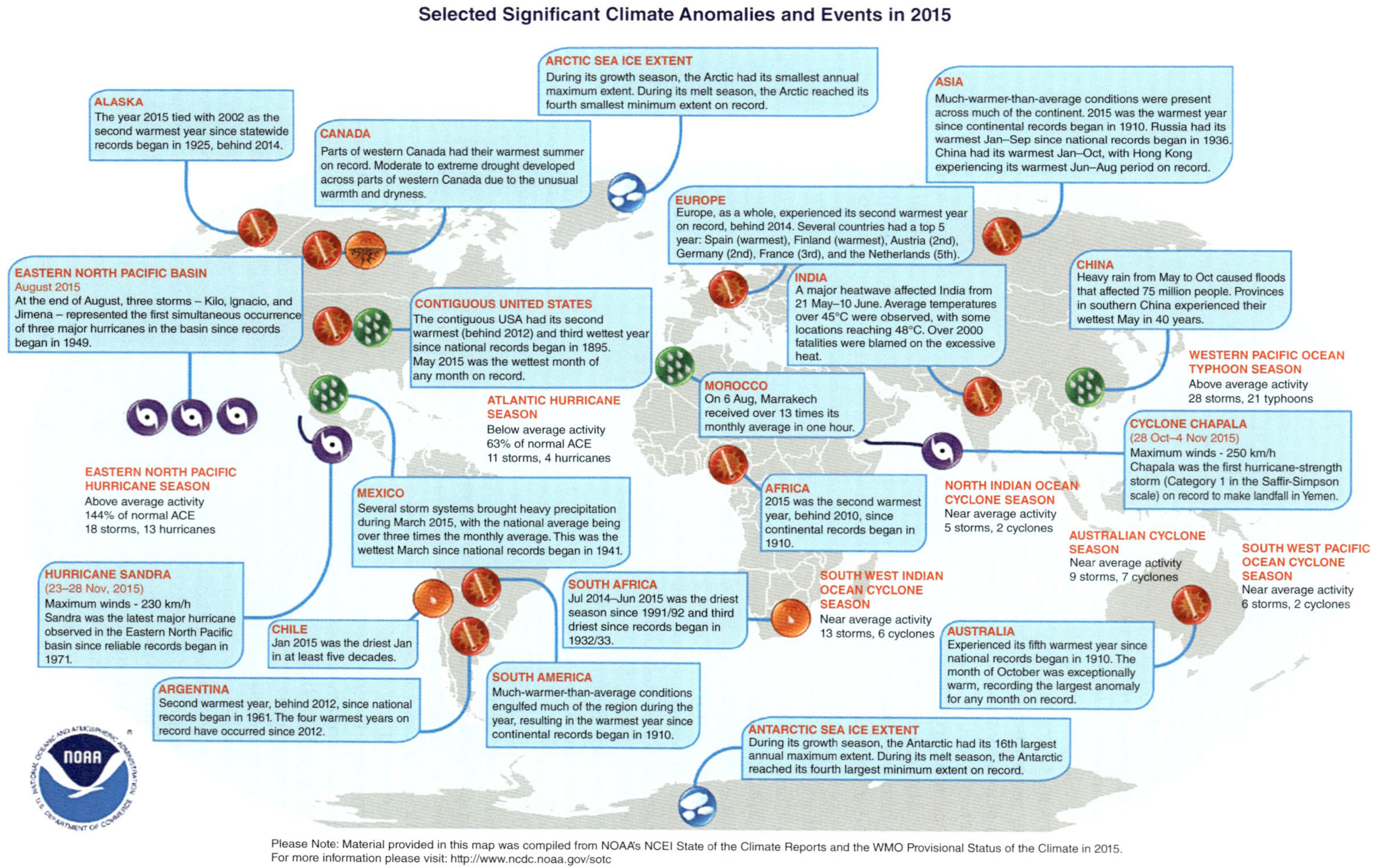

Fig. 2.3. Selected significant climate anomalies and significant events during 2015 (NOAA Global Climate Change portal: Global Analysis – Annual 2015).

is mass extinction, brought about by the inability of the biosphere to adapt so quickly to new climatic conditions. On land, shifts in the timing of the seasons and life-cycle events such as blooming, breeding and hatching are causing mismatches of biological interactions that disrupt patterns of feeding, pollination and other key aspects of food webs (Tripathia *et al.*, 2016). In the oceans, ocean stratification is increasing resulting in less water-mixing between the upper warmer and cooler, deeper waters (Hansen *et al.*, 2016) and linked to an inability for marine life such as phytoplankton found near the surface to access nutrients from below. Phytoplankton constitutes the very foundation of the ocean food web. Readers should refer to additional sources of information for additional impacts such as ocean acidification, coral bleaching and increased incidence of vector-borne diseases.

Adaptive governance requires adequate information on the nature of these impacts in order to respond effectively to related risks. Significant improvements have been made in quantifying future temperature trends on the basis of the current and future emission scenarios of GHGs, and are now helping policy and decision makers to act accordingly. From this adaptation standpoint, the main requirement needed is the evaluation of expected risks by taking into account a range of possible future climate conditions and their associated impacts, and whether the cost of limiting the harm is acceptable. However, due to the current limitations in technology and knowledge of the climate system it is still not possible to predict the precise timing, magnitude and location of these events from decades in advance; there is instead considerable improvement in the equally useful knowledge concerning their nature, risk and social vulnerability.

A timely publication by National Research Council (2013) describes how the initial impact of a warming climate takes the form of disconnected clusters of extreme events. However, with time these sporadic extreme impacts will start taking their toll on integrated socio-economic and environmental systems, which can potentially trigger an internal cascade reaction of negative events with yet unknown consequences if their exposure and susceptibility are sufficiently high and repetitive, and our response is inadequate. Under such situations, national security might be better off if it applies a scenario-based approach of altered integrated systems rather than focus on *isolated, individual extreme* events, which perhaps are more predictable. In this context, experts argue that it would now be more appropriate to consider security risks that are able to disrupt, even at low but chronic levels, the internal linkages of global systems rather than simply prioritize risk on the basis of the magnitude of these extreme events. Therefore understanding how integrated systems are internally connected and how vulnerable they are to impacts of climate change is without a shadow of doubt the most important aspect of climate change from a national security point of view. Incidentally, this approach has so far received very little scientific attention (National Research Council, 2013).

Current Quick Fixes

Despite warnings of impending climatic extreme events, fossil fuels remain the world's primary energy source. Economic and technological progress continue to determine significantly CO_2 emissions (Kais and Sami, 2016). Solace in minimizing the effects of climate change is being sought in the slow but continuous application of mitigation measures and by the use of advanced technologies that are able to harness energy from renewable sources. This strategy started since UNFCCC came into force in 1994 and is periodically revisited at the Conferences of the Parties (COP) to the UNFCCC by means of lengthy negotiations aimed at keeping to the pre-determined emissions goals till 2100. Such targets are indeed very important and are considered as tangible steps towards shifting towards clean technologies, energy efficient processes and a low carbon economy.

However, this strategy fails to recognize the nature of the threat posed, the recognition of which many top climatologists ascribe as being too little, too late. Clark *et al.* (2016) argue that assigning emission targets and introducing new technologies is short-sighted and is unlikely to have a major impact on emissions, especially if the general 'political' perspective of anthropogenic climate change continues to limit itself till 2100.

Climatologists view this perspective as a political and strategic gridlock, which in turn myopically identifies the climatic problem as being only relevant to the next 100 years, and that the expected negative impacts of a changing climate can be reversed if cuts (which some consider to be not overly drastic) are made to current rates of emissions. Consequently, the current proposals and agreements to drastically curb emissions remain limited in scope and effectiveness simply because they are short-term in nature and do not take into account the lag of the climate system to respond to the *current* levels of GHGs in the atmosphere (Clark *et al.*, 2016). Even if we were to entertain the idea of keeping to the current rates of GHG emissions,[9] the climate would still continue to warm up well beyond the 0.87°C already observed and capable of drastically changing its dynamics for the next 10,000 years.

With the help of historical data going back hundreds of thousands of years and the latest climate models, scientists are now able to obtain highly probable long-term scenarios even if their precise timing remains uncertain. Their findings point towards the need for humankind to act fast and bring about the necessary change by 2040.

An interesting study that is hot off the press is a study made by Hansen and 18 other co-authors published this year in the journal *Atmospheric Chemistry and Physics*. They used climate simulation models, palaeo-climatic data from the past 120,000 years and modern observations to study the effect of growing ice melt from Antarctica and Greenland. Continued high GHG emissions until 2100 are predicted to yield an imbalance in the ocean heat budget and to the global thermohaline circulation in such a manner as would result in increasingly powerful storms and an exponential increase in sea level rise. Similar conditions have prevailed some 120,000 years ago, during which huge swaths of polar ice melted down and resulted in a sea level rise of from 6 to 9 m (Hansen *et al.*, 2016). Nothing new so far; sea level rise is a sure thing that is already ongoing. However, this study presents a sea level rise scenario that is diametrically opposed to what we have been hearing so far – one which will only require 50 years to unfold and not centuries. Results also show that this expected melting will be exponential. The resultant costs incurred by coastal communities and megacities are yet unknown.

Realistic visualizations of the impact of sea level rise can be derived from inundation maps modelled for important, low-lying coastal areas. Venice lagoon, for example, represents a focal area of immense art and cultural value because of its unique location, landscape, as well as its cultural history. One can process satellite-derived topographic data (ASTER GDEM) to simulate the projected impact of sea level rise within the lagoon (Fig. 2.4). With a total sea level rise of 5 m (as modelled under RCP 8.5 conditions), much of the current coastline within the lagoon area will already be under water.

Huge tracts of low-lying countries (some of which below sea level) such as Belgium may not survive climate change. On 24 March 2016, Nicholas Kristof from the *New York Times*[10] argued to having a political system that is as sensitive to such risks as it is to homeland security. Clark *et al.* (2016) illustrated how 50–75% of the people living in Indonesia and residing in low-lying areas will eventually be submerged (Fig. 2.5).

Short- Versus Long-term Actions

It is important that policy makers now place what we have learnt so far about the climate system, i.e. since the last Ice Age till the next 10,000 years, into an appropriate climate action. The sole consideration of a 250-year period since pre-industrialization is inappropriate since it is unable to account for all possible perturbations of the climate system as a response to increased rates of CO_2. Climatologists instead are forcing policy makers to

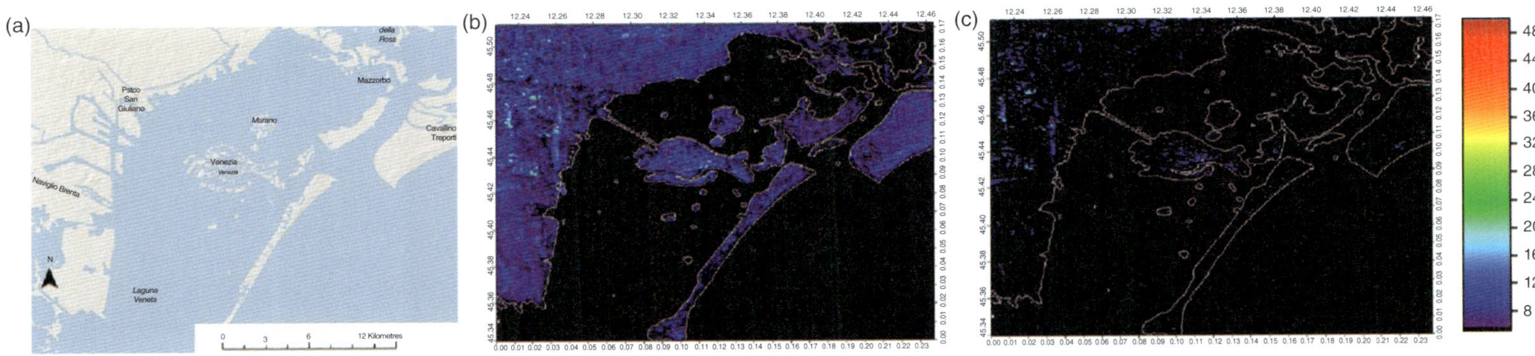

Fig. 2.4. Inundation maps of part of the Venetian Lagoon at varying sea level rise. Legend shows height above sea level in metres. Topographic data are based on ASTER GDEM data (ASTER GDEM is a product of METI and NASA). (a) Current coastline of the Venetian Lagoon, Italy; (b) 5 m above current sea level; and (c) 10 m above current sea level.

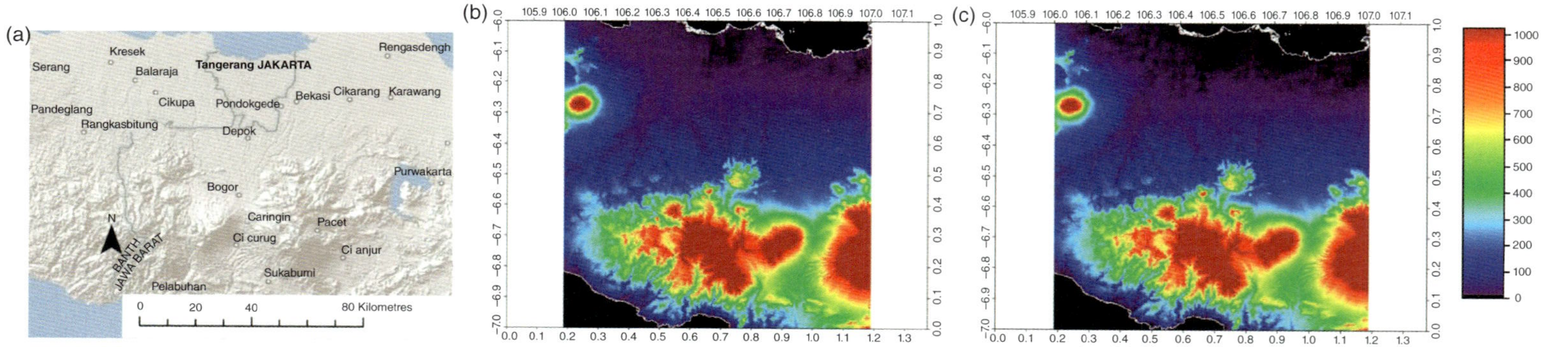

Fig. 2.5. Inundation maps of part of Jakarta area (Indonesia) at varying sea level rise. Legend shows height above sea level in metres. Topographic data are based on ASTER GDEM data (ASTER GDEM is a product of METI and NASA). (a) Current coastline of Jakarta area, Indonesia; (b) 5 m above current sea level; and (c) 10 m above current sea level.

opt for complete de-carbonization of the world's energy systems as soon as possible, which is different from IPCC's efforts to reduce global GHG emissions by at least 50% until 2050 (Clark *et al.*, 2016; Zhou and Wang, 2016). Irrespective of the time frame, a 'carbon quota' on future cumulative CO_2 emissions should already be in place in order to keep the global temperature to 2°C from pre-industrialized temperatures, which would correspond to a global mitigation rate of over 5% per year with a 50% probability of success (Raupach *et al.*, 2014). In quantitative terms, this amounts to a remaining emission quota of 1500 $GtCO_2$ starting from 2015, or of 1300 $GtCO_2$ as of 2020 (Friedlingstein *et al.*, 2014). This quota implies that near future cumulative CO_2 emissions consistent with a given warming limit will be a common and finite resource that must be somehow shared among countries on the basis of equity, and institutionally guided by an international policy, economics and financing.

There are several key opportunities to achieve the expected high mitigation rate. These range from land-based bio-sequestration or bioenergy with carbon capture and storage to the use of low- and zero-carbon economy. Interim opportunities that limit demand for goods and services that require energy as well as improving energy efficiency should be sought, such as: (i) changes in abatement technologies, fuel quality and fuel switching; (ii) changes in the structure and efficiency of the energy systems; and (iii) changes of the total economic activity (Andreoni and Galmarini, 2016). Governments have a pivotal role to play in influencing key stakeholders through effective policies and incentives addressing both private sector investments and consumer behaviour, backed by complementary policies.

The 2015 Paris COP 21 merits a special mention. For the first time, producers of 20% of the world's oil and gas (including BP, Shell, Saudi Aramco and Total, among others) expressed their willingness to assist in the 2°C temperature limit by committing themselves to reduce the GHG intensity of the global energy mix. This will be done through improved efficiency of their own production by giving preference to natural gas over coal, by investing in carbon capture and storage as well as through the harvesting of renewable energy.

A paradigm shift is needed in order to implement the COP 21 agreement's ambition (Obergassel *et al.*, 2016) and its fate is linked to the decoupling of economic growth and the use of fossil fuels as early as possible. However, critics of the Paris Agreement expressed that the financial part of this agreement is weak and does not go beyond what has already been agreed at the Copenhagen Summit in 2009 to mobilize an annual US$100 billion of financial flows by 2020 and beyond. Its effectiveness depends on whether the momentum gained during the December 2015 Conference can be translated into a major political force. This depends entirely on the willingness rapidity of nations to fulfil their declared pledges.

Conclusion

So far the journey from the 1997 Kyoto Protocol to the Paris 2015 Agreement has been problematic and more work remains on the implementation of binding agreements. Decoupling economic growth from continued carbon emissions is an ambitious objective at all three national, international and supranational levels (OECD, 2013). This chapter discussed the reason why there should be a drastic reduction of GHG emissions. Environmental economists support a system of 'negative emissions' up till 2050 in order to limit warming to 2°C over pre-industrial levels by the end of the century (Zhou and Wang, 2016). Huge capital investment in large-scale technologies such as carbon sequestration and storage, assisted by massive harnessing of renewable energy, could make this possible. At the same time, another important fact has to be considered in all major decisions: an ever-growing global population that needs food, energy, access to basic necessities, and an environment that promotes economic growth and social justice.

Notes

[1] GHGs are referred to as positive forcing agents because of their ability to shift the planet's energy balance towards the higher side of internal energy. The forcing power of every GHG can be calculated on the basis of their atmospheric levels over time and on their energy transfer through the atmosphere.

[2] The level of atmospheric CO_2 was higher in the distant past, with higher global temperatures and sea level. The CO_2 level in the atmosphere reached today's levels some 3–5 million years ago, a period when global average temperature is estimated to have been about 2–3.5°C higher than in the pre-industrial period. The level of CO_2 may have even reached 1000 ppm around 50 million years ago, linked to a global average temperature of around 10°C higher than the current one. Under those conditions, the sea level was at least 60 m higher than it is today (Clark *et al.*, 2016).

[3] Currently the temperature anomaly for the period 1880–2015 stands at 0.87°C when compared to the 1951–1980 baseline reference. http://climate.nasa.gov/system/internal_resources/details/original/647_Global_Temperature_Data_File.txt (accessed 17 April 2016).

[4] The pathway with the highest GHG emission (i.e. RCP 8.5) represents 'business as usual' – strong economic development for the rest of this century, driven primarily by dependence on fossil fuels, where the concentration of airborne CO_2 exceeds 1000 ppm CO_2 eq. (Riahi *et al.*, 2011).

[5] CMIP5 is an international climate model inter-comparison exercise involving more than 20 modelling groups and over 40 global models. It was set up to provide a basis for coordinated climate change experiments for IPCC's assessment reports through the World Climate Research Programme (WRCP). Its objective is to provide projections of future climate change on a time scale up until 2100 and beyond, and in doing so, understand some of the factors responsible for differences in model projects. http://regclim.coas.oregonstate.edu/visualization/gccv/cmip5-global-climate-change-viewer/index.html (accessed 17 April 2016).

[6] www.esrl.noaa.gov/gmd/ccgg/trends/global.html (accessed 17 April 2016).

[7] Here it is important to keep in mind the uncertainties embedded in the techniques used to understand our ancient climate. A high CO_2 level could well have been one of the other key factors in controlling the palaeo-climate.

[8] The Pacific Decadal Oscillation (PDO) is a pattern of Pacific climate variability similar to the El Niño–Southern Oscillation (ENSO) in character, but with a duration of 20 to 30 years, in contrast to ENSO's phase of around 6 to 18 months. The PDO consists of a warm and cool phase, which alters upper level atmospheric winds. It has been found that shifts in the PDO phase can have significant implications for global climate and weather. Experts also believe that the PDO can intensify or diminish the impacts of ENSO (Xie, 2016).

[9] An estimated 374 billion t of carbon has been released to the atmosphere since 1751. Half of the CO_2 emitted by fossil fuels has occurred since the mid-1980s. The 2011 global fossil-fuel carbon emission was 9449 million t of carbon. This represented so far a record high and was 3.4% higher than the global 2010 emissions. These are the latest figures released in the public domain (Boden *et al.*, 2015).

[10] www.nytimes.com/2016/03/24/opinion/terrorists-bathtubs-and-snakes.html?ref=opinion (accessed 17 April 2016).

References

Alder, J.R. and Hostetler, S.W. (2013) CMIP5 Global Climate Change Viewer. US Geological Survey. Available at: http://regclim.coas.oregonstate.edu/visualization/gccv/cmip5-global-climate-change-viewer/index.html (accessed 5 June 2017).

Alder, J.R., Hostetler, S.W. and Williams, D. (2013) An interactive web application for visualizing climate data. *Eos Transactions American Geophysical Union* 94, 197–198. DOI: 10.1002/2013EO220001.

Andreoni, V. and Galmarini, S. (2016) Drivers in CO_2 emissions variation: a decomposition analysis for 33 world countries. *Energy* 103, 27–37.

Archer, D. (2005) Fate of fossil fuel CO_2 in geologic time. *Journal of Geophysical Research* 110, C09505. DOI:10.1029/2004JC002625.

Boden, T.A., Marland, G. and Andres, R.J. (2015) *Global, Regional, and National Fossil-Fuel CO_2 Emissions.* Carbon Dioxide Information Analysis Center, Oak Ridge National Laboratory, US Department of Energy, Oak Ridge, Tennessee. DOI: 10.3334/CDIAC/00001_V2015.

Clark, P.U., Shakun, J.D., Marcott, S.A., Mix, A.C., Eby, M. *et al.* (2016) Consequences of twenty-first-century policy for multi-millennial climate and sea-level change. Perspective. *Nature Climate Change*. DOI: 10.1038/NCLIMATE2923.

EPA (2016) Available at: www3.epa.gov/climatechange/science/causes.html (accessed 03 April 2016).

Folland, C.K., Karl, T.R. and Vinnikov, K.Y.A. (1990) Observed climate variations and change. In: Houghton, J.T., Jenkins, G.J. and Ephraums, J.J. (eds) *Climate Change: The IPCC Scientific Assessment*. Cambridge University Press, Cambridge.

Friedlingstein, P., Andrew, R.M., Rogel, J., Peters, G.P., Canadell, J.G. *et al.* (2014) Persistent growth of CO_2 emissions and implications for reaching climate targets. *Nature Geoscience* 7, 709–715.

Fyfe, J.C., Meehl, G.A., England, M.H., Mann, M.E., Santer, B.D. *et al.* (2016) Making sense of the early-2000s warming slowdown. *Nature Climate Change* 6, 224–228. Available at: http://dx.doi.org/10.1038/nclimate2938 (accessed 5 June 2017).

Galdies, C. (2012) Temperature trends in Malta (central Mediterranean) from 1951 to 2010. *Meteorology and Atmospheric Physics*. DOI: 10.1007/s00703-012-0187-7.

Galdies, C. (2015) Potential future climatic conditions on tourists: a case study focusing on Malta and Venice. *Xjenza Online* 3, 6–25.

Hansen, J., Sato, M., Hearty, P., Ruedy, R., Kelley, M. *et al.* (2016) Ice melt, sea level rise and superstorms: evidence from paleoclimate data, climate modeling, and modern observations that 2°C global warming could be dangerous. *Atmospheric Chemistry and Physics* 16, 3761–3812. DOI: 10.5194/acp-16-3761-2016.

IPCC (2013) *Summary for Policymakers in Climate Change. The Physical Science Basis* (eds Stocker, T.F. *et al.*). Cambridge University Press, Cambridge, pp. 3–29.

IPCC (2014) *Climate Change 2014: Impacts, Adaptation, and Vulnerability*. Available at: www.ipcc.ch/pdf/assessment-report/ar5/wg2/WGIIAR5-TS_FINAL.pdf (accessed 14 April 2016).

IPCC (2016) *Climate Change 2007: Working Group II: Impacts, Adaptation and Vulnerability. Glossary*. Available at: www.ipcc.ch/publications_and_data/ar4/wg2/en/annexessglossary-a-d.html (accessed 7 April 2016).

Kais, S. and Sami, H. (2016) An econometric study of the impact of economic growth and energy use of carbon emissions: panel data evidence from 58 countries. *Renewable and Sustainable Energy Reviews* 59, 1101–1110.

Mellor, J.E., Levy, K., Zimmerman, J., Elliott, M., Bartram, J. *et al.* (2016) Planning for climate change: the need for mechanistic systems-based approaches to study climate change impacts on diarrheal diseases. *Science of the Total Environment* 548–549, 82–90.

NASA Global Climate Change Portal (2016) Available at: http://climate.nasa.gov (accessed 14 April 2016).

National Research Council (2013) Climate and social stress: implications for security analysis. Committee on assessing the impacts of climate change on social and political stresses. In: Steinbruner, J.D., Stern, P.C. and Husbands, J.L. (eds) *Board on Environmental Change and Society, Division of Behavioral and Social Sciences and Education*. The National Academies Press, Washington, DC, p. 252.

National Research Council (2015) *Climate Intervention: Carbon Dioxide Removal and Reliable Sequestration*. The National Academies Press, Washington, DC. DOI: 10.17226/18805.

NOAA (2016) National Centers for Environmental Information, State of the Climate: Global Analysis for Annual 2015. Published online January 2016. Available at: www.ncdc.noaa.gov/sotc/global/201513 (accessed 26 October 2016).

Obergassel, W., Arens, C., Hermwille, L., Kreibich, N., Mersmann, F. *et al.* (2016) *Phoenix from the Ashes – An Analysis of the Paris Agreement to the United Nations Framework Convention on Climate Change*. Wuppertal Institute for Climate, Environment and Energy, p. 54.

OECD (2013) *Climate and Carbon Aligning Prices and Policies*. OECD Environment Policy Papers. ISSN2309-7841 (online), p. 57. DOI: 10.1787/23097841.

Raupach, M.R., Davis, S.J., Peters, G.P., Andrew, R.M., Canadell, J.G. *et al.* (2014) Sharing a quota on cumulative carbon emissions. *Nature Climate Change*. Perspective. Published Online: 21 September 2014. DOI: 10.1038/NCLIMATE2384.

Riahi, K., Rao, S., Krey, V., Cho, C., Chirkov, V. *et al.* (2011) RCP 8.5 – A scenario of comparatively high greenhouse gas emission. *Climate Change* 109, 33–57.

Solomon, S., Daniel, J.S., Sanford, T.J., Murphy, D.M., Plattner, G.K. *et al.* (2010) Persistence of climate changes due to a range of greenhouse gases. *Proceedings of the National Academy of Sciences USA* 107, 18354–18359.

Tripathia, A., Tripathi, D.K., Chauhana, D.K., Kumarc, N. and Singh, G.S. (2016) Review Paradigms of climate change impacts on some major food sources of the world: a review on current knowledge and future prospects. *Agriculture, Ecosystems and Environment* 216 (2016), 356–373.

Tripati, A.K., Roberts, C.D. and Eagle, R.A. (2009) Coupling of CO_2 and ice sheet stability over major climate transitions of the last 20 million years. *Science* 326(5958), 1394–1397. DOI: 10.1126/science.1178296.

van Vuuren, D.P., Edmonds, J., Kainuma, M., Riahi, K., Thomson, A. *et al.* (2011) The representative concentration pathways: an overview. *Climatic Change* 109, 5–31. DOI: 10.1007/s10584-011-0148-z.
Xie, S.-P. (2016) Leading the hiatus research surge. *Nature Climate Change. News and Views* 6. Available at: www.nature.com/nclimate/journal/v6/n4/full/nclimate2973.html (accessed 5 June 2017).
Zhou, P. and Wang, M. (2016) Carbon dioxide emissions allocation: a review. *Ecological Economics* 125, 47–59.

3 Integrated Coastal Zone Management: Policy Evolution and Effective Implementation?

M.R. Phillips*
Coastal and Marine Research Group, University of Wales Trinity Saint David, Swansea, Wales, UK

Abstract

Although progress has been made with Integrated Coastal Zone Management (ICZM), the main challenges to implementation are still inadequate capacity and finance. Effective ICZM provides an evidence-based framework for decision makers to ensure activities such as coastal tourism are protected against the consequences of climate change. In England and Wales, Shoreline Management Plans (SMP) rely on assessments that are large scale along whole stretches of coastline, while smaller scale studies are often not undertaken. Consequently, shoreline management policies of 'managed retreat' and 'no active intervention' in response to climate change at Fairbourne will eventually result in the loss of approximately 400 homes at a conservative cost of £60 million. This does not include business and infrastructure losses and damage to the coastal tourism industry. Therefore, before making decisions that could have significant adverse effects on coastal communities, detailed measurements and analyses over an agreed time frame should be undertaken along threatened frontages. Change can then be linked to sea levels and wind/wave data with sea level rise and storm surges superimposed for more accurate predictions to give a more realistic timescale of critical change (or may even result in a change of decision). Although evidence gathering has financial implications, costs are little in comparison to relocation.

Introduction

Over the years coastal zones have become increasingly important, mainly because they encompass the range of anthropogenic activities. Ketchum (1972) identified six major spheres of human activity in the coastal zone, and these still hold true: residency and recreation; industrial and commercial; waste disposal; agricultural, aquaculture and fishing; conservation; and military and strategic. Although coastal zones are home to a range of activities that are sources of national income, due to inadequate planning in some regions these activities have contributed to a degradation of built and natural coastal environments. Shorelines are also highly vulnerable to natural disasters and consequences of climate change, while anthropogenic

*E-mail: mike.phillips@uwtsd.ac.uk

activities have negatively affected coastal zones. Coastal zones provide food for more than 1 billion people (McGinn, 2002), are barriers to natural hazards (Varghese *et al.*, 2008) and provide recreational activities for local inhabitants to earn a livelihood (Lau, 2005). Costanza *et al.* (2014) estimated coastal zones give a combined global annual value of goods and services of approximately US$22 trillion and, consequently, conflicting stakeholder interests, concerns, ecosystem limitations and resource constraints are prevalent (Ahmed *et al.*, 2014). Therefore, authorities are faced with managing highly complex environments that are subject to natural and anthropogenic pressures. Management strategies are complicated by population increases and economic migration to coastal areas as a result of increased tourism, recreation, residential and industrial development and urban expansion (Goble *et al.*, 2014). With coastal pressures increasing worldwide (Lopez y Royo *et al.*, 2009), and populations and development placing significant stresses on coastal resources (Leatherman, 2001), there is a need to reconsider how coastal zones are managed, with policy makers defining policies and legislative instruments to manage effectively this complex space (Nobre, 2011). Erosion and climate change impacts on tourism infrastructure were identified by Phillips and Jones (2006) and Jones and Phillips (2011), who highlighted issues that needed to be addressed. Although these are now widely recognized, there is still no clear strategy for protecting and managing coastal resources while accommodating socio-economic activities, such as tourism, against the consequences of climate change and sea level rise. Consequently, this chapter will consider climate change, coastal processes and shoreline management and, using international examples, show the evolution of ICZM. ICZM strategies from high level to operational implementation will then be focused through the lens of a case study in Fairbourne, Wales, UK, which is a popular tourist destination and where there have been unintended consequences of a regional ICZM strategy. The chapter concludes with a discussion of lessons learned and recommendations for effective ICZM strategies in response to climate change.

Climate Change, Coastal Processes and Shoreline Management

Climate change, weather patterns, impacts on coastal defences and the fate of coastal communities highlights the need for effective use of resources. One of the major physical impacts of sea-level rise is erosion, particularly along the open coast and this will leave coastal infrastructure even more vulnerable to storm waves in future. Therefore, sea-level rise coupled with storm surges and high tides will pose severe problems for shoreline management and the economy, which will require effective management for resolution. Obviously, policy and implementation will be influenced by available techniques for defending the coast with due consideration being given to their cost and sustainability.

Through natural phenomena such as tides, waves and winds, and socio-economic activities, coastal areas undergo transformation. Coastal erosion and the disappearance of beaches as a result of wrong management decisions reflect a lack of effective legislation and often require a series of engineering solutions (van der Salm and Unal, 2003). Because of the problems associated with hard engineering in the near-shore zone, not least cost and maintenance, alternative soft engineering techniques that work in conjunction with natural processes are increasingly being used. These include the construction of multifunctional artificial reefs (Ng *et al.*, 2014), which reduce effective depth offshore, thereby reducing wave power and erosion of the shoreline, and beach nourishment, a soft engineering solution, both of which are often used for the benefit of the tourism industry. Multifunctional artificial reefs provide surfing amenity whilst protecting the shoreline from erosion, and following beach nourishment projects, the new wider beach serves as shore protection from the impacts of storms (Benassai *et al.*, 2001). Both techniques address problems associated with climate change while at the same time provide recreational benefits and new tourism opportunities.

Anthropogenic modification of natural coastal areas is not new, but advances in engineering capability have significantly increased consequences on a global scale (Kay and Alder, 2005). Human activities are also increasingly in conflict with one another as well as with longer-term natural processes. In 2007, a fifth of the enlarged EU's coastline was already severely affected by erosion, with coastlines retreating by between 0.5 and 2 m year^{-1} and by 15 m in a few dramatic cases (Europa, 2016). A worldwide tendency to coastal erosion (Cipriani *et al.*, 2004) has been locally aggravated by some of the very strategies implemented to reverse the pattern (Gillie, 1997; Weerakkody, 1997), while Europa (2016) reported that for the 875 km of European coastlines that had started to erode within the previous 20 years, 63% were located <30 km from coastal areas altered by recent engineering works.

An evaluation of the present-day status of a coastline is fundamental in deciding whether to actively manage or to refrain from intervention. Granja (2001) highlighted that the lack of coastal zone management has irreversibly contributed to the gradual degradation of Portugal's natural heritage, including the landscape that is one of its key elements. She further argued that promoting conservation implies an understanding of geoforms, their generation and longevity and their relationship with the ecosystems dependent on them. Therefore, to promote conservation and management of the coastal zone, it is important to understand not only present-day factors and processes, but also those that have been active in the recent geological past. Granja and Carvalho (2000) highlighted an objection to fundamental conservation and implied that it was unrealistic to believe that all coastlines can be protected from impending sea level rise and subsequent coastal erosion. They suggested managed retreat and selective conservation of parts of the coast that are important to society and to use technological developments, where possible, to halt inland beach migration. Measures should be taken to attenuate or reduce the effects of erosion, but where irreversible loss is evident, society must be persuaded to undertake a gradual and planned retreat from the waterfront. Beach management and sea defences can be justified on socio-economic grounds for a particular region and community, which depend on the beach, even though these may cause the acceleration of coastal erosion further along the coastline. Consequently, the use of risk assessment tools in conjunction with benefit–cost analyses could be used to justify policies of 'managed retreat' or 'no active intervention' where irreversible loss is evident (Granja and Carvalho, 2000).

Reeve and Spivack (2001) argued that strategic planning for coastal management relies implicitly upon understanding the physical processes responsible for shoreline change. Sorensen (1997) argued that it was necessary to determine and apply coastal management lessons derived from cross-national comparisons and that lessons learned are vital for effective shoreline protection strategies. As the coastal zone has evolved from many natural and anthropocentric factors and processes, it is logical that management strategies focused on these should be integrated. ICZM is a cyclic process of problem recognition, planning, implementation and monitoring, which includes all stakeholders and institutional mandate, capacity, commitment and available finance. It is important to accept there is potential conflict when deciding who has most rights and, therefore, policy and implementation will be influenced by strategic and socio-economic aspects in conjunction with available techniques for defending the coastline.

Integrated Coastal Zone Management

Coastal management is constantly evolving (Kay and Alder, 2005) and there is a long history of attempting to create a political setting in which coastal communities can collaboratively participate in the decision-making process (Bremer and Glavovic, 2013). Hence the need for ICZM, which is a dynamic, continuous process designed to promote sustainable management of coastal zones within limits set by natural dynamics and carrying capacity (EC, 1999). ICZM became a globally utilized approach for management of the coast and its resources, predominantly emerging in response to failures of sectoral management (Cicin-Sain and Knecht, 1998), and is considered essential for sustainably managing coastal environments (EU, 2001). As well as socio-economic benefits, however, environmental considerations are important and without an element of sustainability and measurable indicators within an integrated coastal management strategy, deterioration of the coastal environment is likely to persist (Charlier and Bologa, 2003). Therefore, it is important to identify whether a coastline has high environmental and/or socio-economic value to enable assessment of change and, consequently, there is a need for public involvement in coastal management strategies. This is especially the case when assessing consequences of climate change on coastal tourism.

The World Bank's 1996 guidelines for effective ICZM outlined roles and responsibilities for governments, the general public and other coastal stakeholders. They were intended to assist countries initiating ICZM legislation and showed how plans could be implemented, monitored, evaluated and enforced (Post and Lundin, 1996). In 1999, the Intergovernmental Oceanographic Commission (IOC) identified the emergence of coastal megacities and human migration to coastal areas as issues of concern. It recognized the need to increase education, training and public awareness in order to understand interactions between coastal and urban environments (UNESCO, 2000). The European Commission's communication evaluating ICZM found that coastal management policies, legislation and implementation mechanisms had been independently developed on a sectoral basis (CEC, 2007). As a result, the EU ICZM recommendation emphasizes integration across sectors and governance levels, and based on these principles, EU coastal Member States were invited to develop national strategies (Calado *et al.*, 2009). The Commission noted that ICZM tended to focus more on the environment instead of sustainable socio-economic development and concluded this needed more prominence (Defra, 2010). Subsequently, in 2010, Member States were invited to provide a progress update on ICZM implementation between 2006 and 2010 and the national reports were important as a basis to follow up the ICZM recommendation and impact assessment (EC, 2011).

Although many countries have dedicated coastal management legislation, they are faced with implementation and enforcement challenges in response to destruction and degradation of coastal ecosystems. While Glavovic (2006) thought it to be an elusive concept at best, Beeharry *et al.* (2014) recognized that over time ICZM has evolved. This has led to the development of specific policies and legislation to balance the coastal demands of socio-economic activities with ecosystem functionality. Despite many international instruments related to global coastal management and collaboration, and progress in formulation and adoption of these instruments, implementation has been primarily hindered by inadequate human resource capacity and financial constraints (Hewawasam, 2002). According to Goble *et al.* (2014), coastal management has largely been fragmented in South Africa, with different departments and spheres of government often having various overlapping roles and responsibilities. This often leads to conflict and lack of action and there is a need to change the way that coasts are managed, finding innovative ways to support coastal decision making. In Portugal, ICZM stimulated a more integrative and participative approach, with the government enacting legislation for coastal zone protection, recovery, management and governance. Its principles were incorporated into sectoral policies and strategies, which facilitated an ICZM framework (Alves *et al.*, 2011). There was also an increase in funding to help deliver ICZM and this is seen as critical for its success and clearly shows the national and regional importance of the coastal zone. ICZM is seen as a means of reducing risk to people and assets whilst improving the coastal environment and socio-economic development.

ICZM in England and Wales is implemented through the production of Shoreline Management Plans (SMP) (EA, 2015). These plans recognize that natural processes do not follow the human-defined land-based boundaries of administrative authorities, e.g. county councils, but are based on natural boundaries such as sediment cells. Within the length of these sediment cells, coastal processes are similar. The Welsh government has an ICZM policy that encourages all organizations with an interest in the coastline of Wales to work together to formulate policies and plans that will lead to effective economic and sustainable management of the Welsh coastline. SMPs are a vital component of this process in providing key information to all the other organizations. Earlier SMPs, for example Bullen (1993) and SBCEG (1999), have been refined to produce current versions such as the West of Wales SMP 2 (Royal Haskoning, 2011).

SMPs consider how a section of coastline is to be managed over epochs, e.g. 25 years, 50 years and 100 years, to address issues such as flooding and/or erosion. Decisions consider predictions of climate change consequences which are linked to socio-economic use, e.g. coastal tourism, etc. (EA, 2015). Therefore, the coastline is divided into units for which management options are assessed under four criteria:

- No active intervention – No planned investment to defend coastline at a particular location against flooding or erosion. This even includes areas where there are existing coastal defences.
- Hold the line – This section of coastline will be defended, or continue to be defended, by constructing artificial defences to maintain the shoreline position. This includes modifying types of coastal defence.
- Managed realignment or managed retreat – This section of the coastline will be abandoned over a period of time and socio-economic assets will be relocated or lost to the sea.
- Advance the line – New coastal defences are constructed on the seaward side of the existing shoreline or coastal defences. This can include beach nourishment or hard engineering.

Management options are developed by local authorities and the Welsh government (in Wales)/Environment Agency (in England) working together in Coastal Groups. SMPs are agreed following engagement with all interested stakeholders, but even then, funding has to be secured and may require contributions from those who benefit. It is possible for the management strategy of small sections of coastline within a management unit to differ from the overall strategy, e.g. due to tourism infrastructure a policy of 'hold the line' is applied whilst the policy for the remainder of the coastline remains as 'no active intervention'. Similarly, policy options may change through the epochs due to climate change effects such as sea level rise, and a coastline goes from 'managed retreat' to 'no active intervention' (EA, 2015).

Case Study: Fairbourne, Wales, UK

Following severe infrastructure damage from storms in January 2014, the Welsh government increased its flood and coastal erosion budgets (Davies, 2014). Despite it being a time of austerity with cutbacks throughout the public sector, the Welsh government set aside £240 million for flood and coastal defences over a 5-year period. Davies (2014) also allocated more than £50 million from the European Regional Development Fund to reduce the risk to over 7000 homes and businesses across Wales. However, with the UK voting to leave the European Union, it is not clear how certain it is that this funding allocation will remain. Sea level rise and climate change will mean that such severe storms will become more frequent and have increasing shoreline impacts.

In Beatty *et al.*'s (2010) report, the coastal tourism industry in England and Wales directly supports some 210,000 jobs, an average year-round figure spread across six sectors of the local economy. Figures were determined from an analysis of 121 resorts, including principal seaside resorts, smaller seaside towns, sub-parts of some larger towns and important holiday parks. They further argued that additional jobs are supported through the supply chain and via multiplier effects, making the total employment directly or indirectly supported far greater. The report concluded that seaside tourism in England and Wales is by any standards a large industry and in terms of employment is comparable to the telecommunications sector and larger than the motor industry, aerospace, pharmaceuticals or steel. Beatty *et al.*'s (2010) results agree with those of Houston (2008), who similarly compared the tourism industry with other US industries and he further showed that coastal tourism represented 80% of the global industry (Houston, 2008). Therefore, coastal tourism makes a major contribution not only to seaside towns, but to the economy as a whole.

With SMPs published and agreed management strategies in place, the next stage is their implementation. However, whilst the over-arching strategy based on four management options is valid, some decisions may not have been thought through. Although managing climate change impacts on coastal communities will on occasions require tough decisions to be taken, then if that is the case, evidence underpinning the decision should be robust. The implementation of ICZM decisions based on future climate change scenarios can have a devastating impact on coastal communities. This is highlighted by the case study of Fairbourne, North Wales (Fig. 3.1). To the north and south are the towns of Barmouth and Tywyn, respectively, which were included in Beatty *et al.*'s (2010) assessment of the seaside tourist industry in England and Wales.

Following publication of the second generation SMP and adoption by Gwynedd County Council, the long-term management policy for Fairbourne (Fig. 3.1) will be 'no active intervention' from 2050 (Royal Haskoning, 2011). From 2025, the SMP recommends 'managed retreat' as the first implementation stage of not defending the Fairbourne shoreline. Defending the shoreline is costly, as shown to the south at Borth, where recently completed multifunctional artificial reef sea defences and ancillary works cost £29 million (Ng *et al.*, 2014). Since publication of the SMP, much media attention has been devoted to the decision of managed retreat at Fairbourne. Indeed, the first time the population of Fairbourne realized this decision had been taken was when a BBC documentary was broadcast in February 2014 (BBC Wales, 2014). The impact was an immediate blight on the community with falling property prices and problems for residents obtaining mortgages and insurance. Defending the coastline has significant political and social ramifications and it is clear that 'managed retreat' and 'no active intervention' decisions will have significant consequences for residents. Therefore, the question that has to be answered is how robust was the underpinning data to support the policies of 'managed retreat' and 'no active intervention' at Fairbourne?

Measuring beach profiles is the methodology adopted by Shoreline Management Partnerships to monitor coastal change. Based upon annual survey regimes, measurements are taken along widely spaced beach profiles that extend from the backshore to mean low water. Research carried out by Thomas *et al.* (2010, 2011a,b, 2012) showed that these annually averaged profiles can provide insights into the long-term health of a beach and results did show correlation between similarly averaged forcing such as wind and waves along with their directional components. However, widely spaced beach profiles, often 1 km or further apart, cannot provide enough detail to enable overall planform change to be accurately calculated, or for assessment of erosive impacts and recovery from storms. Therefore, decisions such as 'managed retreat' and 'no active intervention' are based on coarse-scale data and not the fine-scale robust dataset that should be collected to support such policies. Present coastal defences and strategic management have not provided the best solutions, and an absence of accurate data has resulted in poor investment decisions. With sea level rise and storms becoming more unpredictable, there is a need to focus more effective use of limited resources, but only if decisions are evidence based. For Fairbourne, the socio-economic impacts on the community are devastating.

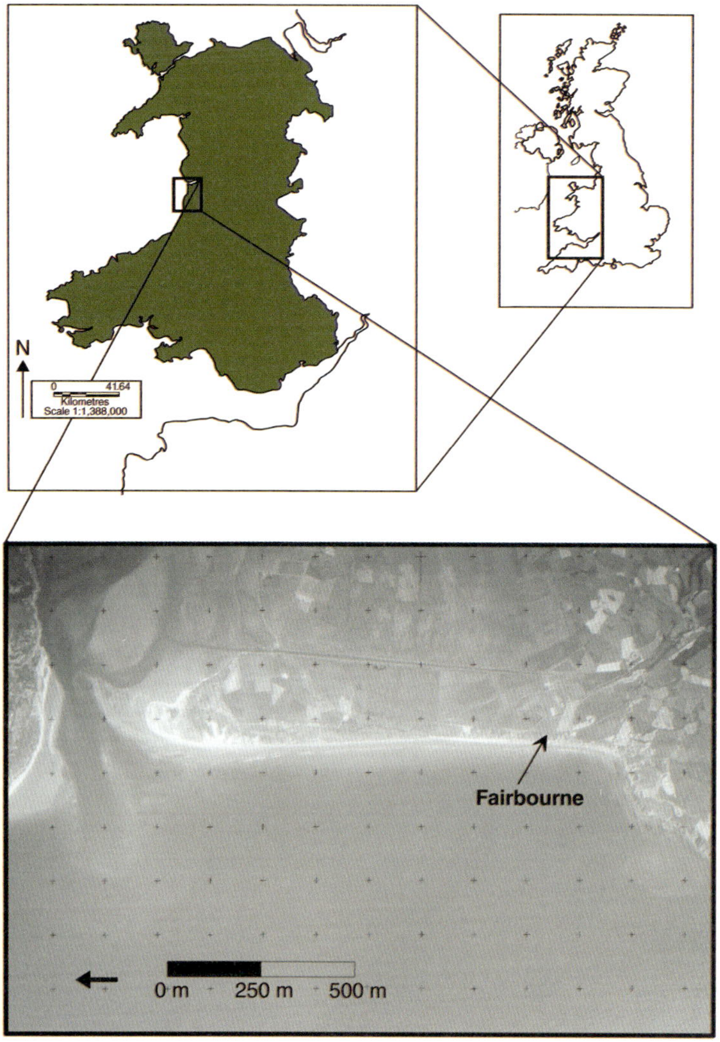

Fig. 3.1. Location of Fairbourne.

Discussion

Alves *et al.* (2013) argued that successful implementation of ICZM needs strong governance models to improve integration of coastal and maritime planning and management. To achieve success, there must be coordination and cooperation between relevant actors and portfolios. An example of how this can be achieved has been demonstrated in Portugal, where cooperation between ministries responsible for ICZM and National Sea Strategies made it possible to share initiatives such as developing a strategic reference frame for high-value economic activities and developing a national programme to monitor coastal systems, biotic communities and environmental quality (Alves *et al.*, 2013). This has addressed concerns such as those raised by Granja and Carvalho (2000) and Granja (2001).

Although over recent years there have been more technical studies and research projects which have increased scientific knowledge and best practice at various national and international levels, further work is needed on coastal behaviour to minimize risk, reduce vulnerability, increase resilience and climate change adaptation. Consequently, ICZM helps identify necessary trade-offs among many competing policy objectives enabling incentives to be introduced for more environmentally sustainable practices. Driven by climate change, coastal development and infrastructure is at risk from flooding and displacement, coastal erosion, increased storm surges and changes in tidal regime and sediments. Therefore, scientific knowledge should be developed to minimize risk and improve climate change adaptation, including the integration of ecosystem services within an ICZM framework to show economic and sustainability benefits. Implementation of ICZM should be based on an action plan where funding sources and lead organizations are identified and progress monitored.

Managers should seek to understand and work with natural processes if ICZM plans are to be effective and long lasting. Identification of local characteristics and/or problems and the socio-economic value of resources are vital for effective management. Therefore, as it has spatial and temporal dimensions, ICZM should be considered through the lens of socio-economic pressures and physical environmental constraints. Spatial measures with respect to ICZM should attempt to identify developments that can influence natural shoreline dynamics, whilst temporal analysis should address timeline variation and causal change, for example, climate change. Data acquisition is often expensive and time consuming but there is a need to collect long-term quantitative data (Phillips, 2008), especially when manging climate change impacts on coastal tourism (Phillips and Jones, 2006).

Coastal management initiatives often fail as local populations have limited understanding of the economic, cultural and aesthetic values that a sustainably managed coast can provide. This is made worse because of the increasing rate at which development impacts are occurring with long-term management strategies not being perceived as relevant in the short-term (Goble *et al.*, 2014). Therefore, ICZM initiatives are needed that create and empower institutions with the best available knowledge, by integrating natural and social sciences into the political processes of decision-making (Bremer and Glavovic, 2013). Managing a coastal zone that is subject to a range of unique natural and anthropogenic pressures requires management intervention (Goble *et al.*, 2014). For ICZM to be effective, capacity and understanding is required by government at all levels and all stakeholders, something that was definitely lacking in the case of Fairbourne. The OECD (2009) recommended that governments need to develop a regulatory framework to enable economic and social development within the coastal environment and this also applies at a regional government level, whereby operational capacity in coastal management is vital for its success.

ICZM strategies agree with those generally suggested for coping with the negative effects of climate change. Difficult decisions will need to be taken irrespective of stakeholder interests and there will be economic consequences (Leatherman, 2001). Policies will therefore need to reflect the complex inter-relationships of all activities. Strategic management involves three stages: (i) analysis, where one seeks to understand the content of a system, the status of existing management, if any, and whether one should take any action; (ii) establishment of what different courses of action are available; and (iii) implementation, where the chosen action is put into effect (Johnson and Scholes, 1988). The coastal zone has evolved from many natural and anthropocentric factors and processes and it is logical that management concepts inherent to these factors should be integrated, i.e. ICZM. Therefore, in the case of Fairbourne, the evidence justifying the decision needs to be made available.

Factors that influence sustainability in coastal management span social, economic, institutional, bio-physical and legal conditions (Christie, 2005). Clear trends of continued human migration to the coast and major growth in coastal tourism have resulted in escalating investment in coastal locations (JHCS, 2000). Global warming and sea level rise are coastal hazards difficult to quantify, but indirect costs will include falling property values and loss or transfer of tourism revenues. As the coastal zone is naturally dynamic and with progress being made in sustainable management practices, there will be a tendency to adopt more soft engineering solutions to work alongside natural coastal processes. It must be remembered, however, that as advocated by van der Weide *et al.* (2001), to best design appropriate mitigating measures the causes of change should be properly analysed and the technical and economic feasibility of mitigation should be evaluated. Furthermore, resort areas depend on facilities and infrastructure and, as argued by JHCS (2000), if tourist perception is negatively affected, then the resorts themselves may be severely affected for many years. Garcia *et al.* (2003) highlighted the importance of the coast as a resource for both visitors and residents and considered the need for integrated management to ensure: erosion control, physical protection of the landscape and implementation of effective urban planning measures to ameliorate tourism pressures. Without effective ICZM prospects for controlling costs and losses would not be good.

Davies (2014) recognized that it was neither affordable nor sustainable to defend the entire Welsh coast and difficult decisions would have to be taken. He promoted a risk-based approach to flood and coastal erosion management, including building coastal resilience and working with others to resolve problems. SMPs are based on limited datasets, which it is argued are insufficient to support strategic decisions such as 'managed retreat'. Many economic activities take place in the coastal zone and it is often where significant and high value infrastructure is located. Economic well-being of coastal communities is critical and there needs to be greater knowledge and understanding of problems informed by robust data. This is especially true when tourism income is threatened. Policies of 'managed retreat' and 'no active intervention' at Fairbourne will result in the loss of approximately 400 homes at a conservative cost of £60 million, without taking into consideration business and infrastructure losses and damage to the seaside tourism industry. A final consideration, if the Cambrian railway which passes behind the town of Fairbourne was to become threatened, would 'hold the line' be the SMP response? If so, 'managed retreat' will only continue to a land-based position where economic importance will change it to 'hold the line'.

Conclusions and Recommendations

Definite progress has been made with ICZM, although there is a difference between countries and the main challenges to implementation include capacity and finance. Effective ICZM will be critical for ensuring coastal tourism is buffered against future challenges of climate change by providing an evidence-based framework for decision makers. In England and Wales, SMPs need an enabling strategy that provides evidence to underpin decision making as current assessment is coarse and not fine scaled. Therefore, detailed analysis should be undertaken over shorter lengths on threatened coastal frontages and should include seasonal beach measurements over preferably 5 years, as the longer the dataset, the more valid the interpretation. Coastal change should be linked to sea levels and wind/wave data over the same period to understand fully coastal processes and shoreline response at a particular location. Subsequently, sea level rise and storm surges can be superimposed for more accurate predictions to be made, with consequences modelled and visualized and this would identify a more realistic timescale of critical change (or may even result in a change of decision). Gathering of such evidence has financial implications, but the costs are little in comparison to relocation costs and loss of equity, as has happened in Fairbourne. It should be noted that although managed retreat will be allowed at specific locations, if erosion continues there will ultimately be a position where the decision will change to 'hold the line', i.e. not allow further loss. That will normally be when critical infrastructure is threatened. Consequently, recommendations for developing a regional ICZM strategy include:

- Understand regional coastal dynamics (wave climate, sediment transport, etc.);
- Assess impacts of climate change (sea level rise, storms, etc.);
- Establish vulnerability;
- Establish realistic buffer zone between development and the sea;
- Use soft engineering solutions working with natural processes;
- Assess economic consequences of decisions such as damage to coastal tourism;
- Educate and include all stakeholders in the decision-making process;
- Learn from mistakes; and
- Understand there will be a cost, whatever the decision.

References

Ahmed, F., Perry, E.C. and Bob, U. (2014) Africa's coasts: natural resource management and conflict-sensitive adaptation. In Bob, U. and Bronkhorst, S. (eds) *Conflict-sensitive Adaptation to Climate Change in Africa*. Climate Diplomacy Series African Centre for the Constructive Resolution of Disputes (ACCORD), Berliner Wissenschafts-Verlag, Berlin.

Alves, F.L., Silva, J.V., Pereira, C.A. and Sousa L.P. (2011) Ten years assessment of ICZM principles applied at a local scale: Ria de Aveiro case study. *Journal of Coastal Research* SI 64, 1311–1315.

Alves, F.L., Sousa, L.P., Almodovar, M. and Phillips, M.R. (2013) Integrated Coastal Zone Management (ICZM): a review of progress in Portuguese implementation. *Regional Environmental Change* 13, 1031–1042.

BBC Wales (2014) *Week In, Week Out: The Last Resort*. BBC TV Documentary, 11 February 2014.

Beatty, C., Fothergill, S., Gore, T. and Wilson, I. (2010) *The Seaside Tourist Industry in England and Wales: Employment, Economic Output, Location and Trends*. Report by Centre for Regional Economic and Social Research, Sheffield Hallam University, 69 pp. ISBN 978-1-84387-324-2.

Beeharry, Y., Makoondlall-Chadee, T. and Bokhoree, C. (2014) Policy analysis for performance assessment of integrated coastal zone management initiatives for coastal sustainability. *APCBEE Procedia*, pp. 30–35.

Benassai, E., Calabrese, M. and Uberti, G.S.D. (2001) A probabilistic prediction of beach nourishment evolution. In: Ozhan, E. (ed.) *Medcoast 01: Proceedings of the Fifth International Conference on the Mediterranean Coastal Environment*. Medcoast, Ankara, 1, pp. 1323–1332.

Bremer, S. and Glavovic, B. (2013) Mobilizing knowledge for coastal governance: reframing the science policy interface for integrated coastal management. *Coastal Management* 41, 39–56.

Bullen (1993) *Coastline Response Study: Worms Head to Penarth Head: Final Report: Vol. 1*. Bullen and Partners, 285 pp.

Calado, H., Ng, K. and Alves, F. (2009) *Would Regional Strategy on Integrated Coastal Zone Management (ICZM) Benefit Autonomous Regions of Azores and Madeira?* 15th Congress of the Portuguese Association for Regional Development.

CEC (2007) *Communication from the Commission 'Report to the European Parliament and the Council: An Evaluation of Integrated Coastal Zone Management (ICZM) in Europe'*. Commission of the European Communities, Brussels, COM(2007) 308 final.

Charlier, R.H. and Bologa, A.S. (2003) Philosophical approach to a sustainable integrated coastal zone management. *Journal of Coastal Research* 19(4), 867–869.

Christie, P. (2005) Is integrated coastal management sustainable? *Ocean and Coastal Management* 48, 208–232.

Cicin-Sain, B. and Knecht, R.W. (1998) *Integrated Coastal and Ocean Management: Concepts and Practices*. Island Press, Washington, DC.

Cipriani, L.E., Wetzel, L., Aminti, D.L. and Pranzini, E. (2004) Converting seawalls into gravel beaches. In: Micallef, A. and Vassallo, A. (eds) *Management of Coastal Recreational Resources – Beaches, Yacht Marinas and Coastal Ecotourism*. ICoD, Malta, pp. 3–12.

Costanza, R., De Groot, R., Sutton, P., van der Ploeg, S., Anderson, S.J. *et al.* (2014) Changes in the global value of ecosystem services. *Global Environmental Change* 26, 152–158.

Davies, A. (2014) Written Statement on Coastal Flooding. Alun Davies (AM) Welsh Government, January, 2014. Available at: http://gov.wales/about/cabinet/cabinetstatements/previous-administration/2014/coastalflooding/?lang=en (accessed 15 August 2016).

Defra (2010) *United Kingdom Report to the European Commission. Implementation of the ICZM Recommendation 2006–2010*. Department for Environment, Food and Rural Affairs, London.
EA (Environment Agency) (2015) *Shoreline Management Plan Policies – What Do They Mean?* Environment Agency. Available at: http://apps.environment-agency.gov.uk/wiyby/134834.aspx (accessed 15 August 2016).
EC (1999) *Towards a European Integrated Coastal Zone Management (ICZM) Strategy: General Principles and Policy Options*. Office for Official Publications of the European Communities, Luxembourg.
EC (2011) *Public Hearing on Integrated Coastal Zone Management 30 May 2011 – Hearing Report*. European Commission, Directorate-General Environment.
EU (2001) EU Focus on Coastal Zones: Turning the Tide for Europe's Coastal Zone. Available at: http://ec.europa.eu/environment/iczm/pdf/2000brochure_en.pdf (accessed 2 August 2016).
Europa (2016) Coastal Zone Policy. Available at: http://ec.europa.eu/environment/iczm/prop_iczm.htm (accessed 2 August 2016).
Garcia, G.M., Pollard, J. and Rodriquez, R.D. (2003) The planning and practice of coastal zone management in Southern Spain. *Journal of Sustainable Tourism* 11(2/3), 204–223.
Gillie, R.D. (1997) Causes of coastal erosion in Pacific Island nations. *Journal of Coastal Research* SI 24, 173–204.
Glavovic, B.C. (2006) The evolution of coastal management in South Africa: why blood is thicker than water. *Ocean Coastal Management* 49(12), 889–904.
Goble, B.J., Lewis, M., Hill, T.R. and Phillips, M.R. (2014) Coastal management in South Africa: historical perspectives and setting the stage of a new era. *Ocean and Coastal Management* 91, 32–40.
Granja, H.M. (2001) Paleoenvironmental indicators from the recent past: a contribution to CZM purposes. In: Ozhan, E. (ed.) *Medcoast 01: Proceedings of the Fifth International Conference on the Mediterranean Coastal Environment*. Medcoast, Ankara 1, pp. 59–70.
Granja, H.M. and Carvalho, G.S. (2000) Inland beach migration (beach erosion) and the coastal zone management (The experience of the Northwest coastal zone of Portugal). *Periodicum Biologorum* 102(1), 413–424.
Hewawasam, I. (2002) *Managing the Marine and Coastal Environment of Subsaharan Africa, Strategic Directions for Sustainable Development*. The World Bank, Washington, DC.
Houston, J.R. (2008). The economic value of beaches – A 2008 update. *Shore & Beach*, 76(3), 22–26.
JHCS (John Heinz III Centre for Science) (2000) *The Hidden Costs of Coastal Hazards: Implications for Risk Assessment and Mitigation*. Island Press, Washington, DC.
Johnson, G. and Scholes, K. (1988) *Exploring Corporate Strategy*. Prentice-Hall, Hemel Hempstead, UK.
Jones, A.L. and Phillips, M.R. (eds) (2011) *Disappearing Destinations: Climate Change and Future Challenges for Coastal Tourism*. CAB International, Wallingford, UK.
Kay, R.C. and Alder, J. (2005) *Coastal Planning and Management*. Taylor and Francis, New York.
Ketchum, B.H. (ed.) (1972) *The Waters Edge*. MIT Press, Cambridge, Massachusetts.
Lau, M. (2005) ICZM in the People's Republic of China – An assessment of structural impacts on decision-making processes. *Ocean and Coastal Management* 48(2), 115–159.
Leatherman, S.P. (2001) Social and economic costs of sea level rise. In: Douglas, B.C., Kearney, M.S. and Leatherman, S.P. (eds) *Sea Level Rise – History and Consequences*. Academic Press, San Diego, California, pp. 181–223.
Lopez y Royo, C., Silvestri, C., Pergent, G. and Cassazza, G. (2009) Assessing human-induced pressures on coastal areas with publicly available data. *Journal of Environmental Management* 90(3), 1494–1501.
McGinn, A.P. (2002) The Importance of Healthy Oceans in the Fight Against Poverty. Available at: www.worldwatch.org.br/alertas/oceanos.htm (accessed 23 August 2016).
Ng, K., Phillips, M.R., Borges, P., Thomas, T., August, P. *et al.* (2014) Maintaining a way of life for São Miguel Island (the Azores Archipelago, Portugal): an assessment of environmental forcing and coastal protection. *Science of the Total Environment* 481C, 142–156.
Nobre, A. (2011) Scientific approaches to address challenges in coastal management. *Marine Ecology Progress Series* 434, 279–289.
OECD (2009) *Assessing Environmental Management Capacity: Towards a Common Reference Framework*. OECD environment working papers, OECD Publishing, Paris.
Phillips, M.R. (2008) Beach erosion and marine aggregate dredging: a question of evidence? *The Geographical Journal* 174(4), 332–343.
Phillips, M.R. and Jones, A.L. (2006) Erosion and tourism infrastructure in the coastal zone: problems, consequences and management. *Tourism Management* 27(3), 517–524.

Post, J.C. and Lundin, C.G. (1996) Guidelines for Integrated Coastal Zone Management. In: *Environmentally Sustainable Development Studies and Monographs Series* No. 9. World Bank, Washington, DC.

Reeve, D.E. and Spivack, M. (2001) Stochastic prediction of long-term coastal evolution. In: Brebbia, C.A. (ed.) *Coastal Engineering V*. WIT Press, Southampton, UK, pp. 55–64.

Royal Haskoning (2011) *West of Wales Shoreline Management Plan 2, Section 4. Coastal Area D.* Report No. 9T9001/RSection 4CADv4/303908/PBor.

SBCEG (Swansea Bay Coastal Engineering Group) (1999) *Swansea Bay Shoreline Management Plan. Sub-Cell 8b: Worms Head to Lavernock Point*. Shoreline Management Partnership and H.R. Wallingford, UK.

Sorensen, J. (1997) National and international efforts at integrated coastal management: definitions, achievements and lessons. *Coastal Management* 25(1), 3–41.

Thomas, T., Phillips, M.R. and Williams, A.T. (2010) Mesoscale evolution of a headland bay: beach rotation process. *Geomorphology* 123, 129–141.

Thomas, T., Phillips, M.R., Williams, A.T. and Jenkins, R.E. (2011a) A multi-century record of linked nearshore and coastal change. *Earth Surface Processes and Landforms* 36, 995–1006.

Thomas, T., Phillips, M.R., Williams, A.T. and Jenkins, R.E. (2011b) Short-term beach rotation, wave climate and the North Atlantic Oscillation (NAO). *Progress in Physical Geography* 35(3), 333–352.

Thomas, T., Phillips, M.R., Williams, A.T. and Jenkins, R.E. (2012) Medium time-scale behaviour of adjacent embayed beaches: influence of low energy external forcing. *Applied Geography* 32, 265–280.

UNESCO (2000) IOC-SOA International workshop on coastal megacities: challenges of growing urbanisation of the World's coastal areas. In: *Intergovernmental Oceanographic Commission Workshop Report* No. 166. Hangzhou, China.

van der Salm, J. and Unal, O. (2003) Towards a common Mediterranean framework for beach nourishment projects. *Journal of Coastal Conservation* 9(1), 35–42.

van der Weide, J., de Vroeg, H. and Sanyang, F. (2001) Guidelines for coastal erosion management. In: Ozhan, E. (ed.) *Medcoast 01: Proceedings of the Fifth International Conference on the Mediterranean Coastal Environment*. Medcoast, Ankara, 3, pp. 1399–1414.

Varghese, K., Ganesh, L.S., Mani, M., Anilkumar P.P., Murthy, R. and Subramaniam, B.R. (2008) Identifying critical variables for coastal profiling in ICZM planning – a systems approach. *Ocean and Coastal Management* 51(1), 73–94.

Weerakkody, U. (1997) Potential impact of accelerated sea-level rise on beaches of Sri Lanka. *Journal of Coastal Research* SI 24, 225–242.

4 Climate Change and Tourism Sustainability – The Red Queen Theory: Tourists as Climate Refugees

Ian Jenkins*
University of Iceland, Iceland

Introduction

More than 1 billion people live within low-lying areas close to the coast and it is estimated that over 200 million more live on land less than 5 m above sea level; this includes 13 million Europeans who will be at risk by just a 1 m rise in sea levels. The world's coastlines are like many other things on this planet, somewhat impermanent and constantly subject to change. This is not without existential support from a casual walk around a coastal area; certainly so within the UK where the residual effects of the last Ice Age such as glaciation are somewhat ubiquitous, with vestiges of relic cliffs, raised beaches, fossilized trees and periglacial deposits on cliff lines. Hence to assume that current coastal locations of tourist destinations will permanently remain as they are is a rather imprudent and ill-prepared treatise. Preparation for impending coastal change can be seen in low-lying countries such as the Netherlands, which is preparing plans for resilience to climate change that will perhaps include floating, rather than terrestrial cities. Not too implausible, if one considers Venice, which is built on a lagoon, and which seems to be in a constant state of flux between sinking and floating (Laskowski, 2014). Even cities such as London, many kilometres from the coast will be affected and already many millions have been spent on flood defences such as the Thames Barrier. There are also examples of sea level mapping in different countries, for example the USA where some scientists have mapped US major cities and the areas that might disappear (Copeland *et al.*, 2012) through sea level rises. Therefore there is an increasing risk of 'climate change' refugees appearing in the 21st century, if predicted sea level rises become an actuality.

Climate change purported by climate science is said to be 'beyond doubt', with predictions not looking propitious for tourist locations situated at low lying milieus. Portents of this global crisis have been heralded by countries with close associations to the sea. For example, the President of Kiribati is actually planning to abandon his country if the need arises and migrate to Fiji! This highlights the seriousness of current forecasts for climate change if it continues at its current rate. This chapter will encapsulate the climate change debate within the context of coastal tourism destinations throughout the world, providing examples of some of the threats and advantages for such significant changes.

*E-mail: jenkins@hi.is or dr.isjenkins@gmail.com

It further attempts to demonstrate how a sustainable tourism product can exist and be viable, whilst the world's coastlines change dramatically due to anthropological pollution, especially atmospheric emissions of greenhouse gases (GHGs).

Although climate change is now considered inevitable, the actual rate of change is a critical factor relating to rising sea levels and how this will affect the tourism industry and destinations (Scott *et al.*, 2012). However, it is not just rising sea levels that tourism destinations need to be concerned about, but also changes to weather patterns such as the increased risk of flooding and more violent storms. The World Meteorological Organization (WMO, 2015) has listed the types of threats that will happen as the following:

> Rise in sea levels – will eventually inundate small islands and coastal regions. Causing coastal erosion and loss of land. Many regions depending on tourism are under threat.
>
> Ocean warming and acidification – will damage and potentially kill coral reefs, a major tourist attraction for many regions.
>
> Desertification and water scarcity – can make regions less hospitable for both local communities and tourists and may lead to an increased risk of outbreaks and epidemics of infectious diseases.
>
> Deforestation and the harm to biodiversity – affecting both the ecosystem and discouraging demand for such destinations. Deforestation also directly reduces the global carbon sink.
>
> Melting of snow and glaciers – affecting mountains, ski resorts, and winter activities. This can lead to a shift in destination demands. The melting of glaciers is also one of the causes behind rising sea levels.

These changes will inevitably disrupt tourism flows; for example the 2010 eruption of Eyjafajallajokull in Iceland caused serious disruption to air traffic in the northern hemisphere. Whilst not part of climate change per se, it illustrates the force and effect of nature and the impacts it can have on global transport networks, key features to any tourism product. Destinations will need to learn how to adapt and evolve in order to manage the effects of more turbulent weather patterns and rises in sea levels.

Sustainability and Climate Change

Sustainability's fit with climate change

The floods in Cumbria UK in December 2015 have been attributed to climate change; more rain fell in 24 hours than at any time in this area since records began and the last serious flooding in 2005 created severe damage and cost billions of pounds. Prophetically the floods in 2005 were said to be a 1 in 100-year chance, yet within 10 years an even greater deluge in this area overran the flood defences (perhaps this is the 1:100 flood?). Furthermore, within the UK this has not been a one-off event or an isolated occurrence; 2014 saw similar floods on the Somerset Levels. These cases illustrate what is to come in the very near future with the predicted rise in global temperature by 2°C, which will increase the frequency of these types of events, possibly with even greater magnitude. Additionally, for those not familiar with these particular UK flood locations, both are on or close to coastal areas. It is also interesting to note that the Minster for the Environment spoke of resilience and future planning for such floods and proposed the concept of adaptation, a concept that will be further explored later in the chapter.

Although this narrative seems rather mundane, it is asserted that it illustrates precisely the theme of the book and the challenges that economies and societies face. It is argued that the most favourable model projections of the IPCC would not have prevented the floods of Cumbria. The prediction is that with the best intentions in the world, $C0_2$ will remain at 2000 levels, which are reflected in the UK events that happened in 2015, particularly, extremities of weather. So how does this relate to sustainability?

Climate change and sustainability are human concepts, devised to explain and manage outcomes on society and economies, but they are also linked to policies that are enacted by organizations and individuals. They are concepts of the Anthropocene, whereby *Homo sapiens* are trying to control environments, or rather trying to affect environmental systems which in turn affect economic, social and cultural systems. Both concepts are closely linked and one might also say that sustainability is the foundation of what has now become the climate change lobby (Yohe *et al.*, 2007). If sustainability is enacted correctly then we will be able to mitigate climate change and perhaps ensure a reduction in extreme weather events such as occurred in the UK, Cumbria (2015) and Somerset Levels (2014). Sustainability has been on the tourist agenda for some considerable time. Since the Brundtland Report 1987 (UNECE, 2005) global markets have been aware of the importance of *Sustainable Development* even if they took little action or tried to ignore it. But the growth of global warming, and what is now better known as climate change, as highlighted by the Stern Report 2006 (The Guardian, 2006), took longer to be taken seriously. Climate change and sustainability are concepts which are not interchangeable but are on the same continuum. Their aims and objectives also have similar purposes in terms of how to control environmental issues which are closely associated with human societies, including economic systems.

It is suggested that climate change and its associated policies are simply environmental aspects of sustainability. It is asserted that nature (however this is socially constructed) is far more resilient and adaptable than humans and regardless of whether we manage to reduce CO_2 emissions the natural world will survive and adapt; there have been five mass extinctions, perhaps this will be the sixth (Barnosky *et al.*, 2011). This polemic suggests that humans are '*apart from*' nature a rationale not agreed with by this author. It is asserted that we are '*a part of*' nature but have now become *apart from* nature!

Sustainability and climate change are inextricably linked. Certainly, any company that is trying to follow ethical aspects of sustainability needs also to be pursuing climate change goals. To ignore the policies of climate change is not a sustainable option.

The above discussion raises many questions relating to the tourism industry. Taking as an example package holidays; most western tourists use a form of transport, air travel, that contributes to between 2 and 5% of global carbon emissions. Yet it was noticeable that aircraft emissions were not included in agreements at the Paris COP21 summit (ATAG, 2015). Was this because tourism provides 10% of global GDP and is regarded as an important economic resource for many developing countries and a staple of many developed countries' economies? Some destinations have also developed their tourism industries purely by using air transport, especially island tourism; for example, 97% of Iceland's tourists arrive by air. This raises a number of important questions for the tourism industry. What are the carbon implications for the Icelandic tourism industry and how does Iceland mitigate these carbon emissions which are contributing to climate change? Who should pay for this pollution? Should the tourist pay for offsets or should Iceland, or maybe both? Furthermore, how is this to be done? Carbon sinks or other methods to capture carbon? Ironically, at the time of writing (2015), Reykjavik has just experienced two extreme weather events that seem linked to climate change; first one of the largest snow falls in living memory and the following week a mini-hurricane.

What is sustainability for tourism?

Sustainability has many different definitions and has evolved in terms of its aims and objectives together with its four pillars (Environment, Economic, Social and Cultural). But as stated above, sustainability's premise is simple, that of 'ensuring future generations have resources to survive in the same way as is currently enjoyed'. So for tourism the main aspect is to ensure that tourism survives on the basis of today's market developments. A simple assertion but one plagued with anomalies, especially as many of the tourism developments of the last 40 years have probably contributed significantly to climate change.

The GSTC (Global Sustainability Tourism Criteria) (GSTC, 2016) represents a global adaptation policy for the tourism market and is regarded as being the very minimum standard to which tourism companies need to adhere. Yet it is asserted that on close examination, many major tourism companies fail to have this type of validation. There is also an assumption that tourism companies are either sustainable or working towards sustainable principles. However, this also raises further questions relating to conflicts between sustainability and profitability (not necessarily exclusive). For example, can package tour companies be considered to be sustainable and climate friendly? It is interesting to note that Virgin Atlantic heralds sustainable measures and even sponsors Responsible Tourism awards, yet as discussed above surely there is conflict of interest, especially with the increasing growth of air transport and the need to reduce CO_2 emissions?

Reinforcing circles

It seems evident that sustainable measures and climate change practices are very similar and are working towards similar ends. In many cases it can be argued that if a company has a climate change policy it necessarily follows that it also has a sustainability policy. That said, sustainability has more elements to be acted upon than perhaps climate change. However, Figs 4.1 and 4.2 illustrate the similarities of the measures that are taken by both. In many ways, both represent a circular process, as both are trying to

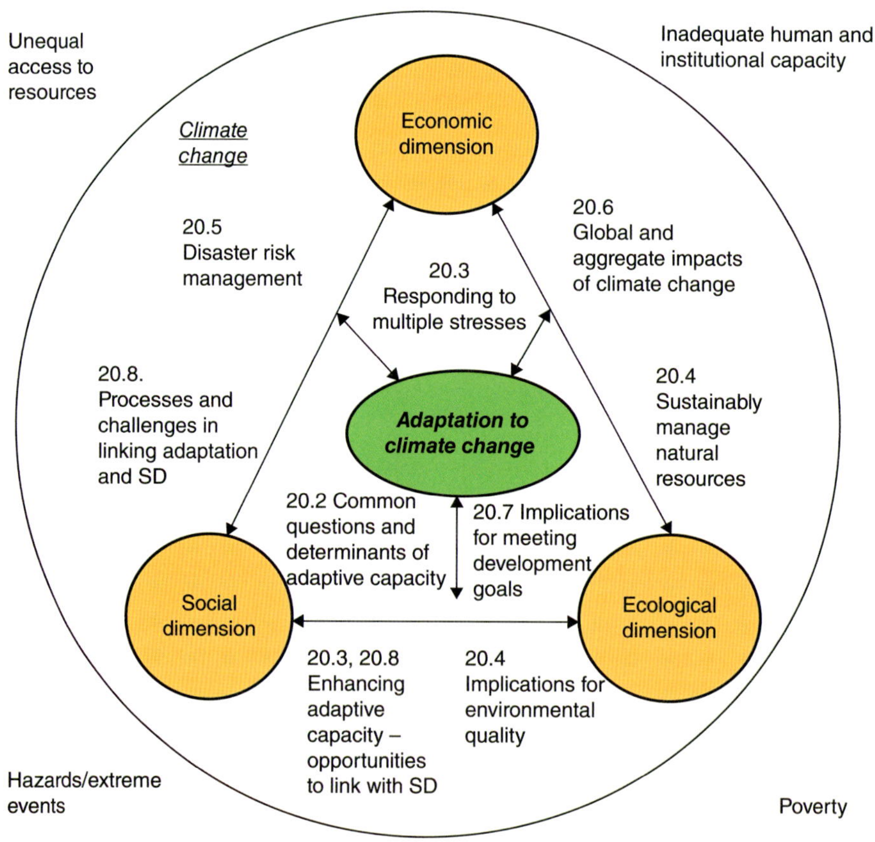

Fig. 4.1. Sustainable development and adaptation to climate change (Fig. 20.1, Fig. 20.2 in Yohe *et al.*, 2007).

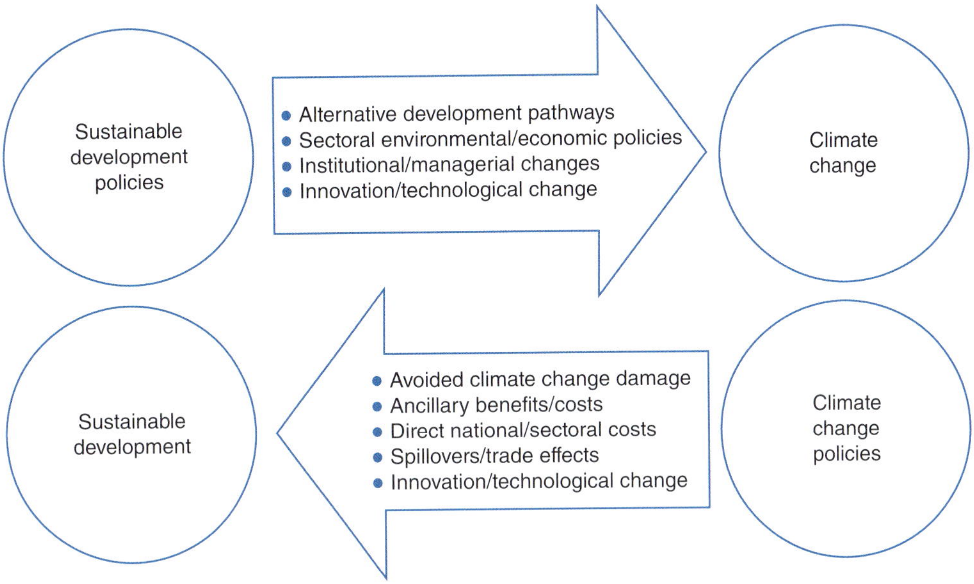

Fig. 4.2. Linkages between climate and sustainable development (Fig. 20.1, Fig. 20.2 in Yohe *et al.*, 2007).

ensure that resources are available for future generations to use.

Tourist Refugees

Currently there is a great deal of media attention concerning climate refugees. This also has to be connected with tourism and the notion of tourist refugees; tourist refugees are not a phenomenon of climate alone but have been part of the tourism market and tourism development for some time. Narratives and case studies of local populations being removed and exploited by tourism developments are fairly common and some notable examples can be found in places such as Cancun, Costa Rica, Brazil, Barbados, Kenya and Gambia (Page and Connell, 2006; Mowforth and Munt, 2009). These are situations where thousands of people have been removed from their homes and displaced to other locations in order for tourism destinations to be developed. This has happened over the last century and is still an issue with new destination developments, especially in less developed countries.

However, climate change is now creating new types of refugees related to tourism. First there is the refugee who will be displaced from destinations that are at, or close to sea levels. Most notably those in the Indian and Pacific Oceans, mainly small island chains with coral reefs, such as Kiribati (Milman, 2015). Although many of these populations are relatively small many may have to move or migrate. However, that said, many of these peoples are opposed to the term 'climate refugee' and at the Paris talks (2015) were determined to challenge this concept.

Then there is the potential tourist refugees or perhaps better termed 'pseudo refugees' or 'demand refugee' that are actually tourists. According to the IPCC some destinations are likely to be submerged in the future and will disappear. Potential tourists will therefore change their choice of destination and transfer demand to other locations, so in essence these will become Climate Benefiting Destinations (CBDs). Looking at the IPCC predictions this seems inevitable and these market-demand 'refugees' may well transfer to more northern destinations offering different types of products than those found in the equatorial regions; or perhaps there will be increased demand for destinations that are located in the equatorial regions, but with better *resilience* to climate change. Certainly, this is an area of research that needs to be more fully investigated in terms of where the transfer of demand will go, if many of the current Indian and Pacific Ocean coastal resorts disappear (IPCC, 2012).

Who should be responsible for them?

The tourism refugee debate raises certain questions relating to ethical aspects of who should be responsible for the likely migration, who needs to support them and where will they go. This predicted tourism diaspora is also another area that needs further research and Fig. 4.3 shows the 'hot spots' where refugees might be generated due to climate change. There is indubitably a need to assess where these will be and who should be responsible for them. Perhaps there is a case for UNWTO to consider and suggest funding platforms and transfer plans for these destinations, if the predictions for climate change materialize. What is evident is that these changes will be incremental and that some form of adaptation before migration is possible. It also raises the issue at what point should a destination's population migrate or evacuate. Clearly tourists will have stopped arriving well before a destination reaches this stage of evacuation or perhaps these destinations could adapt to new forms of tourism based on water-world concepts.

Climate refugees

If anthropogenic emissions continue at their present rate then there will inevitably be market changes to the global economics of tourism and the destinations visited. It will inevitably create a new migration of people. The worst predictions are that the sea will rise by at least 1 m by 2100 (Gregory, 2013).

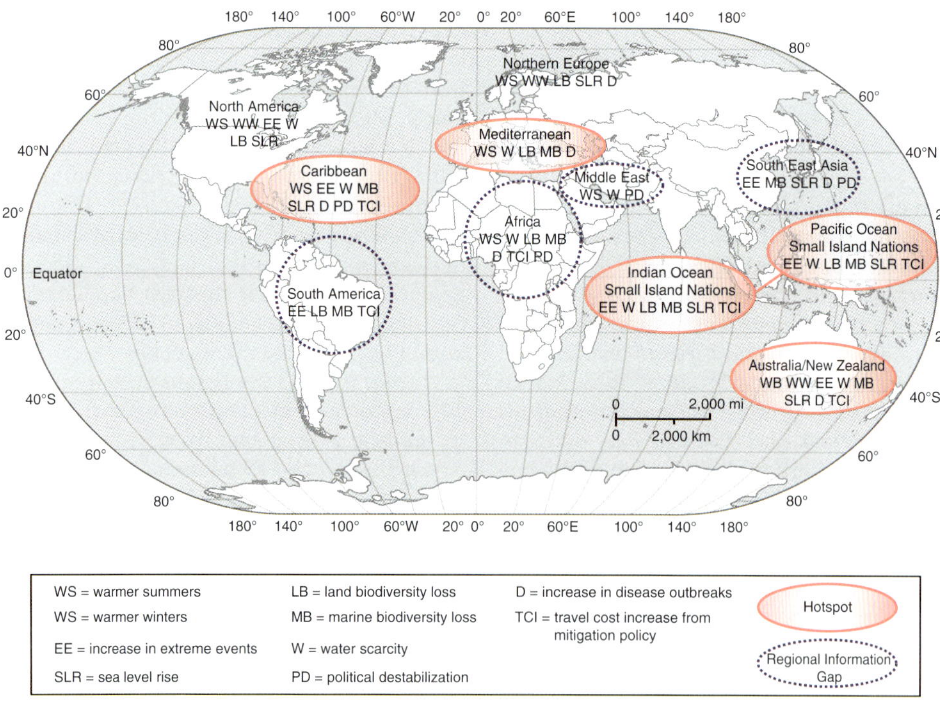

*Key destination vulnerabilities are identified at the sub-regional scale in the full technical report

Fig. 4.3. Geographic distribution of major climate change impacts affecting tourism destinations (WTO and UNEP, 2008).

Based upon IPCC global warming models this means that some tourist destinations as we now know them may well disappear. Adaptation and resilience have become essential strategies of climate change. Tourists currently visit many of these threatened destinations, and many destinations will have to adapt if they are to survive. That said, there are examples around the world where people and tourist locations do exist below or close to sea level and adaptation has meant their survival. The Netherlands (Van Aalst, 2006) must be seen as a nation state that has effectively dealt with living at low and below sea levels; certainly they do not plan to become climate refugees as might happen to Kiribati in the Pacific and other similar islands (IPCC, 2012; Warne, 2015). Moving a population of 15 million would be an incredible task.

As mentioned earlier, Venice is a classic example of adaptation and resilience even though it is sinking and sea levels continue to rise (Laskowski, 2014) and it is perhaps a useful model for other tourist destinations. It has adopted the MOSE system (Vergano *et al.*, 2010), which is a flood-gate assembly similar to the Thames Barrier. If this is effective then Venice will be saved for another 100 years. Given technological developments and the existence of urban areas currently below sea level, Venice may well survive the worst effects of climate change.

Perhaps the tourism industry needs to be thinking about two questions. What will be the type of climate tourist refugee and how many climate tourist refugees will there be? One answer would be to work backwards and look at destinations under threat and the volume of tourists who go there. It could then be estimated how many tourists would change one destination for another. But this raises another question, where are they likely to go if the destinations they usually frequent start to disappear? One must also be aware that this process will be slow and not happen overnight. It also assumes that coastal destinations are unable to adapt or ameliorate the changes, unlike Venice.

As argued earlier it may well be that certain countries will benefit from climate change and Iceland may now be exhibiting that new trend. Perhaps the rise in Icelandic tourism can be seen as a change in markets and demand towards an alternative aspect of tourism in the light of climate change. Clearly the sun, sea and sand elements for tourism seem to have been a 20th-century phenomenon where the beach was an all-encompassing milieu for hedonism and escape. Conceivably, the tourist has now become jaded with this mass tourism product, although it is still popular and remains economically successful. Yet there appears, as always, to be a change in antecedents relating to the choice of tourist products. Iceland offers a new product form, which is not sun and beach tourism. It is more ‘edgy’ and ‘on the edge of the world’ in terms of its position and products. It may well be that Iceland will be a receiver of climate change tourists and currently Iceland has exceeded the 1 million tourist mark (2015), with a huge exponential growth since 2010.

Perhaps what is being seen in Iceland is the start of ‘climate refugee tourism’. Certainly the Icelandic market has hit an exponential growth market and the reasons for the increases are unclear. One might well predict movements to different destinations away from the sun, sea and sand market, and there is growing evidence that this market (hot climates near the equator) will be affected most by climate change (National Geographic, 2015). Notwithstanding, climate change will also affect the extremes of the northern hemisphere.

It is also interesting to note that Iceland may well have grown its tourism market from naturally risky phenomena such as volcanic eruptions, glacial walking, etc. Perhaps, too, climate change affects different destinations and may also be engendering a new kind of climate risk tourism to new areas with extremes of climate. Inevitably there will be winners and losers of climate change and perhaps this has already started with scares of skin cancers, an inevitability of beach tourism and its followers, together with sinking destinations located in sunny climes. Iceland too has a geographical location that is subject to extreme weather conditions, yet this does not seem to have deterred tourism. Although its temperatures are constant, rarely ‘arctic’ (but still cold), the island is relentlessly battered by weather and its aberrations, with tourists every so often having to be rescued.

Destinations and Coastal Regions

The current climate change debate seems to be focused upon three concepts relating to tourism destinations: Resilience, Vulnerability and Adaptation; these are now common themes in dealing with change (Haimes, 2009). In particular the key word is that of resilience and how effective it can be against climate change. The major risk factor of coastal destinations is that of sea level rise and this will be coupled with more extreme weather conditions. But sea level rises will be incremental, which means that there is clearly time to consider how to adapt and be resilient to climate change.

Vulnerability

It should be remembered that these concepts are dynamic not static and a destination or system is vulnerable to only given factors that are not acted upon. There is a large body of work on aspects of vulnerability (Surugiu *et al.*, 2010), much relating to the natural environment but also issues relating to economic systems. A useful definition of vulnerability is:

> Vulnerability is the pre-event, inherent characteristics or qualities of social systems that create the potential for harm. Vulnerability is a function of the exposure (who or what is at risk) and sensitivity of system (the degree to which people and places can be harmed).
>
> Cutter *et al.*, 2008, p. 599.

This might also include reviewing river catchment areas that feed into coastal plains which add to a destination's vulnerability, especially with high rainfall and a surging storm tide. It is suggested that current coastal destinations are vulnerable depending upon the physiognomy of the location and also the preparation and planning for climate change effects and the predicted changes in weather. It is suggested that some destinations will be more vulnerable than others, depending upon the height above sea level, but also aspects of sea defences and what planning has been put in place to resist/adapt to increasing changes in extreme weather conditions.

UNEP (United Nations Environmental Programme) defines vulnerability as:

> Vulnerability: the conditions determined by physical, social, economic and environmental factors or processes, which increase the susceptibility of a community to the impact of hazards.
>
> UNEP, 2008, p. 8.

It is evident that vulnerability relates to destinations remaining in static positions relating to climatic changes, and without considerations of actions to adapt or change then the destination is clearly increasing its vulnerability by cumulative future risk factors (UNEP, 2008, p. 20; UNEP, 2008b).

As can be seen from the UNEP map (Fig. 4.3) there are clear zones of vulnerable destinations and many are coastal island chains, found towards the equatorial zones. They also happen to be countries that have little in the way of economic resources and with a higher reliance on tourism for their economic survival. This will therefore compound the impacts of climate change, as the evidence suggests that the world's richer economies are better able to deal with natural disasters, such as extreme weather conditions. This indicates that there is a need for resource assistance to help reduce a destination's vulnerability. The COP21 talks spoke of helping less-developed countries in terms of carbon emissions and perhaps there is a need to also assist in a reduction of vulnerability. It is not only coastal zones that will experience extremes of weather but other inland destinations too.

Resilience

Currently, resilience is a popular concept, being used as a means to respond to increasing extremes of weather linked to climate change. Resilience has a number of perspectives related to hazards and hazard management.

Yet systems cannot be totally resilient and resilience has its limits; it must be emphasized that tourist destinations cannot withstand extensive, unremitting pressure of very high magnitude. It is inevitable that at some point relocation and abandonment of a destination will be an outcome of climate change, especially for those located at sea level.

A useful definition for understanding resilience is as follows:

> Resilience is the ability of a social system to respond and recover from disasters and includes those inherent conditions that allow the system to absorb impacts and cope with an event, as well as post-event, adaptive processes that facilitate the ability of the social system to re-organize, change, and learn in response to a threat.
>
> Cutter *et al.*, 2008, p. 599.

UNEP also defines resilience as:

> The capacity of a system, community or society potentially exposed to hazards to adapt, by resisting or changing in order to reach and maintain an acceptable level of functioning and structure. This is determined by the degree to which the social system is capable of organizing itself to increase its capacity for learning from past disasters for better future protection and to improve risk reduction measures.
>
> UNEP, 2008, p. 8.

The key aspect here of resilience is that of preparation in terms of extreme events. Unfortunately, humans tend to 'put off' what they could do today 'until tomorrow' and therefore preparation for extreme events is currently not a priority. Perhaps increasing media reports and news items linking climate change to extreme weather has started to prompt destinations to respond to climate change; planning and adaptation would increase a destination's resilience in the short- to medium-term. There is a need to risk-assess destinations and put in place plans and resources to try and adapt to the future predicted changes; however, as with other concepts, there is a large array of adaptation models that perhaps may confuse which approach to take (Dickinson, 2007).

Destination survival depends on adaptation to changes. Destinations that have the greatest ability to adapt appear to be the ones who will survive any moderate changes to the systems in which they live. This is clearly the case with tourism and it can be seen that over the past 250 years, tourism has changed alongside changes to the type of tourist and systems. An excellent example is perhaps the European northern seaside resorts which have struggled to compete with the large global markets of beach tourism locations of the 1960s and 1970s. The UK seaside resorts had their 'heyday' when transport was limited to regions of a country and the aspect of sun (as a resource) was a minor consideration for a holiday. Yet many seaside resorts in the UK still exist but have adapted and evolved into different types of tourist destination.

A definition of adaptation is: 'an adjustment in ecological, social or economic systems in response to observed or expected changes in climatic stimuli and their effects and impacts in order to alleviate adverse impacts of change or take advantage of new opportunities' (Adger *et al.*, 2005, p. 78).

Tourists will certainly react and adapt to changes – they always have and will continue to do so in the future. Tourism is littered with heritage products related to tourism's bygone ages. For example, The Grand Tour and the development of Spa Towns, all of which still exist but in a different form today, were important to the tourist developments of their day. Yet clearly these developments were affected by changes in the market very similar to what is being predicted today.

Adaptation perhaps cannot be seen as a singular response but must be viewed through the lens of tourist structures and agencies at destinations. Adaptation must be seen as Adger *et al.* (2005) note 'in a spatial and temporal aspect' also raising issues of success and effective adaptation; both are normative assessments, depending upon the criteria used to measure them.

In addition, adaptation seems to occur only when extreme events are reported as being outside the normal averages (Füssel, 2007), and adaptation involves a large range of organizations and factors such as technology, planning, education, research, economics and cultural integrity to name but a few.

Clearly, there are recommendations for adaptation strategies but tangible examples appear hard to find, particularly where climate change has effected a new direction for tourist destinations and where the superstructure and infrastructures have been changed to encompass climate change.

The Maldives have started to think about what needs to be done given their location and the potential devastation that the next 100 years of predicted temperature changes could produce. In essence the Maldives could disappear completely, but is adaptation a possibility or will it simply be abandonment? The same can be said too of many low-lying Pacific island states. Clearly, early planning is needed if adaptation is going to be successful, as can be seen with the barrier in Venice and the planning of floating cities for the Netherlands. Aspects of preparing for more violent weather is possibly easier to adapt to, compared to sea level rises, especially if the highest point of a destination is only 3 m above sea level.

Adaptation, then, is a response to changes in the environment and the ability of the tourism industry to absorb changes to the tourist system. It will appear in a variety of forms, not just that of environmental change, but also resource allocation relating to economic and social modifications.

The Red Queen Theory

First proposed by Van Valen in 1973 relating to the Law of Constant Extinction, this theory has been used by biologists to explain the extinction of species over the earth's time frame (Hoffman, 1991). The emphasis upon the importance of the environment was adapted by Barnosky (2001) and perhaps would seem more relevant to the aspect of climate change in the Anthropocene and its effects on tourism. The theory was developed using the analogy of the Red Queen in Alice in Wonderland when Alice tries to catch the Red Queen but is unable to do so:

> 'Well, in our country,' said Alice, still panting a little, 'you'd generally get to somewhere else – if you run very fast for a long time, as we've been doing.'
>
> 'A slow sort of country!' said the Queen. 'Now, here, you see, it takes all the running you can do, to keep in the same place. If you want to get somewhere else, you must run at least twice as fast as that!'
>
> Caroll, 2009, p. 33.

The Red Queen Theory is in essence a competition between predator and prey and the ability of both to adapt to the other's evolving capacities. True, climate is not considered a living organism in the true sense of the word, but it is an agent of change, with qualities that affect the environment and the resources available for living species.

The premise is that this theory could be adapted to fit current aspects of anthropogenic climate change and its effects on man. Man has affected the climate in a negative way through emissions of GHGs and nature has responded by using climate as a 'predatorily factor against man', vis-à-vis the climate is now threatening man's created environments such as tourist destinations.

The theory is seen in biological terms as an arms race between natural enemies. In the case of the tourist, in particular the destination (host), this raises conflict with the pressures of climate change (predator) on resources which will be in short supply if carbon emissions continue at the current rate. It relates to how man will adapt to the current changing environment and this could be seen in a number of modes, such as reducing carbon emissions or resilience through adaptation to the changing conditions. The Paris Accord of 2015 seems to have heralded a means to reduce the emissions of carbon to below 2°C, which can be seen as a form of adaptation by man to reducing the predatory aspects of climate as an agency of negative effects on resources.

Alice was unable to keep up with the Red Queen, which seems an appropriate analogy of man's inability to reduce emissions against accelerating climate change. The predicted rate of climate change is forecast to accelerate, which means that even current changes to carbon emissions will not be enough to prevent the processes of increasing extreme weather conditions prevailing on future tourist destinations. So the

Paris climate talks could be seen as Alice (man) simply just 'running on the spot', not really going anywhere in terms of emissions and carbon reductions, while climate change (predator) accelerates. The conflict will be about whether man can adapt to climate change before climate change disastrously affects man's resource systems.

Actions and Policy Changes at Destinations

So what should be the policies and actions at destinations, especially coastal destinations that are going to be subject to the weather changes that the IPCC are predicting? There need to be definite actions and policies, together with planning for, and identification of, the best and worst case scenarios. The following action chart (Fig. 4.4) is a useful example of management approaches to climate change.

The evidence suggests that there is a need to ameliorate opinions but with a firm process that does not allow the procedure to be halted by superfluous argument and comments. However, given the predictions of serious sea level rises, low-lying countries need to act now to reduce the risks. There should be a coordinated international effort to support these countries, together with planning for the possible climate refugees who may have to leave their homelands to settle in new locations. This should be supported by UNWTO, which is an important agency in effecting the outcome of this eventuality. Targeted funding and appropriate actions are therefore required.

Conclusion

The IPCC predictions have a tangible message for many coastal destinations around the world. Some of these are not obvious places, such as capital cities situated many miles from the coast, especially those situated near river estuaries. Clearly, there is an immediate need to evaluate the risks to coastal tourist destinations and to develop resilience through adaptation strategies.

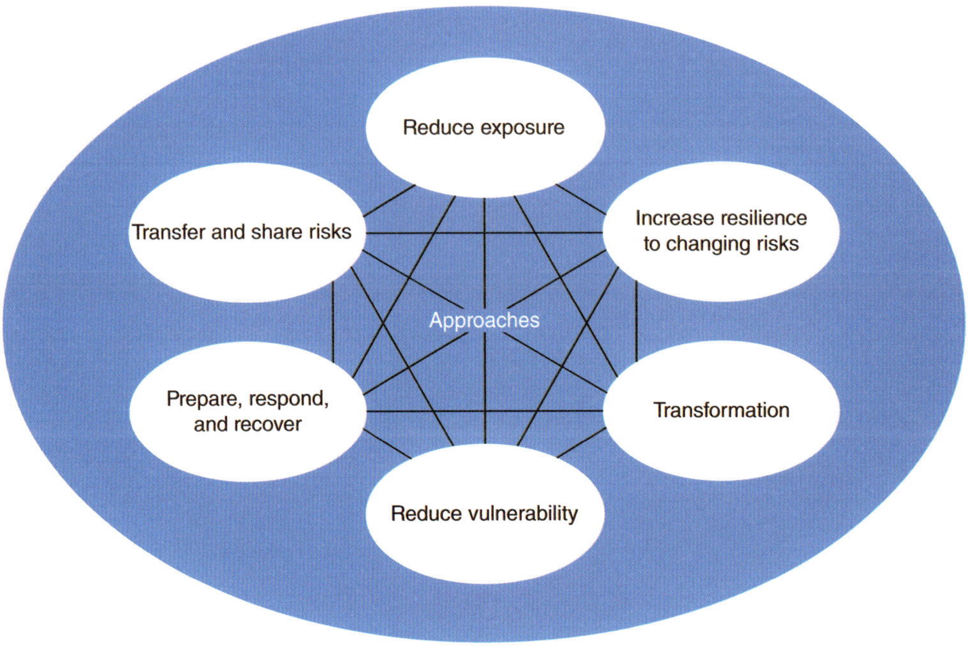

Fig. 4.4. Adaptation and disaster management approaches to climate change (Fig. SPM.2 from IPCC, 2012, p.6).

Although the process of sea level rises will be slow and incremental, action needs to begin immediately. Time is a relative concept and a hundred years may seem a long way off, but delaying until climate change starts to seriously affect destinations will be too late for many coastal areas. Perhaps a key to climate change adaptation is to revisit sustainable development polices and to incorporate climate change action and planning to mitigate and increase resilience to climate change.

The Red Queen Theory seems suited to the current debates on climate change and tourism destinations. It appears that climate change is the new predator for humans and humans are trying to outrun the predator by simply 'running on the spot' (reducing GHG emissions). Presently, climate change predictions are disconcerting and attempts to cut emissions are too little and too late, especially with some sectors being ignored, in particular air transport.

There will be climate refugees emerging from tourist destinations, affected by extreme weather and sea level rises and there need to be policies in place to deal with how the global community will help relocate these refugees. In addition, there will be changes in demand to the tourism market and these changes need to be planned for and properly resourced.

It seems inevitable that some coastal destinations will disappear, but this will also herald new locations. As with all aspects of existence things change, but currently it is the rate of change and what exactly will be the legacy of the tourist industry in the Anthropocene that needs to be catechized and answered. 'Alice needs to run a lot faster to keep up with the Red Queen!' The simple question is can Alice (man) do it?

References

Adger, W.N., Arnell, N.W. and Tompkins, E.L. (2005) Successful adaptation to climate change across scales. *Global Environmental Change* 15, 77–86.

Adger, W.N., Brown, K., Nelson, D.R., Berkes, F., Eakin, H., *et al.* (2011) Resilience implications of policy responses to climate change. *WIREs Climate Change* 2, 757–766.

Askja Energy (2012) The Independent Icelandic Energy Portal, Blog. Available at: http://askjaenergy.org/2012/11 (accessed 5 October 2015).

ATAG (Air Transport Action Group) (2015) Aviation CO_2 Emissions to be Dealt with Next Year at ICAO. Available at: http://aviationbenefits.org/newswire/2015/12/aviation-co2-emissions-to-be-dealt-with-next-year-at-icao (accessed 9 February 2016).

Barnosky, A.D. (2001) Distinguishing the effects of the red queen and court jester on Miocene mammal evolution in the Northern Rocky mountains. *Journal of Vertebrate Paleontology* 21(1), 172–185. Available at: www.ucmp.berkeley.edu/miomap/RESULTS-MIOMAP/barnoskyjvp2001.pdf (accessed 14 December 2015).

Barnosky, A.D., Matzke, N., Tomiya, S., Wogan, G.O.U., Swartz, B., *et al.* (2011) Has the earth's sixth mass extinction already arrived? *Nature* 471, 51–57.

Caroll, L. (2009) *Through the Looking-Glass, and What Alice Found There*. Evertype, Éire.

Copeland, B., Keller, J. and Marsh, B. (2012) What could disappear. *New York Times, Sunday Review*, Opinion pages. Available at: www.nytimes.com/interactive/2012/11/24/opinion/sunday/what-could-disappear.html?_r=0 (accessed 30 November 2015).

Cutter, S.L., Barnes, L., Berry, M., Burton, C., Evans, E. *et al.* (2008) A place-based model for understanding community resilience to natural disasters. *Global Environmental Change* 18, 598–606.

Dickinson, T. (2007) *The Compendium of Adaptation Models for Climate Change*, 1st edn. Adaptation and Impacts Research Division, Environment Canada.

EU (2016) Food Nutrition and Health Claims. Available at: http://ec.europa.eu/food/safety/labelling_nutrition/claims/index_en.htm (accessed 9 February 2016).

Fairtrade (2016) What is Fairtrade? Available at: www.fairtrade.net/about-fairtrade/what-is-fairtrade.html (accessed 9 February 2016).

Füssel, H.-M. (2007) Adaptation planning for climate change: concepts, assessment approaches, and key lessons. Review Article *Sustainability Science* 2, 265–275. DOI: 10.1007/s11625-007-0032-y.

Gregory, J. (2013) Projections of Sea Level Rise, Chapter 13, Sea Level Change. Available at: www.ipcc.ch/pdf/unfccc/cop19/3_gregory13sbsta.pdf (accessed 9 February 2016).

GSTC (2016) Global Sustainable Tourism Council Criteria. Available at: www.gstcouncil.org/en/gstc-criteria/sustainable-tourism-gstc-criteria.html (accessed 9 August 2017).

Haimes, Y.Y. (2009) On the definition of resilience in systems. *Risk Analysis* 29(4), 498–501. DOI: 10.1111/j.1539-6924.2009.01216.x (accessed 5 October 2015).

Hoffman, A. (1991) Mini-review testing the red queen hypothesis. *Journal of Evolutionary Biology* 4, 1–7.

IPCC (2012) *Managing the Risks of Extreme Events and Disasters to Advance Climate Change Adaptation. Special Report of the Intergovernmental Panel On Climate Change*. Cambridge University Press, Cambridge. Available at: www.ipcc.ch/pdf/special-reports/srex/SREX_Full_Report.pdf (accessed 9 February 2016).

Laskowski, A. (2014) Lessons from Venice. *Bostonia*, Boston University. Available at: www.bu.edu/bostonia/fall14/lessons-from-venice (accessed 16 February 2016).

Milman, O. (2015) Pacific nations beg for help for islanders when 'calamity' of climate change hits. *The Guardian*, 13 October 2015. Available at: www.theguardian.com/environment/2015/oct/14/pacific-nations-beg-for-help-for-islanders-when-calamity-of-climate-change-hits (accessed 9 February 2016).

Mowforth, M. and Munt, I. (2009) *Tourism and Sustainability: Development, Globalisation and New Tourism in the Third World*. Routledge, Abingdon, UK.

National Geographic (2015) Here's the challenge. *National Geographic* 228(5), 16–17.

Page, S. and Connell, J. (2006) *Tourism: A Modern Synthesis*. Thomson, London.

Scott, D., Hall, C.M. and Gössling, S. (2012) *Tourism and Climate Change: Impacts, Adaptation and Mitigation*. Routledge, Abingdon, UK.

Surugiu, C., Dincă, A.-I. and Micu, D. (2010) Tourism destinations vulnerable to climate changes: An econometric approach on Predeal resort. *BULETINUL Universităţii Petrol – Gaze din Ploieşti* Vol. LXII No. 1/2010 111–120, Seria Ştiinţe Economice.

The Guardian (2006) Stern Report: The key points. *The Guardian*, 30 October 2006. Available at: www.theguardian.com/politics/2006/oct/30/economy.uk (accessed 9 February 2016).

UNECE (United Nations Economics for Europe) (2005) Sustainable Development – Concept and Action. Available at: www.unece.org/oes/nutshell/2004-2005/focus_sustainable_development.html (accessed 9 February 2016).

UNEP (2008a) *Climate Change Adaptation and Mitigation in the Tourism Sector: Frameworks, Tools and Practices*. United Nations Environment Programme.

UNEP (2008b) *Disaster Risk Management for Coastal Tourism Destinations Responding to Climate Change – A Practical Guide for Decision Makers*. United Nations Environment Programme, Paris.

UNEP (2011) Tourism Investing in Energy and Resource Efficiency. Available at: www.unep.org/resourceefficiency/Portals/24147/scp/business/tourism/greeneconomy_tourism.pdf.

UNEP (2012) United Nations Environment Programme and (UNWTO) World Tourism Organization Tourism in the Green Economy. Background Report. Available at: www.unep.org/greeneconomy/Portals/88/documents/ger/ger_final_dec_2011/Tourism%20in%20the%20green_economy%20unwto_unep.pdf.

Van Aalst, M.K. (2006) *The Impacts of Climate Change on the Risk of Natural Disasters*. Climate Specialist, Red Cross/Red Crescent Centre on climate change and Disaster Preparedness, the Netherlands.

Vergano, L., Umgiesser, G. and Nunes, P.A.L.D. (2010) An economic assessment of the impacts of the MOSE barriers on Venice port activities. *Transportation Research Part D: Transport and Environment* 15(6), 343–349.

Warne, K. (2015) How to live with it against the tide. *National Geographic*, November, pp. 120–135.

WMO (World Meteorological Organization) (2015) Tourism Sector. Available at: www.wmo.int/pages/themes/climate/applications_tourism.php (accessed 30 November 2015).

WTO (World Tourism Organization) and UNEP (United Nations Environment Programme) (2008) Climate Change and Tourism – Responding to Global Challenges. Available at: http://sdt.unwto.org/sites/all/files/docpdf/climate2008.pdf (accessed 18 February 2016).

Yohe, G.W., Lasco, R.D., Ahmad, Q.K., Arnell, N.W., Cohen, S.J., Hope, C., Janetos, A.C. and Perez, R.T. (2007) *Perspectives on Climate Change and Sustainability. Climate Change 2007: Impacts, Adaptation and Vulnerability. Contribution of Working Group II to the Fourth Assessment Report of the Intergovernmental Panel on Climate Change*. Cambridge University Press, Cambridge, pp. 811–841.

5 Climate Change and its Impacts on Coastal Tourism: Regional Assessments, Gaps and Issues

C. Michael Hall*

University of Canterbury, New Zealand; University of Oulu, Finland; and Linneaus University, Kalmar, Sweden

Introduction

Tourism is a 'climate-sensitive human activity' with a vulnerability to climate change that is 'generally greater in certain high-risk locations, particularly coastal and riverine areas' (Wilbanks *et al.*, 2007, p. 359). Tourism in coastal regions, along with alpine and island tourism, has been the focus of much of the research on climate change and tourism relationships (Hall, 2008; Scott *et al.*, 2012a). This situation is also reflected in much of the more general work on climate change where coastal areas are regarded as particularly vulnerable to the impacts of sea-level rise (SLR) as well as high-magnitude storm events (Moser *et al.*, 2012; Wilby and Keenan, 2012). For example, with respect to coastal systems and low-lying areas the IPCC AR4 and AR5 have reported with 'very high confidence' that coasts are already experiencing the adverse consequences of climate (Nicholls *et al.*, 2007). The IPCC have also recognized that 'The impact of climate change on coasts is exacerbated by increasing human-induced pressures' (Nicholls *et al.*, 2007, p. 317), including tourism. This chapter addresses some of these issues in a regional context as well as emphasizing some of the factors that make regional and destination level analysis so important yet, at the same time, challenging.

Assessment of Threats

Although coastal regions are recognized as being extremely vulnerable to climate change, there are substantial gaps in knowledge of the impact on coastal tourism both in general and with respect to specific locations. There are very few coastal areas around the world for which baseline conditions have been identified and that have received long-term monitoring in order to chart change over time. Those sites on which long-term research and monitoring have been conducted tend to be in developed countries and/or very particular types of locations, such as within national parks (Hall and Lew, 2009). Furthermore, climate change research itself is often not integrated over various scales of analysis or disciplinary approaches (Scott *et al.*, 2012a; Hall and Ram, 2016). Therefore, while there may be reasonable knowledge of the effects of climate change on coastal areas at a macro-scale, as in the case of IPCC reports, micro-scale, location-specific understandings and forecasts are far more difficult to determine. Table 5.1 identifies the relative level of tourism-specific climate

*E-mail: michael.hall@canterbury.ac.nz

Table 5.1. Relative level of tourism-specific climate change knowledge and estimated impact of climate change on tourism by region (Hall and Higham, 2005; Parry *et al.*, 2007; Hall, 2008; UNWTO *et al.*, 2008; Scott *et al.*, 2012a, 2016; IPCC, 2014a, b, c).

Region	Estimated impact of climate change on tourism	Relative level of tourism specific climate change knowledge	Tourism content (citations) in IPCC Working Group 2: impacts, adaptation and vulnerability		
			TAR 2001	AR4 2007	AR5 2013–2014
Africa	Moderately–strongly negative	Extremely poor	3	28	6
Asia	Weakly–moderately negative	Extremely poor	9	7	1
Australia and New Zealand	Moderately–strongly negative	Poor–moderate (high in Great Barrier Reef)	22	31	27
Europe	Weakly–moderately negative, positive in high latitudes	Moderate (high in alpine areas)	34	40	35
Central and South America	Weakly–moderately negative	Poor	10	10	10
North America	Weakly negative, positive in high latitudes	Moderate (relatively high in coastal areas, high in ski areas)	86	27	4
Polar regions	Weakly negative–weakly positive	Poor	11	7	12
Small islands	Strongly negative	Moderate (although with significant geographic variability)	26	47	40

change knowledge and estimated impact of climate change on tourism by region as a result of analyses of IPCC reports and key literature (see Scott *et al.*, 2016, for a general overview of tourism in the IPCC AR5 reports).

Regional knowledge gaps are substantial. In the case of Africa, Boko *et al.* (2007, p. 450) emphasize that 'very few assessments of projected impacts on tourism and climate change are available' and state:

> There is a need to enhance practical research regarding the vulnerability and impacts of climate change on tourism, as tourism is one of the most important and highly promising economic activities in Africa. Large gaps appear to exist in research on the impacts of climate variability and change on tourism and related matters, such as the impacts of climate change on coral reefs and how these impacts might affect ecotourism.
>
> Boko *et al.*, 2007, p. 459.

These issues remained in the 2014 assessment, although there is a growing recognition that SLR could disrupt tourism as well as other coastal economic activities (Kebede and Nicholls, 2012). This is regarded as a major issue given that 'More than a quarter of Africa's population lives within 100 km of the coast and more than half of Africa's total population living in low-elevation coastal zones is urban, accounting for 11.5% of the total urban population of the continent' (Niang *et al.*, 2014, p. 1225).

Tourism is similarly recognized by the IPCC as one of the most important economic sectors in Asia, although the relative lack of research is recognized (Cruz *et al.*, 2007; Su and Hall, 2014; UNWTO, 2014a). Despite recognition of the potential impacts of climate change and SLR on coastal populations in Asia, the effects on tourism are little studied (Hijioka *et al.*, 2014; Su and Hall, 2014). Even in the case of a developed region such as North America substantial knowledge gaps exist. For example, Field *et al.* (2007, p. 634) noted that, 'Although coastal zones are among the most important recreation resources in

North America, the vulnerability of key tourism areas to sea-level rise has not been comprehensively assessed.' This extremely important issue is only gradually drawing more research to it (Haer *et al.*, 2013; Mitsch and Hernandez, 2013; Martínez *et al.*, 2014).

The understanding of the impacts of climate change on coastal tourism is therefore operating in a relative vacuum at the destination level and with respect to some specific issues, e.g. appropriate planning actions. This situation has occurred in part because of a lack of specific focus on tourism in certain locations by researchers but also reflects a lack of appreciation of the nature of tourism. In particular this relates to not only understanding the effects of climate change in tourism destinations but also the wider changes that are taking place in the tourism system. Figure 5.1 represents some of the issues with respect to coastal tourism.

Tourism is affected by climate change because of the large number of tourism-dependent coastal economies as well as the increasing human-induced pressures it brings to coastal areas (Hall, 2006; Scott *et al.*, 2012a). Nicholls *et al.* (2007) noted a number of physical effects of climate change that impact tourism, including:

- Air and seawater temperature increase;
- Extreme events (storms, waves);
- Sea level rise;
- Erosion (sea level, storms, waves); and
- Biological effects, such as ecosystem change and changes in species composition.

Extreme events, such as cyclones and hurricanes, are regarded as significant because of both their impact on environment, tourism resources and tourism infrastructure, as well as a potentially negative contribution to destination image in both the short and long term (Scott *et al.*, 2012a; Hall *et al.*, 2016). Under climatic change more frequent high-magnitude events may mean there is less time for physical and human systems to recover, meaning that recovery may never be complete therefore resulting in long-term environmental

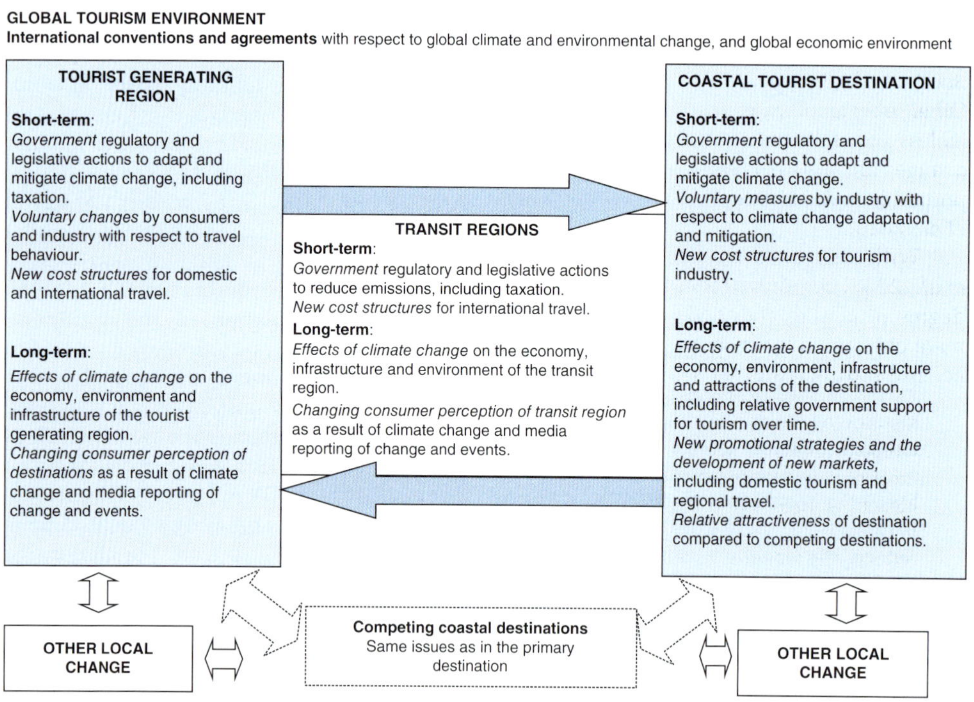

Fig. 5.1. System dimensions of tourism in coastal destinations and the impact of climate change (Hall, 2005).

deterioration with considerable input of resources (Forbes *et al.*, 2013; Hall, 2015).

Given their contribution to biodiversity as well as economic importance via fishing and tourism, coral reefs have been a major focus of research interest and are a coastal ecosystem in which the effects of environmental change has already been felt. Tourism in Australia's Great Barrier Reef, which is a major contributor to the regional and national economy (Zeppel, 2012), is also threatened by coral bleaching (Bohensky *et al.*, 2011), which occurs as a result of a number of factors connected to anthropogenic-induced climate change, including:

- Changes in water temperatures;
- Increased solar irradiance; and
- Ocean acidification and other changes in water chemistry.

SLR is also significant because of the impacts it can have on many low-lying resort and second-home areas. For example, because of the perceived attractiveness of coastal areas there is a strong tendency to concentrate accommodation and hospitality services as close to the beach as possible (Hall, 2006; Scott *et al.*, 2012a, b; Gurran *et al.*, 2013). Other effects of SLR such as rising water tables and saltwater intrusion (also runoff change) are also becoming increasingly important in water-scarce destinations (Gössling *et al.*, 2015).

With respect to small islands, Mimura *et al.* (2007) concluded that there was a high degree of confidence that the effects of climate change on tourism are likely to be direct and indirect, and largely negative:

> Sea-level rise and increased sea water temperature will cause accelerated beach erosion, degradation of coral reefs, and bleaching. In addition, a loss of cultural heritage from inundation and flooding reduces the amenity value for coastal users. Whereas a warmer climate could reduce the number of people visiting small islands in low latitudes, it could have the reverse effect in mid- and high-latitude islands. However, water shortages and increased incidence of vectorborne diseases may also deter tourists.
>
> Mimura *et al.*, 2007, p. 689

Climate change may also have impacts on specific forms of tourism. For example, biodiversity loss has substantial significance for nature-based tourism in marine and coastal environments (Hall, 2010a, b). This can include changes to ecosystems, as in the case of coral reefs noted above, or the loss of iconic or charismatic marine species that are a focal point of ecotourism interest at destinations, such as marine mammals (seals, whales, dolphins), birds (penguins, albatross, gannets), fish (shark), reptiles (marine iguanas, turtles) or fauna (kelp forests) (Hall *et al.*, 2011).

Nevertheless, as Fig. 5.1 indicates, the impacts of climate change on a coastal tourism destination are not just in terms of environmental change but also economic and social change. Regulatory initiatives developed to mitigate climate change will have significant impacts on transport systems and thus the mobility of tourists (Gössling *et al.*, 2013). One possible effect is that those destinations which are the most peripheral and least accessible to markets, which typically are most in need of development initiatives, are potentially the most impacted in relative terms by the costs of mitigation strategies based on distance travelled, size of emissions or energy consumed (Hall and Lew, 2009). Such an observation is important in understanding the potential impacts of international and national climate change mitigation regimes on tourism flows and destinations, as while preferred environmental attributes are significant in destination choice, changes in price structure resulting from climate change policies may have the most immediate and significant impact for tourism development (UNWTO *et al.*, 2008; Gössling *et al.*, 2009a). Importantly for understanding the system effects of mitigation strategies on coastal destinations, it should also be stressed that the price competitiveness of a destination will be influenced by the mitigation policies adopted in tourist generating regions and/or in transit regions. Therefore, the relative price attractiveness of a destination needs to be understood in relation to its source markets, the actions of competitors, as well as the policies in place at the destination itself (Gössling *et al.*, 2016). In such a situation it is therefore possible to envisage scenarios where some destinations may not implement an emissions reduction framework in order to retain cost leadership (Gössling *et al.*, 2013).

In addition to the system-wide complications provided by the implementation of any emissions reduction framework or broader environmental regulations that affect tourism, the determination of future tourism scenarios is complicated by the difficulties of assessing future changes in tourist behaviour (Gössling *et al.*, 2012). Climate change can influence all stages of the travel experience prior to travel, en route, and at the destination as well as in future choice of destination (Hall and Lew, 2009). Higher temperatures may influence summer and winter destination preferences globally, either through direct effects on tourism such as availability of a tourism resource, such as snow in alpine areas, or in terms of making competing destinations more or less attractive in terms of their climatic comfort (Deutsche Bank Research, 2008; UNWTO *et al.*, 2008; Scott *et al.*, 2012a). Changes in temperatures may also lead to new patterns of seasonality as a result of shifts in travel preferences in generating areas as well as product change in destinations, although the capacity of potential tourists to accurately judge the implications of increases in average temperatures or climate overall for a destination is questionable (Gössling *et al.*, 2012).

Despite the complexities of tourist preferences for climate there has been a strong dependence on deterministic models of tourist–climate relationships in trying to forecast future preferences under different climate change scenarios. For example, the Tourism Comfort Index of Amelung and Viner (2006) indicates improved conditions for tourism in northern Europe, while econometric modelling by Hamilton *et al.* (2005) (the only tourist study included in the influential Stern (2006) report) suggested that an arbitrary climate change scenario of 1°C would lead to a gradual shift of tourist destinations further north and up mountains, thereby affecting the preferences of European summer tourists, a finding that was reported in IPCC AR4 (Alcamo *et al.*, 2007) and reinforced by more recent studies (Amelung and Moreno, 2012). Nevertheless, Gössling and Hall (2006a, b, c) argued that there are a number of major weaknesses with respect to predicting changes in travel flows as a result of climate change (see also Scott and Gössling, 2015) and have urged substantial caution in utilizing the results of deterministic approaches to climate change adaptation (Table 5.2). However, despite some of the difficulties in identifying future flows of tourists it is still possible to identify some of the locations that may be most at risk as a result of environmental change. The next section therefore briefly identifies some of the most at risk coastal tourism destinations.

Table 5.2. Major weaknesses of current models in predicting travel flows under climate change forecasts (Gössling and Hall, 2006a, b, c; Hall, 2008; Scott *et al.*, 2012a).

- Validity and structure of statistical databases, including international and domestic statistics
- Temperature assumed as the most important weather parameter
- Importance of other weather parameters (rain, storms, humidity, hours of sunshine, air pollution) largely unknown
- Role of weather extremes unknown
- Role of information in decision making unclear
- Role of non-climatic parameters in influencing travel flows unclear (e.g. perception of security and political instability, risk perceptions, destination perception)
- Existence of fuzzy-variables problematic (terrorism, war, epidemics, natural disasters)
- Assumed linearity of change in behaviour unrealistic
- Future costs of transport and availability of tourism infrastructure uncertain
- Future levels of personal disposable income (economic budget) and availability of leisure time (time budget) that are allocated to travel uncertain

Identifying the Most Vulnerable Coastal Destinations

The capacity for countries and destinations to respond to the impacts of climate change on tourism is a combination of a number of factors, including:

- Direct climatic effects, including high-magnitude events such as storms, floods and droughts;
- Effects of regulatory attempts to mitigate climate change which increase the price of travel;
- Destination substitution effects resulting from climatic or regulatory factors. This includes price substitution whereby a destination that provides similar benefits for customers becomes relatively cheaper, and activity and environmental substitution effects whereby because of the impacts of climate change the capacity to engage in an activity and/or experience a specific environment is greater elsewhere; and
- The actual capacities each destination has to adapt to the effects of climate change.

A further key element in the threat of climate change on tourism is the extent to which a country or region is economically dependent on tourism. The capacity to adapt to climate change depends on a range of factors, including the overall capacity of the local and national economy to respond to change, the national innovation system, and a range of environmental health factors. Countries and destinations that are already under significant environmental, social and/or economic stress will clearly find it harder to adapt than those which are not. Given such a situation it should therefore not be surprising that the coasts of developing countries have less adaptive capacity than the coasts of developed countries (Füssel, 2010; Scott *et al.*, 2012b; Hall, 2015).

Many of the Less Developed Countries (LDCs) that have strongly developed tourism industries are greatly at risk from the effects of climate change. For example, according to the UNWTO (2014b) the number of international tourists visiting Small Island Developing States (SIDS) destinations increased from 28 million in 2000 to 41 million in 2013. In the same period, exports from tourism grew from US$26 billion to US$53 billion. This means that tourism accounts for over one-quarter of the Gross Domestic Product (GDP) in at least seven SIDS and represents 9% of the overall exports (US$61 billion). However, SIDs are highly vulnerable destinations. Table 5.3 details some of the characteristics of SIDS as well as their vulnerability in relation to biodiversity loss, disaster, freshwater availability as well as some other factors. In addition, nearly all SIDS are potentially affected by any changes in the cost of travel because of their status as mid- or long-haul destinations for their major markets as well as having tourism in coastal areas.

The susceptibility of the SIDS identified in Table 5.3 is also reinforced by the assessments of several other studies of climate change, which have consistently identified developing nations in Africa, the Caribbean, Indian Ocean, Oceania, and South and South-east Asia, as the most at-risk tourism destinations from 2020 to the mid-21st century (Table 5.4). Coastal destinations in developing countries that have been frequently identified as being at risk include the Mediterranean as the most at-risk destination because of the cumulative effects of a number of elements of global environmental change while, with respect to specific factors, the Gold Coast (Australia) and Florida were also identified as at risk from sea-level rise. However, the greater capacity of developed countries to adapt to climate change means that forecasts as to the demise of tourism in those countries at least in the medium term may be quite problematic. Indeed, it may even be possible to develop destination management scenarios in the European Mediterranean coastal destinations that will provide more balanced seasonal tourism demand than what exists at present (Hall and Lew, 2009).

Conclusions

There is now a substantial gap between the perspectives of climate change researchers of

Table 5.3. Characteristics of SIDS.

SIDS (UN DESA, 2010)	SIDS (UNCTAD, 2012)	UN-OHRLLS (2014)[a]	LDC[b]	High Environmental Vulnerability[c]	CO_2 per capita (t) (2008)[d]	CO_2 per capita (average annual growth) (1970–2008)[e]	Endangered species (% of all animal species) (2011)[f]	Fresh water withdrawals (% of total renewable water resources) (2003–12)[g]	No. of deaths due to natural disasters (annual average per million people) (2005–11)[h]	Human Development Index (HDI) (rank) (2012)[i]	Medium/long haul destination in relation to main tourist markets	Tourist receipts in country as % of exports[l]
American Samoa		N		NA	10.35[1]	NA	NA	NA	NA	NA	Y	NA
Anguilla		N		NA	NA	NA	NA	NA	NA	NA	Y	87.0
Antigua and Barbuda	Y	UN		Y	5.1	–0.8	8.3	3.3	0	0.760 (67)	Y	55.0
Aruba		N		NA	21.7[a]	NA	NA	NA	NA	NA	Y	69.4
Bahamas	Y	UN		NA	6.5	–2.2	10.0	NA	3	0.794 (49)	Y	61.8
Bahrain		UN		Y	21.4	1.5	7.2	219.8	NA	0.796 (48)	N	4.7
Barbados	Y	UN		Y	5.0	2.7	8.7	76.1	0	0.825 (38)	Y	53.5
Belize		UN			1.3	0.7	6.4	0.8	13	0.702 (96)	Y	33.2
Bermuda		N		NA	NA	NA	NA	NA	NA	NA	Y	31.4
British Virgin Islands		N		NA	NA	NA	NA	NA	NA	NA	Y	NA
Cape Verde	Y	UN		Y	0.6	4.2	12.5	6.8	0	0.586 (132)	Y	64.3
Cayman Islands		N		NA	NA	NA	NA	NA	NA	NA	Y	NA
Comm. of Northern Marianas		N		NA	NA	NA	NA	NA	NA	NA	Y	NA
Comoros	Y	UN	Y	Y	1.0	NA	11.7	0.8	0	0.429 (169)	Y	44.2
Cook Islands		N		NA	12.7[a]	NA	NA	NA	NA	NA	Y	56.2
Cuba		UN		Y	2.8	0.7	18.1	19.8	0	0.780 (59)	Y[k]	13.1
Curacao		N			NA	NA	NA	NA	NA	NA	Y	25.5
Dominica	Y	UN		NA	1.9	4.4	8.6	NA	15	0.745 (72)	Y	41.2
Dominican Republic		UN		Y	2.2	3.1	16.1	16.6	9	0.702 (96)	Y	32.0
Federated States of Micronesia	Y	UN		NA	0.6	NA	13.7	NA	45	0.645 (117)	Y	41
Fiji	Y	UN		Y	1.5	1.0	13.1	0.3	8	0.702 (96)	Y	30.9

French Polynesia				NA	NA	NA	NA	NA	NA	NA	Y	34.4
Grenada	Y	UN		Y	2.4	4.4	10.5	NA	38	0.770 (63)	Y	63.1
Guadeloupe		N		NA	NA	NA	NA	NA	NA	NA	Y	NA
Guam		N		NA	NA	NA	NA	NA	NA	NA	Y	NA
Guinea-Bissau		UN	Y		0.2	1.4	5.7	0.6	1	0.364 (176)	Y	3.2
Guyana		UN			2.0	−0.2	3.8	0.7	4	0.636 (118)	Y	5.0
Haiti		UN	Y	Y	0.3	3.0	19.4	8.6	65	0.456 (161)	Y	36.6
Jamaica	Y	UN		Y	4.5	1.4	15.2	6.2	3	0.730 (85)	Y	47.6
Kiribati	Y	UN	Y	Y	0.3	−1.0	12.4	NA	0	0.629 (121)	Y	NA
Maldives	Y	UN		Y	3.0	NA	9.1	15.7	0	0.688 (104)	Y	75.0
Marshall Islands	Y	UN		Y	1.9	NA	11.0	NA	NA	NA	Y	15.5
Martinique		N		NA	NA	NA	NA	NA	NA	NA	Y	NA
Mauritius	Y	UN		Y	3.1	4.4	15.2	26.4	1	0.737 (80)	Y	21.0
Montserrat		N		NA	NA	NA	NA	NA	NA	NA	Y	41.0
Nauru	Y	UN		Y	3.9	NA	12.1	NA	NA	NA	Y	NA
New Caledonia		N		NA	NA	NA	NA	NA	NA	NA	Y	8.1
Niue		N		NA	NA	NA	NA	NA	NA	NA	Y	NA
Palau	Y	UN		Y	10.5	−0.3	11.4	NA	NA	0.791 (52)	Y	NA
Papua New Guinea	Y	UN		Y	0.3	0.3	11.4	0.0	4	0.466 (156)	Y	0.03
Puerto Rico		N		NA	NA	NA	NA	NA	NA	NA	Y	NA
Samoa	Y	UN	Y	Y	0.9	3.9	10.8	NA	5	0.702 (96)	Y	54.9
Sao Tome and Principe	Y	UN	Y		0.8	3.7	14.9	0.3	NA	0.525 (144)	Y	43.9
Seychelles	Y	UN		Y	7.8	7.3	16.1	NA	0	0.806 (46)	Y	31.7
Singapore		UN		NA	6.7	−0.7	13.7	31.7	NA	0.895 (18)	Y	3.6
Solomon Islands	Y		Y	Y	0.4	1.1	14.8	NA	4	0.530 (143)	Y	12.1
St Kitts and Nevis	Y	UN		Y	4.9	NA	8.6	NA	NA	0.745 (72)	Y	39.5
St Lucia	Y	UN		NA	2.3	3.4	9.4	NA	6	0.725 (88)	Y	60.0
St Vincent and the Grenadines	Y	UN		Y	1.8	4.7	9.0	NA	0	0.733 (83)	Y	48.1
Suriname		UN			4.7	0.2	3.5	0.5	2	0.684 (105)	Y	3.3
Timor-Leste	Y	UN	Y	NA	0.2	NA	5.2	NA	1	0.806 (134)	Y	21
Tonga	Y	UN		Y	1.7	4.6	8.5	NA	0	0.702 (95)	Y	43.8

Continued

Table 5.3. Continued.

SIDS (UN DESA, 2010)	SIDS (UNCTAD, 2012)	UN-OHRLLS (2014)[a]	LDC[b]	High Environmental Vulnerability[c]	CO_2 per capita (t) (2008)[d]	CO_2 per capita (average annual growth) (1970–2008)[e]	Endangered species (% of all animal species) (2011)[f]	Fresh water withdrawals (% of total renewable water resources) (2003–12)[g]	No. of deaths due to natural disasters (annual average per million people) (2005–11)[h]	Human Development Index (HDI) (rank) (2012)[i]	Medium / long haul destination in relation to main tourist markets	Tourist receipts in country as % of exports[l]
Trinidad and Tobago	Y	UN		Y	37.4	3.7	6.8	6.0	0	0.760 (67)	Y	2.3
Turks and Caicos Islands		N		NA	NA	NA	NA	NA	NA	NA	Y	NA
Tuvalu	Y	UN	Y	Y	NA	NA	13.0	NA	NA	NA	Y	NA
US Virgin Islands				NA	NA	NA	NA	NA	NA	NA	Y	NA
Vanuatu	Y	UN	Y	Y	0.4	–0.4	12.0	NA	0	0.626 (124)	Y	69.3
LDCs[j]					0.2	–0.5	7.6	2.8	20	0.449		
SIDS[j]					2.7	1.4	14.9	NA	16	0.648		6.5–8.2
World[j]					4.5	0.4	11.7	7.3	6	0.694		5.1

[a]The United Nations Office of the High Representative for the Least Developed Countries, Landlocked Developing Countries and the Small Island Developing States (UN-OHRLLS) (2014) Country profiles. Online. Available at: http://unohrlls.org/about-sids/country-profiles (accessed 5 January 2014)

[b]Least Developed Country category according to UNCTAD (2012)

[c]Above average environmental vulnerability on UNEP environmental vulnerability index compared with average for all LDCs (UN DESA, 2010). The Environmental Vulnerability Index is based on 50 indicators covering natural/anthropogenic risks, resilience and ecosystem integrity, and covers issues related to climate change, biodiversity, water, agriculture and fisheries, human health, desertification, and exposure to natural disasters.

[d]Anthropogenic carbon dioxide emissions stemming from the burning of fossil fuels, gas flaring and the production of cement, including carbon dioxide emitted by forest biomass through depletion of forest areas; divided by mid-year population (UNDP, 2013). Figure does not include non-CO_2 greenhouse gas emissions. World Bank (2012) in UNDP (2013)

[e]UNDP (2013) calculations based on data from the World Bank

[f]Percentage of animal species classified as critically endangered, endangered or vulnerable by the International Union for the Conservation of Nature (IUCN) (2012) in UNDP (2013)

[g]Total fresh water withdrawn in a given year, expressed as a percentage of total renewable water resources. FAO (2011) in UNDP (2013)

[h]Number of people confirmed as dead and missing and presumed dead as a result of a natural disaster. Natural disasters are classified as climatological, hydrological and meteorological disasters, which include drought, extreme temperature, flood, mass movement, wet storm and wildfire. UNDP (2013) using United Nations Department of Economic and Social Affairs and Centre for Research on the Epidemiology of Disasters data

[i]The Human Development Index (HDI) is a composite index measuring average achievement in three basic dimensions of human development – a long and healthy life, knowledge and a decent standard of living (UNDP, 2013)

[j]Figures from UNDP (2013)

[k]Following the anticipated normalization of Cuban-US relations it is anticipated that the market mix will change to accommodate the very short haul market of the Cuban diaspora in Florida

[l]UNWTO (2014b). Figures are for 2013 or nearest date available

Table 5.4. Estimated climate change vulnerability of coastal tourism in developing and developed country regions (Gössling *et al.*, 2009a; Hall, 2011; Scott *et al.*, 2012a).

Study	Timeframe assessed	Vulnerability categories	Developing Country Regions	Developed Country Regions
Deutsche Bank Research 2008	2030	Negatively affected (slightly or strongly)[a]	South America, Caribbean/Mexico, South-east Asia (including China and India), Middle East, Africa[b]	Mediterranean Europe and the Balkans, Australia
Gössling and Hall, 2006d; Hall, 2008	Mid-21st century	Moderately to strongly negatively impacted[c]	Africa, Asia, especially coastal China, Latin America, Small Island Nations	Countries bordering the Mediterranean, Florida and south-west USA, Queensland (Australia)
Benfield-UCL Hazard Research Centre for Halifax Travel Insurance, 2006/2007	2030	Identifies at risk destinations[d]	Caribbean, North Africa (Morocco and Tunisia), India and the Indian Ocean, Maldives and Seychelles	Alpine areas in Europe and North America, Mediterranean
The Centre for Future Studies for Churchill Travel Insurance, 2009	2020	Top ten most at risk destinations[d]	Goa (India), Maldives	South-east Spain, Everglades (Florida), Athens, Crete (Greece), Amati Coast and Tuscany (Italy)
Hamilton *et al.*, 2005	2025, +4°C warming scenario	Negative impact on tourist arrivals[e]	Caribbean/Mexico, South America (except Chile), Africa (except Zambia and Zimbabwe), Middle East, South-east Asia (except China)	Southern Europe/ Mediterranean, Australia and New Zealand
UNWTO *et al.*, 2008	Mid-21st century	Vulnerability hotspots[f]	Caribbean, Indian and Pacific Ocean Small Island Nations[g]	Countries bordering the Mediterranean, Australia and New Zealand, USA

[a]Impact criteria considered: climatic changes, regulatory burdens, substitution effects, adaptation capacities. Data sources and indicators for these criteria were not identified

[b]Many nations in the Middle East and Africa were 'not examined', however, no rationale was provided regarding availability of information for selected nations

[c]Impact criteria considered: land and marine biodiversity loss, urbanization, water security, sea level rise, regime change, fuel costs, temperature changes, disease potential

[d]Impact criteria not stated although items noted included climate changes, sea level rise, changing weather patterns; no rationale was provided regarding availability of information for selected nations, destinations and attractions

[e]Impact criteria considered: change in annual average temperature

[f]Impact criteria considered: summer and winter climatic change, increases in extreme events, seal level rise, land and marine biodiversity loss, water scarcity, political destabilization, health impacts/disease potential, transportation costs, and relative importance of tourism to the economy (also utilizing Gössling and Hall's (2006c) results)

[g]South America, Africa, Middle East, South-east Asia were identified as potentially vulnerable, but not listed as 'hotspots' due to insufficient information on magnitude of potential impacts

what is required to limit the negative effects of climate change and the actions of government and industry stakeholders around the world (Hall *et al.*, 2015; Scott *et al.*, 2016). In the case of government this includes not only national governments, but also regional and local government as well. However, in some cases local governments are able to take leadership on the planning and legal aspects of coastal climate change adaptation (Reisinger *et al.*, 2014), so long as they are given leeway to do so by central governments.

Nevertheless, initiating adaptive and mitigative strategies is difficult because of the personal costs of such strategies (Monbiot, 2006). For example, application of the 'polluter pays' principle to an area such as aviation would have significant impacts on airline passenger costs and therefore levels of demand (Hall, 2008). In addition, climate change is now part of wider political debates and appears to be emerging as a factor in electoral campaigns (Hall *et al.*, 2015).

There are clearly enormous practical political difficulties in reducing the carbon footprint of tourism yet still achieve the economic benefits that tourism may bring. As discussed above, tourism is proportionately economically more important to LDCs, and it is germane to the present volume that those economies for which visitor expenditure represents a proportion of GDP above 5% are nearly all coastal destinations and island states. However, it is also clear that there are enormous knowledge gaps in understanding tourism and climate change relationships. Although the IPCC has produced excellent macro-accounts of climate change impacts, some regions, such as Asia, Africa and South America, have had very little research conducted specifically on climate change effects and coastal tourism. Such gaps also make it extremely difficult to understand the extent to which adaptation processes can be effectively engaged in by the tourism sector or by government at various levels. Perhaps most importantly though, it needs to be realized that tourism in coastal destinations is often only a small, albeit significant part, of a much larger environmental, economic and political debate. Tourism-specific adaptation and mitigation efforts are important but in the overall scheme of things there is a need for a much more comprehensive and integrative coastal climate change regime.

References

Alcamo, J., Moreno, J., Nováky, B., Bindi, M., Corobov, R. *et al.* (2007) Europe. In: Parry, M., Canziani, O., Palutikof, J., van der Linden, P. and Hanson, C. (eds) *Climate Change 2007: Impacts, Adaptation and Vulnerability*. Cambridge University Press, Cambridge, pp. 541–580.

Amelung, B. and Moreno, A. (2012) Costing the impact of climate change on tourism in Europe: results of the PESETA project. *Climatic Change* 112(1), 83–100.

Amelung, B. and Viner, D. (2006) Mediterranean tourism: exploring the future with the tourism climatic index. *Journal of Sustainable Tourism* 14, 349–366.

Bohensky, E., Butler, J., Costanza, R., Bohnet, I., Delisle, A. *et al.* (2011) Future makers or future takers? A scenario analysis of climate change and the Great Barrier Reef. *Global Environmental Change* 21(3), 876–893.

Boko, M., Niang, I., Nyong, A., Vogel, C., Githekol, A. *et al.* (2007) Africa. In: Parry, M., Canziani, O., Palutikof, J., van der Linden, P. and Hanson, C. (eds) *Climate Change 2007: Impacts, Adaptation and Vulnerability*. Cambridge University Press, Cambridge, pp. 433–467.

Churchill Insurance (2006) The future of travel: The 'disappearing destinations' of 2020. Churchill Insurance, London.

Cruz, R.V., Harasawa, H., Lal, M., Wu, S., Anokhin, Y. *et al.* (2007) Asia. In: Parry, M., Canziani, O., Palutikof, J., van der Linden, P. and Hanson, C. (eds) *Climate Change 2007: Impacts, Adaptation and Vulnerability*. Cambridge University Press, Cambridge, pp. 469–506.

Deutsche Bank Research (2008) *Climate Change and Tourism: Where Will the Journey Take Us?* Deutsche Bank Research, Berlin.

Field, C.B., Mortsch, L., Brklacich, M., Forbes, D., Kovacs, P. *et al.* (2007) North America. In: Parry, M., Canziani, O., Palutikof, J., van der Linden, P. and Hanson, C. (eds) *Climate Change 2007: Impacts, Adaptation and Vulnerability*. Cambridge University Press, Cambridge, pp. 617–652.

Forbes, D., James, T., Sutherland, M. and Nicholls, S. (2013) Physical basis of coastal adaptation on tropical small islands. *Sustainability Science* 8, 327–344.

Füssel, H. (2010) How inequitable is the global distribution of responsibility, capability, and vulnerability to climate change: a comprehensive indicator-based assessment. *Global Environmental Change* 20(4), 597–611.

Gössling, S. and Hall, C.M. (2006a) Uncertainties in predicting tourist flows under scenarios of climate change. *Climatic Change* 79(3–4), 163–173.
Gössling, S. and Hall, C.M. (2006b) An introduction to tourism and global environmental change. In: Gössling, S. and Hall, C.M. (eds) *Tourism and Global Environmental Change*. Routledge, London, pp. 1–34.
Gössling, S. and Hall, C.M. (2006c) Conclusion: wake up... This is serious. In: Gössling, S. and Hall, C.M. (eds) *Tourism and Global Environmental Change*. Routledge, London, pp. 305–320.
Gössling, S. and Hall, C.M. (2006d) Conclusion: wake up... this is serious. In: Gössling, S. and Hall, C.M. (eds) *Tourism and Global Environmental Change. Ecological, Social, Economic and Political Interrelationships*. Routledge, London, pp. 305–320.
Gössling, S., Hall, C.M. and Scott, D. (2009a) The challenges of tourism as a development strategy in an era of global climate change. In: Palosou, E. (ed.) *Rethinking Development in a Carbon-constrained World. Development Cooperation and Climate Change*. Ministry of Foreign Affairs, Helsinki, pp. 100–119.
Gössling, S., Hall, C.M. and Weaver, D. (eds) (2009b) *Sustainable Tourism Futures: Perspectives on Systems, Restructuring and Innovations*. Routledge, New York.
Gössling, S., Scott, D., Hall, C.M., Ceron, J.-P. and Dubois, G. (2012) Consumer behaviour and demand response of tourists to climate change. *Annals of Tourism Research* 39(1), 36–58.
Gössling, S., Scott, D. and Hall, C.M. (2013) Challenges of tourism in a low-carbon economy. *WIRES Climate Change* 4(6), 525–538.
Gössling, S., Hall, C.M. and Scott, D. (2015) *Tourism and Water*. Channel View, Bristol, UK.
Gössling, S., Ring, A., Dwyer, L., Andersson, A.-C. and Hall, C.M. (2016) Optimizing or maximizing growth? A challenge for sustainable tourism. *Journal of Sustainable Tourism* 24(4), 527–548.
Gurran, N., Norman, B. and Hamin, E. (2013) Climate change adaptation in coastal Australia: an audit of planning practice. *Ocean & Coastal Management* 86, 100–109.
Haer, T., Kalnay, E., Kearney, M. and Moll, H. (2013) Relative sea-level rise and the conterminous United States: consequences of potential land inundation in terms of population at risk and GDP loss. *Global Environmental Change* 23(6), 1627–1636.
Halifax Travel Insurance (2006) Holiday 2030: Climate change to drive radical changes in global tourism. Halifax Press Release, 1 September 2006. Available at: www.hbosplc.com/media/pressreleases/articles/halifax/2006-09-01-05.asp (accessed 1 April 2007).
Hall, C.M. (2005) *Tourism: Rethinking the Social Science of Mobility*. Prentice Hall, Harlow, UK.
Hall, C.M. (2006) Tourism urbanization and global environmental change. In: Gössling, S. and Hall, C.M. (eds) *Tourism and Global Environmental Change: Ecological, Economic, Social and Political Interrelationships*. Routledge, London, pp. 142–156.
Hall, C.M. (2008) Tourism and climate change: knowledge gaps and issues. *Tourism Recreation Research* 33, 339–350.
Hall, C.M. (2010a) Tourism and biodiversity: more significant than climate change? *Journal of Heritage Tourism* 5(4), 253–266.
Hall, C.M. (2010b) An island biogeographical approach to island tourism and biodiversity: an exploratory study of the Caribbean and Pacific Islands. *Asia Pacific Journal of Tourism Research* 15(3), 383–399.
Hall, C.M. (2011) Climate change and its impacts on tourism: regional assessments, knowledge gaps and issues. In: Jones, A. and Phillips, M. (eds) *Disappearing Destinations: Climate Change and the Future Challenges of Coastal Tourism*. CAB International, Wallingford, UK, pp. 10–29.
Hall, C.M. (2015) Global change, islands and sustainable development: Islands of sustainability or analogues of the challenge of sustainable development? In: Redclift, M. and Springett, D. (eds) *Routledge International Handbook of Sustainable Development*. Routledge, Abingdon, UK, pp. 55–73.
Hall, C.M. and Higham, J. (eds) (2005) *Tourism, Recreation and Climate Change*. Channel View Publications, Clevedon, UK.
Hall, C.M. and Lew, A. (2009) *Understanding and Managing Tourism Impacts*. Routledge, London.
Hall, C.M. and Ram, Y. (2016) Heritage in the intergovernmental panel on climate change (IPCC) assessment reports: a lexical assessment. *Journal of Heritage Tourism* 11(1), 96–104.
Hall, C.M., James, M. and Baird, T. (2011) Forests and trees as charismatic mega-flora: implications for heritage tourism and conservation. *Journal of Heritage Tourism* 6(4), 309–323.
Hall, C.M., Amelung, B., Cohen, S., Eijgelaar, E., Gössling, S. *et al.* (2015) On climate change skepticism and denial in tourism. *Journal of Sustainable Tourism* 23(1), 4–25.
Hall, C.M., Baird, T., James, M. and Ram, Y. (2016) Climate change and cultural heritage: conservation and heritage tourism in the Anthropocene. *Journal of Heritage Tourism* 11(1), 10–24.

Hamilton, J., Maddison, D. and Tol, R. (2005) Climate change and international tourism: a simulation study. *Global Environmental Change* 15, 253–266.
Hijioka, Y., Lin, E., Pereira, J., Corlett, R., Cui, X. *et al.* (2014) Asia. In: Barros, V., Field, C., Dokken, D., Mastrandrea, M., Mach, K. *et al.* (eds) *Climate Change 2014: Impacts, Adaptation, and Vulnerability. Part B: Regional Aspects. Contribution of Working Group II to the Fifth Assessment Report of the Intergovernmental Panel on Climate Change*. Cambridge University Press, Cambridge, pp. 1327–1370.
IPCC (2014a) *Climate Change 2014: Impacts, Adaptation, and Vulnerability. Part A: Global and Sectoral Aspects. Contribution of Working Group II to the Fifth Assessment Report of the Intergovernmental Panel on Climate Change*. Cambridge University Press, Cambridge.
IPCC (2014b) *Climate Change 2014: Impacts, Adaptation, and Vulnerability. Part B: Regional Aspects. Contribution of Working Group II to the Fifth Assessment Report of the Intergovernmental Panel on Climate Change*. Cambridge University Press, Cambridge.
IPCC (2014c) *Climate Change 2014: Mitigation of Climate Change. Contribution of Working Group III to the Fifth Assessment Report of the Intergovernmental Panel on Climate Change*. Cambridge University Press, Cambridge.
Kebede, A. and Nicholls, R. (2012) Exposure and vulnerability to climate extremes: population and asset exposure to coastal flooding in Dar es Salaam, Tanzania. *Regional Environmental Change* 12(1), 81–94.
Martínez, M., Mendoza-González, G., Silva-Casarín, R. and Mendoza-Baldwin, E. (2014) Land use changes and sea level rise may induce a 'coastal squeeze' on the coasts of Veracruz, Mexico. *Global Environmental Change* 29, 180–188.
Mimura, N., Nurse, L., McLean, R., Agard, J., Briguglio, L. *et al.* (2007) Small islands. In: Parry, M., Canziani, O., Palutikof, J., van der Linden, P. and Hanson, C. (eds) *Climate Change 2007: Impacts, Adaptation and Vulnerability*. Cambridge University Press, Cambridge, pp. 687–716.
Mitsch, W. and Hernandez, M. (2013) Landscape and climate change threats to wetlands of North and Central America. *Aquatic Sciences* 75(1), 133–149.
Monbiot, G. (2006) *Heat: How to Stop the Planet Burning*. Allen Lane, London.
Moser, S., Jeffress Williams, S. and Boesch, D. (2012) Wicked challenges at land's end: managing coastal vulnerability under climate change. *Annual Review of Environment and Resources* 37, 51–78.
Niang, I., Ruppel, O., Abdrabo, M., Essel, A., Lennard, C. *et al.* (2014) Africa. In: Barros, V., Field, C., Dokken, D., Mastrandrea, M., Mach, K. *et al.* (eds) *Climate Change 2014: Impacts, Adaptation, and Vulnerability. Part B: Regional Aspects. Contribution of Working Group II to the Fifth Assessment Report of the Intergovernmental Panel on Climate Change*. Cambridge University Press, Cambridge, pp. 1199–1265.
Nicholls, R.J., Wong, P.P., Burkett, V.R., Codignotto, J.O., Hay, J.E. *et al.* (2007) Coastal systems and low-lying areas. In: Parry, M., Canziani, O., Palutikof, J., van der Linden, P. and Hanson, C. (eds) *Climate Change 2007: Impacts, Adaptation and Vulnerability*. Cambridge University Press, Cambridge, pp. 315–356.
Parry, M.L., Canziani, O.F., Palutikof, J.P., van der Linden, P.J. and Hanson, C.E. (eds) (2007) *Climate Change 2007: Impacts, Adaptation and Vulnerability. Contribution of Working Group II to the Fourth Assessment Report of the Intergovernmental Panel on Climate Change*. Cambridge University Press, Cambridge.
Reisinger, A., Kitching, R., Chiew, F., Hughes, L., Newton, P. *et al.* (2014) Australasia. In: Barros, V., Field, C., Dokken, D., Mastrandrea, M., Mach, K. *et al.* (eds) *Climate Change 2014: Impacts, Adaptation, and Vulnerability. Part B: Regional Aspects. Contribution of Working Group II to the Fifth Assessment Report of the Intergovernmental Panel on Climate Change*. Cambridge University Press, Cambridge, pp. 1371–1438.
Scott, D. and Gössling, S. (2015) Tourism in the future(s): forecasting and scenarios. In: Hall, C.M., Gössling, S. and Scott, D. (eds) *The Routledge Handbook of Tourism and Sustainability*. Routledge, Abingdon, pp. 305–319.
Scott, D., Gössling, S. and Hall, C.M. (2012a) *Tourism and Climate Change: Impacts, Adaptation and Mitigation*. Routledge, London.
Scott, D., Simpson, M. and Sim, R. (2012b) The vulnerability of Caribbean coastal tourism to scenarios of climate change related sea level rise. *Journal of Sustainable Tourism* 20(6), 883–898.
Scott, D., Hall, C.M. and Gössling, S. (2016) A review of the IPCC 5th assessment and implications for tourism sector climate resilience and decarbonization. *Journal of Sustainable Tourism* 24(1), 8–30.
Stern, N. (2006) *The Economics of Climate Change: The Stern Review*. Cambridge University Press, Cambridge.
Su, Y.-P. and Hall, C.M. (2014) Climate change and tourism in Asia: a review. In: *Responding to Climate Change: Tourism Initiatives in Asia and the Pacific*. UNWTO, Madrid, pp. 27–40.

United Nations Conference on Trade and Development (UNCTAD) (2012) *UNCTAD Handbook of Statistics 2012*. UNCTAD, New York.

United Nations Department of Economic and Social Affairs (UN DESA) (2010) *Trends in Sustainable Development: Small Island Developing States*. UN DESA, New York.

United Nations Development Programme (UNDP) (2013) *Human Development Report 2013. The Rise of the South: Human Progress in a Diverse World*. UNDP, New York.

UNWTO (2014a) *Responding to Climate Change: Tourism Initiatives in Asia and the Pacific*. UNWTO, Madrid.

UNWTO (2014b) *Tourism in Small Island Developing States (SIDS): Building a More Sustainable Future for the People of Islands*. UNWTO, Madrid.

UNWTO, UNEP and WMO (2008) *Climate Change and Tourism: Responding to Global Challenges*. United Nations World Tourism Organization, Nations Environment Program, World Meteorological Organization, Madrid.

Wilbanks, T.J., Romero Lankao, P., Bao, M., Berkhout, F., Cairncross, S. *et al.* (2007) Industry, settlement and society. In: Parry, M., Canziani, O., Palutikof, J., van der Linden, P. and Hanson, C. (eds) *Climate Change 2007: Impacts, Adaptation and Vulnerability*. Cambridge University Press, Cambridge, pp. 357–390.

Wilby, R. and Keenan, R. (2012) Adapting to flood risk under climate change. *Progress in Physical Geography* 36(3), 348–378.

Zeppel, H. (2012) Climate change and tourism in the Great Barrier Reef Marine Park. *Current Issues in Tourism* 15(3), 287–292.

6 Assessing the Climate Change Risk of a Coastal-Island Destination

Daniel Scott* and Stephanie Verkoeyen
University of Waterloo, Ontario, Canada

Introduction

Coastal tourism is considered to be one of the largest tourism market segments globally (Hall, 2001; Honey and Krantz, 2007; Phillips and House, 2009) as well as one of the most vulnerable to the impacts of global climate change (Scott *et al.*, 2012a, 2015). As is discussed at length in Chapter 1, sea level rise (SLR), ocean temperatures and ocean acidification are three salient manifestations of climate change that are influencing coastal systems worldwide, such that the Intergovernmental Panel on Climate Change (IPCC) (Wong *et al.*, 2014, p. 14) concluded that '[t]here is no wetland, mangrove, estuary, rocky shore or coral reef which is not exhibiting some degree of impact'. Despite the transformative risk posed by climate change, the multifaceted nature of its impacts remains poorly quantified in the vast majority of coastal destinations. The interactions between impacts and the cumulative effect on coastal destinations remains an even larger knowledge gap.

Coastal tourism is a major market segment in many developed (e.g. the USA, Mediterranean Europe, and Australia) and developing (e.g. Mexico, Brazil, Thailand, South Africa and Egypt) nations, but plays a particularly pivotal role in Small Island Developing States (SIDS). SIDS are also considered to be among the most vulnerable countries to changes in climate (IPCC, 2014), while their often peripheral location heightens vulnerability to changes in transportation costs arising from change in climate change mitigation policy. This perfect storm of climate and carbon risks for SIDS tourism economies is a salient knowledge gap, particularly if tourism is to contribute meaningfully to progress on the 2030 Sustainable Development Goals (UN General Assembly, 2015).

This chapter will begin with a brief review of the coverage on coastal tourism in the most recent IPCC Fifth Assessment Report (AR5) (IPCC, 2014), including the relative state of knowledge regionally and identified knowledge gaps. The discussion will then outline the multiple and interdependent impacts of climate change that must be assessed if climate and carbon risks are to be comprehensively understood at the destination scale; information that is the foundation for informed adaptation decision-making. Finally, the chapter will examine the relative vulnerability of SIDS tourism to climate change using an index approach.

*Corresponding author, e-mail: daniel.scott@uwaterloo.ca

Coastal Tourism in the IPCC Fifth Assessment Report

The IPCC's periodic assessment reports, typically every 7 years, critically assess and synthesize the state of knowledge on climate change, its possible impacts, and options for adaptation and mitigation. A review of the evolving place of tourism in the IPCC assessments provides a useful barometer of progress and key knowledge gaps.

An overarching sentiment of the discussion of tourism within the AR5 is that the potentially significant effects of climate change remain poorly quantified and poorly integrated into local vulnerability studies (Scott *et al.*, 2015). Though the trans-boundary impacts of climate change are recognized as potentially outweighing local impacts in many destinations, such effects remain the least explored. Knowledge gaps related to the limited understanding of tourist response to climate change impacts and climate mitigation policies, climate sensitivity detection and attribution analysis (including analogue studies) and climate adaptation (potential and barriers) are identified as research priorities (Scott *et al.*, 2015).

Specific to coastal tourism, the considerable risk posed by SLR was recognized in the majority of regional chapters, but the long-term implications of this were identified as a knowledge gap in the Australasia, Central & South America and Small Islands chapters. More importantly, the identification of SLR risks to coastal tourism in the Asia, Europe and North America chapters were conspicuous in their absence. While the pace of SLR remains uncertain, that it will continue to occur for centuries is one of the most certain impacts associated with climate change. Scott *et al.*'s (2012a) conclusion that the tourism community has thus far failed to assess its risk to SLR remains valid.

A review of tourism content in the AR5 chapters most relevant to coastal tourism produced mixed results, with a slight decrease in content in the Small Islands chapter, a slight increase in Coastal and Low-lying Areas, and a significant increase in the Oceans chapter. Table 6.1 presents the relative state of knowledge in several key impact areas for Small Islands and Ocean regions. Few of the impact types are evidenced in either chapter; the many shaded areas represent identified or unidentified knowledge gaps. The need to evaluate better the potential impact of SLR and the associated erosion of high-value beach areas, as well as better understand tourists' potential response to degraded beaches and adaptation responses (including beach nourishment, and soft and hard engineering approaches) were identified as research priorities, while potential implications of adaptation by the insurance industry was an overlooked research need.

Multi-dimensionality of Climate Change Impacts on Coastal Tourism

Considering the multiple ways in which climate change will affect the future prospects of tourism is the first step towards an integrated vulnerability assessment and informed considerations of potential adaptation responses.

Figure 6.1 outlines four broad climate change impact pathways for coastal tourism, adapted from the conceptual framework of Scott *et al.* (2012a): (i) direct climate impacts; (ii) indirect climate-induced environmental impacts; (iii) indirect climate-induced socio-economic impacts; and (iv) mitigation and

Table 6.1. Climate impacts identified in IPCC coastal-related regional chapters.

Region	Climate resources and extremes	Ecosystems	Sea level rise	Water resources	Security	Transport cost/access	Adaptation options
Small Islands	I/E	I/E	I/gap	E	?	I/gap	I
Oceans	?	I/E	I/gap	?	?	I/gap	?

gap, knowledge gap specified; I, potential impact identified; E, impact evidenced; ?, impact not identified

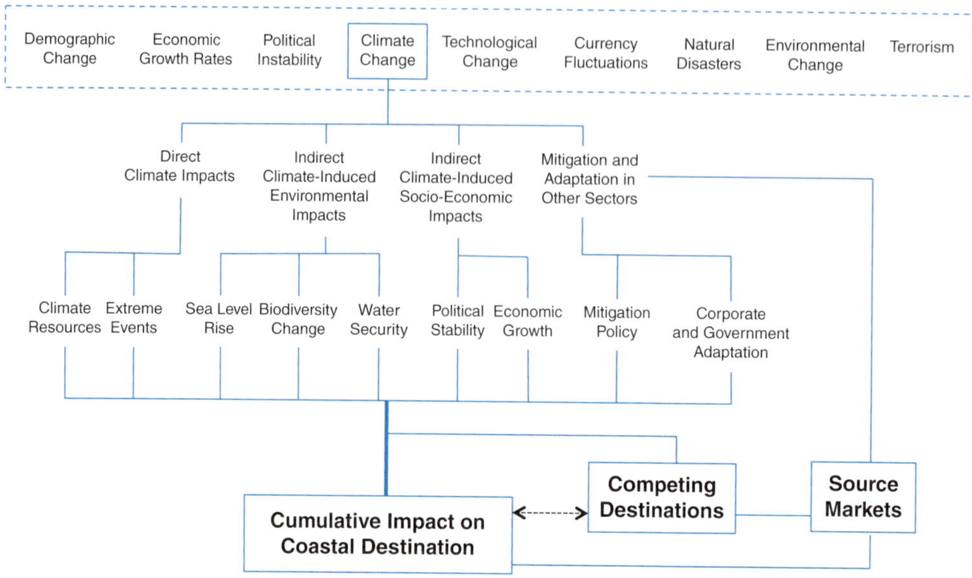

Fig. 6.1. Climate change impact pathways on coastal tourism.

adaptation policy responses of other sectors. These pathways will affect all of the major components of the global tourism system (tourists, source markets, transport systems and destinations), though the aggregate impact will vary substantially among tourism sub-sectors and geographical regions depending on differences in exposure, sensitivity and adaptive capacity. Each of the four pathways is subsequently discussed.

Direct climatic impacts

Climate change is expected to directly affect climate conditions (temperature, precipitation, humidity and winds) and extremes (frequency and severity of storm events or droughts) (IPCC, 2014). Changes in climate affect the length and quality of climate-dependent tourism seasons, operating costs, infrastructure damage and business interruptions, destination attractiveness, and tourist demand and destination choice. Tourism climate indices have been used in multiple studies to examine the redistribution of climate resources for tourism, revealing generally consistent temporal and geographical patterns. In many subtropical and tropical destinations, conditions ideal for tourist activity are expected to degrade, particularly in the peak summer tourism season (Scott *et al.*, 2012a). These impacts are particularly pronounced under warmer scenarios and in the latter half of the century. Consequently, demand for international travel to subtropical and tropical countries is projected to decline, with fewer arrivals from temperate nations (Hamilton *et al.*, 2005). However, coastal destinations may be less affected by temperature-related impacts than urban destinations in tropical and subtropical regions, as the former benefit from moderating influences of the sea and coastal breezes and higher thermal preferences of visitors (Rutty and Scott, 2010, 2014).

Changes in precipitation and water availability are one of the main mechanisms by which the impacts of climate change will be experienced at the destination scale. Tourism is both dependent on freshwater resources and an important factor in freshwater use in many destinations (Gössling *et al.*, 2012, 2015). As population and economic growth increase demand for water in other sectors over the next 20–30 years, water stress and inter-sectoral water competition is expected to intensify in many tourism regions (Gössling *et al.*, 2015). Where tourism will be ranked in the list of priority water users remains largely unasked within the tourism sector. In addition to non-climatic drivers of freshwater use, climate change is expected to become an increasingly important determinant of freshwater resource availability (Jiménez Cisneros *et al.*, 2014). By mid-century, increased water stress from population and economic growth in combination with climate change is expected in several mature tourism markets (Mediterranean, Caribbean, Australia), but also many watersheds where significant growth in international tourism is a key strategy for economic development strategy and poverty alleviation through the 2040s (including north-east and south-east Asia, northern and central Africa, the Middle East and many SIDS) (Gössling *et al.*, 2015). Changes in precipitation and water supply come at a cost to tourist operations. While there has been little research exploring these potential changes in business costs in the tourism sector, the international business community is increasingly turning its collective attention to the issue of water security and price as a strategic operational planning concern (Morrison *et al.*, 2009; Ernst and Young, 2012).

In addition to changes to mean climate conditions, the IPCC has concluded that changes in a number of weather extremes are very probable, including more intense and frequent heatwaves, more intense precipitation events and greater tropical storm intensity (IPCC, 2014). The intensification of tropical storms is of especial concern to SIDS, particularly single-island states, where such events can disrupt ecological systems and most of the economy simultaneously (Anthoff *et al.*, 2010). The Caribbean is one of the regions most affected by extreme weather events (Kreft *et al.*, 2014). The damage incurred by recent hurricanes suggests that the cost to tourism infrastructure can be significant. For example, Hurricane Wilma in 2005 caused US$1.5 billion in damage to the hotel sector in Mexico (Randewich, 2005), plus more than US$70 million to replenish beach assets in Cancun alone (Barrell, 2010). Regardless of the direct cost, there is also an indirect price to pay in terms of business interruption and destination attractiveness (Buzinde *et al.*, 2010).

Indirect climate-induced environmental change

Indirect climate-induced environmental change affects natural assets that define destination image and are critical attractions for tourists. SLR poses one of the most widely recognized climate change threats to low-lying coastal areas (Nicholls and Cazenave, 2010; Church and White, 2011). If emissions continue unmitigated, experts predict a probabilistic SLR ranging from 0.7 to 1.2 m by 2100 (Horton *et al.*, 2014), a scenario that would see the world's coastlines transformed. Beyond erosion and inundation, rising sea levels can impede drainage and increase the risk of riverine flooding, salinity intrusions into freshwater supplies, and coastal habitat loss through the coastal squeeze. Engineering and geomatics-based techniques have rarely been used to quantify the magnitude of impacts on tourism assets (beaches, cultural heritage resources) and infrastructure (resorts, transportation networks) under SLR projections. One exception is a study of over 900 resort properties in 19 Caribbean nations, which found 29% of coastal resorts vulnerable to partial or full inundation by a 1 m SLR, with a substantially higher proportion (49–60%) vulnerable to associated coastal erosion (Scott *et al.*, 2012b). As the visualization of the impact of 1 m SLR in Fig. 6.2 illustrates, when such SLR scenarios are realized the competitiveness of coastal destinations worldwide would be transformed, with implications for property values, pricing structures, insurance, marketing and destination image (Scott *et al.*, 2012a).

In many coastal destinations, coral reefs are a key visitor attraction and major economic asset. The synergistic effects of ocean warming, which has accelerated at a rate 50% faster over the last 4 decades of the 20th century (Durack *et al.*, 2014), and acidification, which has increased 30% (Friedrich *et al.*, 2012), will increase coral mortality, reduce calcification and enhance skeletal

Fig. 6.2. Visualization of 1 m SLR at the resort property level.

dissolution (Manzello *et al.*, 2008), exacerbating other human impacts on the coastal and marine environments within which reefs are situated (e.g. pollution, over-fishing, damage by tourists). By the 2030s, estimates from the World Resources Institute (Burke *et al.*, 2011) predict that more than 90% of the world's reefs will be threatened by local human activities, warming and acidification, with nearly 60% facing high, very high or critical threat levels. Thirty per cent of reefs will shift from low threat to medium or higher threat specifically due to climate or ocean chemistry changes. By 2050, high thermal stress will be ubiquitous. Change in reef health in turn has implications for marine biodiversity, which is both a resource for sport fishing and a supply of fish to tourists in many destinations. The density of reef fish is likely to decrease as a result of increasing post-settlement mortality arising from a lack of hiding places and appropriate food for newly settled juveniles (Wilson *et al.*, 2006). As the spatial scale of disturbances to reefs increases, fish populations will become increasingly fragmented and more susceptible to persistent declines in abundance (Munday, 2004).

Indirect climate-induced socio-economic change

Evidence continues to accumulate that climate change will also take a toll on economic growth, political stability and sustainable development (IPCC, 2014). New evidence suggests climate change could cause ten times as much damage to the global economy as previously estimated, with unmitigated warming reducing average global incomes roughly 23% by 2100 (Burke *et al.*, 2015). Reductions in global GDP in turn reduces the discretionary wealth available for tourism, and is expected to have negative implications for anticipated future growth, though there are no explicit model outputs that provide insight into the future implications of climate change-related economic change for global tourism. The IPCC similarly concluded that climate change would impede the sustainable development of many nations, becoming a security risk that would steadily intensify, particularly under greater warming scenarios (Gemenne *et al.*, 2014). This increased security risk has far-reaching consequences (e.g. reduced access to and quality of certain natural resources), which can lead to large-scale migrations of people, destabilize fragile governments and restrict the freedom of mobility that is fundamental to international tourism. Climate-induced migrations are consistently projected for all SIDS (Smith and Vivekananda, 2007), influencing population densities, intensifying land-use pressures and resource competition, even potentially affecting the availability of human resources for the tourism industry. Contemporary evidence following acts of terrorism, war, or social unrest and political revolutions clearly demonstrates international tourists' aversion to political and social instability (Hall, 2010).

Mitigation and adaptation in other sectors

Climate change mitigation and adaptation policy responses in other sectors will also have important implications for tourism development and operating costs. With respect to the former, discussions have focused almost exclusively on the aviation sector. While currently proposed international aviation policies would have a limited impact on tourism demand (Mayor and Tol, 2007, 2010; Gössling *et al.*, 2008; Pentelow and Scott, 2010, 2011), more significant decreases in tourist arrivals are projected for SIDS under more stringent mitigation policy scenarios (Gössling *et al.*, 2008; Pentelow and Scott, 2011). A review of energy scenarios developed to achieve the 2°C policy target of the international community found that the implications for international tourism have not been explicitly considered, rather the assumption is demand management will play a large role in aviation achieving required emission reductions (Scott and Gössling, 2015). Such a policy framework would be an unsettling prospect for most SIDS and other destinations with significant long-haul tourism markets.

The adaptation response of the insurance industry will also pose a risk to coastal tourism destinations. Tourism infrastructure in high-risk areas (e.g. hurricane-prone coastlines) may become uninsurable or see insurance became unaffordable for many smaller tourism operators. The lack of insurance would restrict reinvestment and further development in some coastal areas. Similarly, under-insurance, as a result of high premiums, would reduce tourism operators' capacity to rebuild or invest in property improvement.

Interdependencies of climate change impact pathways

The diverse impacts associated with the four pathways do not occur in isolation, but rather are interdependent. A vulnerability matrix for coastal destinations is illustrated in Fig. 6.3 using only five major types of impacts, the timing and magnitude of which will vary by location. Achieving a complete understanding of the climate change vulnerability of any coastal destination requires the integration of these multiple dimensions. Scott *et al.* (2015) emphasize that our understanding of the consequences of climate change for destinations is often limited to a single dimension, and '[o]ne of the largest barriers to understanding the implications of climate change for tourism destinations has been the lack of integrated sectoral assessments that analyze the full range of potential impacts and their interactions' (p. 12). The IPCC AR5 similarly concluded that to date there are no tourism studies available that fully 'consider all aspects of the impact of climate change for particular countries or regions' (Arent *et al.*, 2014, p. 679).

The risks of not considering climate change interdependencies and the adaptive responses of other political or business organizations and major competitors for local-scale destinations are further illustrated in Fig. 6.4. Because tourists prefer beach nourishment

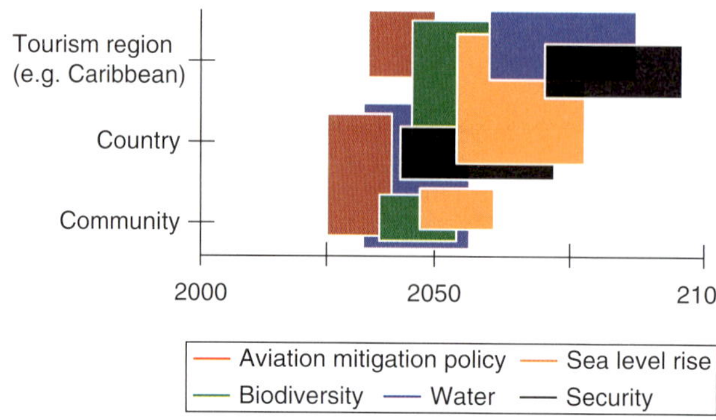

Fig. 6.3. Interdependent Impacts of climate change (Scott *et al.*, 2015).

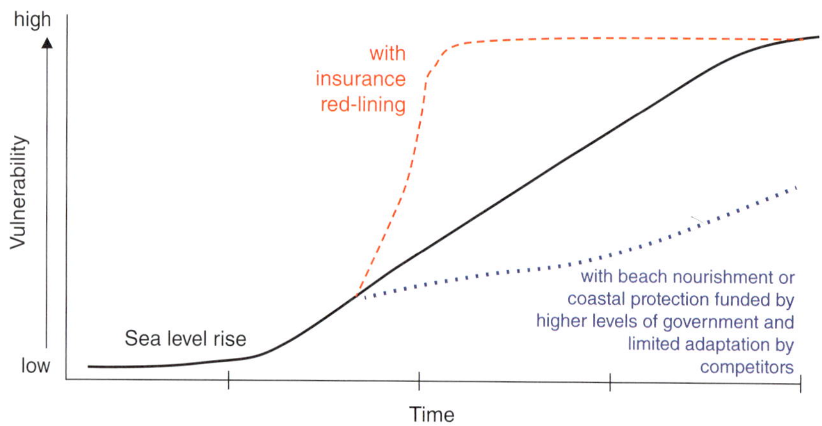

Fig. 6.4. Multifaceted destination vulnerability to sea level rise (Scott *et al.*, 2015).

to artificial engineered coastlines (Hamilton, 2007), if competing destinations cannot implement beach nourishment the relative vulnerability of this destination would decrease and its competitiveness in the coastal tourism marketplace increase. However, if the insurance industry deems coastal risks at this location too high to continue property insurance, regardless of any investment in beach nourishment, the loss of insurability would negate the reduced vulnerability from beach nourishment.

A Climate Change Vulnerability Index for SIDS Tourism

One approach to assess the vulnerability of coastal tourism to the multiple risks posed by climate change is by combining multiple indicators into a vulnerability index. Such an approach has been utilized in a number of studies to compare the relative climate change vulnerability of countries (e.g. Burke *et al.*, 2011; Wheeler, 2011; ND-GAIN, 2012; Kreft *et al.*, 2014). However, few of these climate change vulnerability indices include tourism, and those that do inadequately represent the sector. For example, the DARA index (DARA and the Climate Vulnerability Forum, 2012) ranks the tourism sector vulnerability of 184 countries using two indicators: reef-based tourism and low-elevation ski tourism, despite the fact that neither of these types of tourism may be relevant in many countries. Nor can such a limited approach capture the range of impacts previously discussed.

The index presented here integrates indicators for all four of the impact pathways identified in Fig. 6.1. Variations across these dimensions are used to establish the relative vulnerability of tourism in 28 countries. Focusing on relative vulnerability takes into account that climate change is projected to alter the pattern of tourism flows rather than overall global tourism demand (Hamilton *et al.*, 2005). While SIDS are recognized as having a high overall vulnerability to climate change and the highest economic reliance on tourism, there has been no investigation into the degree to which the vulnerability of the sector varies across island nations. For this reason, country selection was restricted to SIDS, with the exception of Greece and Australia, which were included as comparative developed countries with major coastal tourism markets.

For each of the impact pathways, one or more factors were identified and represented by available indicators (Table 6.2), selection of which was influenced by data availability for a large number of countries, a challenge exacerbated in the case of SIDS, which often have limited data availability. In most cases the selected indicators project risk exposure under some future climate scenario. However, in some instances the indicator reflects current vulnerability, because of the lack of future projections. Present-day indicators are nevertheless valuable in establishing relative vulnerability, e.g. the relative risk exposure of future cyclone intensity can be inferred from a measure of current cyclone damage and disruption risk. Risk exposure is defined in terms of destination attractions (e.g. climate resources, marine biodiversity, coral reefs), tourism demand, operational costs (water, energy) and exposure to nationwide threats posed by extreme weather and SLR. Sensitivity to risk considers both security of the state and its vulnerability to social unrest or conflict, as well as dependence on leisure tourism arrivals. Finally, adaptive capacity is assessed both at the nation-level and specifically for the tourism industry, incorporating indicators of human development and global adaptation, and destination competitiveness, respectively. To enable comparison and index aggregation, all individual indicators were converted to an ordinal scale of 1 and 5

Table 6.2. List of indicators combined in the vulnerability index.

Dimension	Factor	Indicator	Source
Exposure	Attractions	Tourism climate index	Amelung *et al.* (2007)
		Projected change in marine biodiversity (ecosystem service)	ND-GAIN (2012)
		Vulnerability to reef loss	Burke *et al.* (2011)
	Tourism demand	Change in arrivals	Bigano *et al.* (2006)
		Average distance of top 5 markets	D. Scott and S. Verkoeyen (own data)
	Operational costs	Water Stress Index (present)	Reig *et al.* (2013)
		Projected water stress (2050s)	Schlosser *et al.* (2014)
		Electricity from fossil fuel (%)	World Bank (2011)
		Energy cost increase	Raupach *et al.* (2014)
	Country	Climate risk index	Kreft *et al.* (2014)
		Climate extremes index	Wheeler (2011)
		Land area below 5 m sea level (%)	CIESIN (2013)
		SLR risk index	Wheeler (2011)
		Tropical cyclone intensity (present)	Peduzzi *et al.* (2012)
Sensitivity	Security	Fragile state index	Fund For Peace (2014)
		Leisure arrivals (%)	UNWTO (2012)
Adaptive capacity	Tourism	Tourism competitiveness index	Blanke *et al.* (2013)
		Country Brand Index	FutureBrand (2012)
	Country	Global Adaptation Index	ND-GAIN (2012)
		Human Development Index	Malik (2014)

(using quintiles or existing original ranking categories), with lower values representing low vulnerability (low exposure, low sensitivity or high adaptive capacity). The 20 indicators in Table 6.2 were equally weighted in the index, however alternative weighting could be applied to reflect impacts more likely to occur in the near term or based on expert ranking of impact salience.

Table 6.3 identifies the countries most at risk to the direct and indirect impacts of climatic change (climate risk), changes in mitigation policy (carbon risk), and those with the lowest adaptive capacity. When all of the indicators are combined, the top five most at risk tourism sectors are found in Haiti, Dominican Republic, Fiji, Cuba and the Philippines.

Figure 6.5 compares the relative vulnerability of each country's tourism sector against its economic contribution. The countries in the top right quadrant, with high tourism sector vulnerability and high economic dependence on tourism, represent the countries of greatest concern. Those in the top left quadrant, with moderate tourism sector vulnerability but high economic dependence on tourism (e.g. over 50% of GDP in Antigua and Barbuda, Seychelles and Maldives), may experience a greater overall

Table 6.3. Countries where coastal tourism is most at risk (top quartile, with ties).

Climate risk	Carbon risk	Adaptive capacity (low)	Total climate change risk
Haiti	Australia	Haiti	Haiti
Fiji	Dominican Republic	Togo	Dominican Republic
Dominican Republic	Cyprus	Madagascar	Fiji
Solomon Islands	Cuba	Dominican Republic	Cuba
Cuba	Fiji	Vanuatu	Philippines
Philippines	Jamaica	Cuba	Solomon Islands
Vanuatu	Kiribati	Philippines	Madagascar
	Tonga		

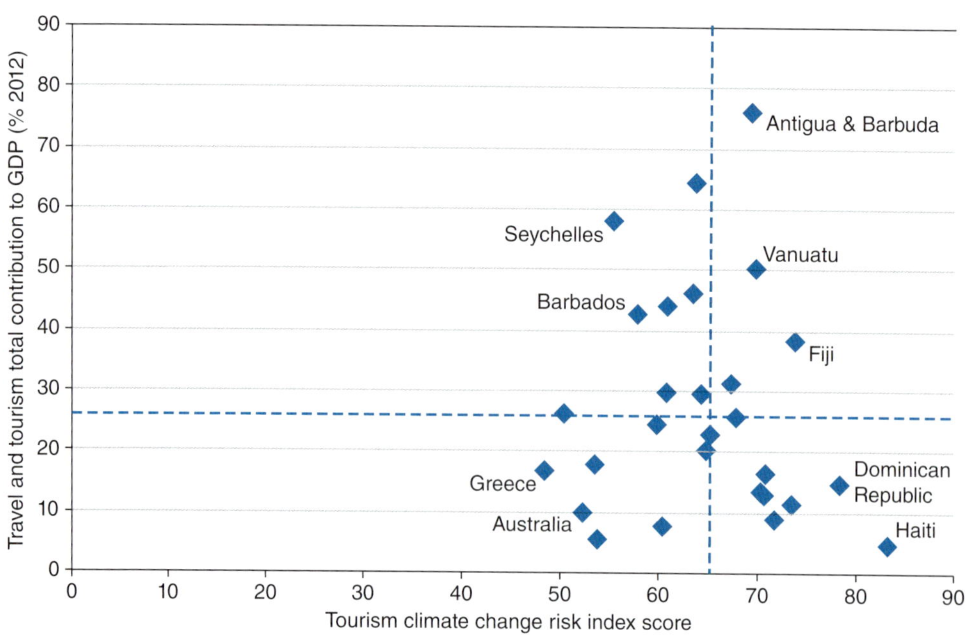

Fig. 6.5. Comparison of climate change risk and economic importance of tourism.

impact than countries that exhibit higher tourism sector vulnerability, but are far less dependent on tourism as a source of income and employment. All of the SIDS in this analysis include tourism growth as part of their development agenda, yet there is no evidence from National Adaptation Programmes of Action (submitted as of 2015) that any of these countries consider the implications of climate change as a potential constraint on tourism development.

Conclusion

Impacts attributable to climate change have already begun. Yet, the vast majority of coastal tourism destinations have a very limited understanding of the potential long-range risks climate change poses to tourism development and sustainable development more broadly. James Michel (2014), the President of the Alliance of Small Island States, was very clear when he stated 'there can be no discourse on security, trade, human rights, and development, unless we are also addressing climate change'. And so it is with sustainable tourism.

Although all tourism destinations will need to adapt to climate change, whether to minimize risks or to capitalize on new opportunities, the adaptation imperative has yet to translate into discernible action within the tourism sector (Scott *et al.*, 2015). The need for innovative vulnerability assessment approaches capable of advancing our understanding of the integrated effect of multiple climate change impacts in coastal destinations is an important research priority.

References

Amelung, B., Nicholls, S. and Viner, D. (2007) Implications of global climate change for tourism flows and seasonality. *Journal of Travel Research* 45(3), 285–296.

Anthoff, D., Nicholls, R.J. and Tol, R.S. (2010) The economic impact of substantial sea-level rise. *Mitigation and Adaptation Strategies for Global Change* 15(4), 321–335.

Arent, D.J., Tol, R., Faust, E., Hella, J., Kumar, S. *et al.* (2014) Key economic sectors and services. In: Field, C.B., Barros, V., Dokken, D., Mach, K., Mastrandrea, M. *et al.* (eds) *Climate Change 2014: Impacts, Adaptation, and Vulnerability*. Cambridge University Press, Cambridge, pp. 659–708.

Barrell, S. (2010) The battle for the beaches of Cancun. *Independent*, 8 May 2010. Available at: http://www.independent.co.uk/travel/americas/the-battle-for-the-beaches-of-cancun-1968917.html (accessed 14 December 2015).

Bigano, A., Hamilton, J.M. and Tol, R.S. (2006) The impact of climate on holiday destination choice. *Climatic Change* 76(3–4), 389–406.

Blanke, J., Chiesa, T. and Crotti, R. (2013) The travel and tourism competitiveness index 2013: contributing to national growth and employment. In: Blanke, J. and Chiesa, T. (eds) *The Travel and Tourism Competitiveness Report 2013*. World Economic Forum, Geneva, Switzerland, pp. 3–42.

Burke, L.M., Reytar, K., Spalding, M. and Perry, A. (2011) *Reefs at Risk Revisited*. World Resources Institute, Washington, DC, 114 pp.

Burke, M., Dykema, J., Lobell, D.B., Miguel, E. and Satyanath, S. (2015) Incorporating climate uncertainty into estimates of climate change impacts. *Review of Economics and Statistics* 97(2), 461–471.

Buzinde, C.N., Manuel-Navarrete, D., Yoo, E.E. and Morais, D. (2010) Tourists' perceptions in a climate of change: Eroding destinations. *Annals of Tourism Research* 37(2), 333–354.

Center for International Earth Science Information Network (CIESIN) (2013) *Sea Level Rise Impacts on Ramsar Wetlands of International Importance*. NASA Socioeconomic Data and Applications Center (SEDAC), Palisades, New York.

Church, J.A. and White, N.J. (2011) Sea-level rise from the late 19th to the early 21st century. *Surveys in Geophysics* 32(4–5), 585–602.

DARA and the Climate Vulnerable Forum (2012) *Climate Vulnerability Monitor: A Guide to the Cold Calculus of a Hot Planet*. Fundación DARA Internacional, Madrid.

Durack, P.J., Gleckler, P.J., Landerer, F.W. and Taylor, K.E. (2014) Quantifying underestimates of long-term upper-ocean warming. *Nature Climate Change* 4(11), 999–1005.

Ernst and Young (2012) *Water Resources at the Corporate Level: Moving from a Risk Based Approach to Active Management*. Ernst and Young, New York.

Friedrich, T., Timmermann, A., Abe-Ouchi, A., Bates, N.R., Chikamoto, M.O. *et al.* (2012) Detecting regional anthropogenic trends in ocean acidification against natural variability. *Nature Climate Change* 2(3), 167–171.

Fund for Peace (2014) Fragile States Index. Available at: http://fundforpeace.org/fsi/data (accessed 14 December 2015).

FutureBrand (2012) Country Brand Index 2012–2013. Available at: www.slideshare.net/BTOEducational/country-brand-index-2012-2013-futurebrand (accessed 14 December 2015).

Gemenne, F., Barnett, J., Adger, W.N. and Dabelko, G.D. (2014) Climate and security: evidence, emerging risks, and a new agenda. *Climatic Change* 123(1), 1–9.

Gössling, S., Peeters, P. and Scott, D. (2008) Consequences of climate policy for international tourist arrivals in developing countries. *Third World Quarterly* 29(5), 873–901.

Gössling, S., Peeters, P., Hall, C.M., Ceron, J.P., Dubois, G. and Scott, D. (2012) Tourism and water use: supply, demand, and security. An international review. *Tourism Management* 33(1), 1–15.

Gössling, S., Hall, C.M. and Scott, D. (2015) *Tourism and Water*, Vol. 2. Channel View Publications, Bristol, UK.

Hall, C.M. (2001) Trends in ocean and coastal tourism: the end of the last frontier? *Ocean and Coastal Management* 44(9), 601–618.

Hall, C.M. (2010) Crisis events in tourism: subjects of crisis in tourism. *Current Issues in Tourism* 13(5), 401–417.

Hamilton, J.M. (2007) Coastal landscape and the hedonic price of accommodation. *Ecological Economics* 62(3), 594–602.

Hamilton, J.M., Maddison, D.J. and Tol, R.S. (2005) Climate change and international tourism: a simulation study. *Global Environmental Change* 15(3), 253–266.

Honey, M. and Krantz, D. (2007) *Global Trends in Coastal Tourism*. Center on Ecotourism and Sustainable Development, Stanford, California.

Horton, B.P., Rahmstorf, S., Engelhart, S.E. and Kemp, A.C. (2014) Expert assessment of sea-level rise by AD 2100 and AD 2300. *Quaternary Science Reviews* 84, 1–6.

IPCC (2014) *Climate Change 2014: Impacts, Adaptation, and Vulnerability. Part A: Global and Sectoral Aspects. Contribution of Working Group II to the Fifth Assessment Report of the Intergovernmental Panel on Climate Change* (Field, C.B., Barros, V.R., Dokken, D.J., Mach, K.J., Mastrandrea, M.D. *et al.* (eds)). Cambridge University Press, Cambridge, p. 1132.

Jiménez Cisneros, B.E., Oki, T., Arnell, N.W., Benito, G., Cogley, J.G. *et al.* (2014) Freshwater resources. In: Field, C.B., Barros, V., Dokken, D., Mach, K., Mastrandrea, M. *et al.* (eds) *Climate Change 2014: Impacts, Adaptation, and Vulnerability*. Cambridge University Press, Cambridge, pp. 229–269.

Kreft, S., Eckstein, D., Junghans, L., Kerestan, C. and Hagen, U. (2014) *Global Climate Risk Index 2015*. Germanwatch, Bonn, Germany.

Malik, K. (2014) *Human Development Report 2014. Sustaining Human Progress: Reducing Vulnerabilities and Building Resilience*. United Nations Development Programme, New York.

Manzello, D.P., Kleypas, J.A., Budd, D.A., Eakin, C.M., Glynn, P.W. and Langdon, C. (2008) Poorly cemented coral reefs of the eastern tropical Pacific: possible insights into reef development in a high-CO_2 world. *Proceedings of the National Academy of Sciences USA* 105(30), 10450–10455.

Mayor, K. and Tol, R.S. (2007) The impact of the UK aviation tax on carbon dioxide emissions and visitor numbers. *Transport Policy* 14(6), 507–513.

Mayor, K. and Tol, R.S. (2010) The impact of European climate change regulations on international tourist markets. *Transportation Research Part D: Transport and Environment* 15(1), 26–36.

Michel, J. (2014) President Michel: AOSIS is the Conscience of the Climate Change Negotiations. Available at: http://www.mfa.gov.sc/static.php?content_id=36&news_id=913 (accessed 14 December 2015).

Morrison, J., Morikawa, M., Murphy, M. and Schulte, P. (2009) *Water Scarcity and Climate Change: Growing Risks for Businesses and Investors*. CERES, Boston, Massachusetts.

Munday, P.L. (2004) Habitat loss, resource specialization, and extinction on coral reefs. *Global Change Biology* 10(10), 1642–1647.

ND-GAIN (2012) Notre Dame Global Adaptation Index. Available at: http://index.gain.org (accessed 14 December 2015).

Nicholls, R.J. and Cazenave, A. (2010) Sea-level rise and its impact on coastal zones. *Science* 328(5985), 1517–1520.

Peduzzi, P., Chatenoux, B., Dao, H., De Bono, A., Herold, C. *et al.* (2012) Global trends in tropical cyclone risk. *Nature Climate Change* 2(4), 289–294.

Pentelow, L. and Scott, D. (2010) The implications of climate change mitigation policy and oil price volatility for tourism arrivals to the Caribbean. *Tourism and Hospitality Planning and Development* 7(3), 301–315.

Pentelow, L. and Scott, D.J. (2011) Aviation's inclusion in international climate policy regimes: implications for the Caribbean tourism industry. *Journal of Air Transport Management* 17(3), 199–205.

Phillips, M.R. and House, C. (2009) An evaluation of priorities for beach tourism: case studies from South Wales, UK. *Tourism Management* 30(2), 176–183.

Randewich, N. (2005) Wilma batters Mexico's Caribbean tourism industry. *The Star Online*, 25 October 2005. Available at: www.thestar.com.my/news/world/2005/10/22/hurricane-wilma-batters-mexicos-caribbean-resorts (accessed 14 December 2015).

Raupach, M.R., Davis, S.J., Peters, G.P., Andrew, R.M., Canadell, J.G. *et al.* (2014) Sharing a quota on cumulative carbon emissions. *Nature Climate Change* 4(10), 873–879.

Reig, P., Shiao, T. and Gassert, F. (2013) *Aqueduct Water Risk Framework*. Working Paper. World Resources Institute, Washington, DC. Available at: http://pdf.wri.org/aqueduct_water_risk_framework.pdf (accessed 14 December 2015).

Rutty, M. and Scott, D. (2010) Will the Mediterranean become 'too hot' for tourism? A reassessment. *Tourism and Hospitality Planning and Development* 7(3), 267–281.

Rutty, M. and Scott, D. (2014) Thermal range of coastal tourism resort microclimates. *Tourism Geographies* 16(3), 346–363.

Schlosser, C.A., Strzepek, K., Gao, X., Fant, C., Blanc, É. *et al.* (2014) The future of global water stress: an integrated assessment. *Earth's Future* 2(8), 341–361.

Scott, D. and Gössling, S. (2015) What could the next 40 years hold for global tourism? *Tourism Recreation Research* 40(3), 269–285.

Scott, D., Hall, C.M. and Gössling, S. (2012a) *Tourism and Climate Change: Impacts, Adaptation and Mitigation*. Routledge, New York.

Scott, D., Simpson, M.C. and Sim, R. (2012b) The vulnerability of Caribbean coastal tourism to scenarios of climate change related sea level rise. *Journal of Sustainable Tourism* 20(6), 883–898.

Scott, D., Hall, C.M. and Gössling, S. (2015) A review of the IPCC fifth assessment and implications for tourism sector climate resilience and decarbonization. *Journal of Sustainable Tourism* 24(1), 8–30.

Smith, D. and Vivekananda, J. (2007) *A Climate of Conflict: The Links between Climate Change, Peace and War*. International Alert, London.

UN General Assembly (2015) Transforming our world: The 2030 Agenda for Sustainable Development. *UN Doc. A/70/L, 1.*

UNWTO (2012) *Tourism Highlights: 2012 Edition*. World Tourism Organization, Madrid.

Wheeler, D. (2011) *Quantifying Vulnerability to Climate Change: Implications for Adaptation Assistance*, Working Paper 240. Center for Global Development, Washington, DC.

Wilson, S.K., Graham, N.A., Pratchett, M.S., Jones, G.P. and Polunin, N.V. (2006) Multiple disturbances and the global degradation of coral reefs: are reef fishes at risk or resilient? *Global Change Biology* 12(11), 2220–2234.

Wong, P.P., Losada, I.J., Gattuso, J.P., Hinkel, J., Khattabi, A. *et al.* (2014) Coastal systems and low-lying areas. In: Field, C.B., Barros, V., Dokken, D., Mach, K., Mastrandrea, M. *et al.* (eds) *Climate Change 2014: Impacts, Adaptation, and Vulnerability*. Cambridge University Press, Cambridge, pp. 361–409.

World Bank (2011) Fossil Fuel Energy Consumption. Available at: http://data.worldbank.org/indicator/EG.USE.COMM.FO.ZS (accessed 14 December 2015).

7 Climate Change Governance and Trade Policy: Challenges for Travel and Tourism in Small Island Developing States

Keith Nurse,[1]* Danielle Edwards[2] and Denyse Dookie[3]

[1]*University of the West Indies, Barbados;* [2]*The University of the West Indies Open Campus Dominica, Dominica;* [3]*School of International and Public Affairs, Columbia University, New York*

Introduction

Climate change policies and governance have a significant impact on the sustainability of trade in key global industries like tourism. The promotion of sound climate policies is considered necessary for the continued sustainability of the travel and tourism sector, given the significant global economic and social value of these industries and their close links with climate (UNWTO and UNEP, 2008, p. 13). Climate is an important resource for tourism, as it influences the perceptions of suitability of locations for a wide range of tourist activities, for example suitability for a beach holiday or a ski trip (UNWTO and UNEP, 2008, p. 28). Climate also influences seasonality in tourism demand, and by extension the competitiveness of destinations and the profitability of tourism enterprises. However, the tourism industry is 'a non-negligible contributor to climate change through greenhouse gas emissions derived especially from the transport and accommodation of tourists' (UNWTO and UNEP, 2008, p. 27).

The travel and tourism sector is a major contributor to greenhouse gas (GHG) emissions and consequently, there is a potential economic threat from the implementation of climate change policies in tourism source countries in terms of emission and taxation schemes. Shifts in consumer preferences on account of the carbon labelling may also deter holiday makers from long-haul travel. The intersection of these factors makes for a critical scenario for Small Island Developing States (SIDS) in the evolving context of climate change and trade in international services (Nurse, 2010).

Therefore, the challenge of climate change for SIDS in the context of the travel and tourism industry is twofold: the sector is both a victim of climate change due to factors like sea-level rise, coral bleaching and storm surges; and an indirect vector of climate change through the contribution of the airline and cruise ship industry to GHG emissions. This raises concerns regarding the potential adverse impact of prospective climate regulation by OECD countries on international aviation and bunker fuels as

*Corresponding author, e-mail: keithnurse@me.com

well as the shift in corporate policies and consumer preferences towards long-haul destinations. The implications for SIDS are considerable given the varied policy measures, such as border carbon adjustments that are being implemented or contemplated by OECD countries.

For SIDS the implications of climate change policies are very concerning, notwithstanding the universal need for international climate policies and the impact of climate change on all states. Indeed, international tourism accounts for the largest share of gross domestic product (GDP), exports and employment in most SIDS, and consequently these small island nations and territories are highly vulnerable to carbon border measures that would impact on international trade, travel and tourism. International tourism and travel and its associated services have been the main means by which the economies of SIDS have diversified in the last several decades with the growth of mass tourism. These countries are highly dependent on the aviation and cruise ship industries that are considered to be major emitters and so are vulnerable to the climate change policies.

The travel and tourism sector is also captured in broader economic policy perspectives such as the 2007 Davos Declaration (UNWTO, 2007), which identifies the tourism sector as an industry that must respond rapidly to climate change, urging governments to face climate change as one of the greatest challenges to sustainable development and implement climate policies for effective adaptation.[1] The Davos Declaration also urges consumers to consider the climate, economic and environmental impacts of their journeys prior to making decisions to travel, and where possible to reduce their carbon footprint.[2]

From this perspective, it is argued that the global community should create frameworks that shield SIDS from the potentially devastating economic impact of travel-related climate policies, as these states are heavily dependent on international tourism and travel. This theme was central to the debate and discussions of the Third International Conference on Small Island Developing States in Samoa, September 2014, where the Heads of State and government officials affirmed that 'sustainable tourism represents an important driver of SIDS sustainable economic growth and creation of decent jobs'.[3]

SIDS may be significantly disadvantaged by the introduction of tourism-related climate policies due to their heavy dependence on international tourism and travel. For these reasons this chapter explores the intersection between climate change and trade governance with specific consideration for tourism industry policies based on the recognition that it is not sufficient to cope solely with the challenges of environmental change. Strategies are required to identify new market spaces and opportunities and develop an implementation agenda for adaptation. The chapter first provides a brief overview of the climate change impact on SIDS and then focuses on the intersection between trade policy, tourism and climate change governance. Specific attention is given to the policy dimension in the aviation and maritime sectors given their importance to the tourism sector in SIDS.

Climate Change Governance and Trade Policies

Climate change governance has rapidly grown in importance as a trade policy, international relations and development challenge in recent years, given the implications for the global commons of increased GHG emissions. The intersection of climate change and international trade is a particularly critical dimension of this scenario because global efforts to mitigate emissions and to promote adaptation are likely to impact on cross-border economic activities as well as production and consumption patterns. Global trade has grown significantly within the last several decades, outpacing the rate of growth in industrial production. This is exemplified by the fact that the share in world GDP for international trade has increased from 5.5% in 1950 to 21% in 2007. This rapid growth in global trade underpins the importance of the nexus between climate change and international trade and the urgency of the international community to find a solution to curtail GHG emissions emitted by international trade (WTO, 2009).[4]

There is increasing tension between the objectives of open trading systems and the reduction of GHGs. As such, the evolving global context is fraught with potential conflict between climate regimes like the UNFCCC and the Kyoto Protocol and the trade rules embodied in the World Trade Organization. There is also a well-founded concern that mitigation policies will impact on international trade. For instance, Pascal Lamy (2009), former Director-General of the WTO, has stated that 'since the main thrust of climate change policies is to increase the price of carbon, trade costs are impacted, with implications for the global pattern of production and consumption and thus for trade patterns'. Climate policies designed to control carbon emissions often involve strategies to increase the price of carbon, thereby leading to heightened costs for end consumers. Increased air transport fares may also lead to reduced tourist flows and revenues, as price is an important factor in tourists' choices (UNCTAD, 2014).[5]

The climate policies introduced in the aviation and marine transport sectors often fall under the scope of market-based instruments (MBIs) for the reduction of GHG emissions at the multilateral climate change negotiations (ICTSD, 2010). To date, the most controversial proposed cross-border regime for global efforts to mitigate emissions in the travel sector has been the European Union Emissions Trading Scheme (ETS) Directive 2008/101, designed to support the European Union's efforts to achieve its Kyoto commitments and the international goal of keeping the planet's temperature increase to 2°C. The ETS imposes absolute limits on CO_2 emissions through a cap-and-trade programme. Under the EU ETS companies are required to give annual reports on their GHG emissions to an independent authorized body. Those businesses exceeding their established limit are subject to penalties.

Generally, MBIs are developed to encourage the use of more carbon-efficient uses of energy resources (WTO, 2009). Even though these mechanisms have been lauded as comprehensive proposals to aid in the reduction of GHG emissions the implications for international trade in terms of rising transportation costs need to be considered, especially with regard to the Caribbean context. This principle is embraced by the UNFCCC, which states that 'measures taken to combat climate change, including unilateral ones, should not constitute a means of arbitrary or unjustifiable discrimination or a disguised restriction on international trade. Therefore states should ensure that climate change initiatives do not create barriers to international trade' (Nurse *et al.*, 2009).

A country which suffers a hindrance in its ability to trade as a result of an environmentally related measure adopted by another state such as a cap-and-trade emissions programme may seek recourse before the WTO Dispute Settlement Body, provided that both states are WTO members. In *Venezuela, Brazil v US: Gasoline* (WTO, 1996), Venezuela and Brazil challenged a regulation instituted by the US Environmental Protection Agency on the grounds that it was inconsistent, *inter alia*, with the General Agreement on Tariffs and Trade (GATT) Articles III and XX on National Treatment on Internal Taxation and Regulation and General Exceptions to the GATT, respectively. The regulation in question was introduced by the USA with the objective of motivating trading partners to improve the quality of gasoline used, thereby leading to a decrease in emissions within the USA.

The USA argued that the Gasoline Rule complied with Article XX (g), which allows WTO members to take measures 'relating to the conservation of exhaustible natural resources if such measures are made effective in conjunction with restrictions on domestic production and consumption'. The USA further contended that clean air is an exhaustible natural resource which the Gasoline Rule aimed to conserve. The WTO dispute settlement panel held that the Gasoline Rule was inconsistent with the GATT Article III, as the regulations for domestic and imported gasoline differed. Although the Appellate Body (AB) supported the US argument that clear air is an exhaustible natural resource, the AB upheld the panel's conclusions.[6] In accordance with this ruling, states which are desirous of adopting climate mitigation policies should refrain from influencing the mode of production in other countries through unilateral trade measures.

Another significant WTO decision which involved environmental conservation measures is the *Shrimp-Turtle* case (WTO, 1998). The dispute arose over measures introduced by the USA to prohibit the importation of shrimp from countries that did not have a turtle-conservation programme comparable to that of the USA. In the *Shrimp-Turtle* case, the WTO AB ruled that the measures introduced by the USA satisfied the GATT Article XX(g) exception for conservation of exhaustible natural resources. The AB also noted that such measures should be applied in a manner that conforms to the chapeau of Article XX: that is, the application of such measures should neither result in unjustified nor arbitrary discrimination among States in which similar circumstances prevail. Additionally, the application of such measures should not generate hidden restrictions on international trade. The AB's *Shrimp-Turtle* ruling 'has been interpreted as justifying environmental policies with extraterritorial effect provided certain safeguards are met, including the consideration of comparative regulations in third countries and the conducting of prior negotiations with the aim of concluding an international agreement' (CISDL, 2012).

Climate Change Governance and Travel and Tourism Regulation

The discourse on the issue of climate change, travel, tourism and trade has been very diverse and has manifested itself in many forms. These discussions have been facilitated by several key international agreements and institutions, and influenced by a plethora of other governmental and non-governmental international organizations. The result of this dialogue has been a series of international climate conferences, policy tools, instruments ranging from declarations, to conventions, to international agreements all geared towards addressing the global challenges related to climate change. However, only a handful of these measures to address climate change propose policies to safeguard the interests of the tourism industry in SIDS at least directly. In this regard, the third International Conference on Small Island Developing States and the conference document, the SIDS Accelerated Modalities of Action (SAMOA) Pathways must be celebrated as its declaratory provisions are a milestone for SIDS. The SAMOA document stresses the economic importance of sustainable tourism and travel for SIDS. Article 30 of the SAMOA document also declares strong support by SIDS in taking the following actions:

1. Developing and implementing policies that promote responsive, responsible, resilient and sustainable tourism inclusive of all peoples in SIDS.

2. Diversifying sustainable tourism through products and services, including large scale tourism projects with positive economic, social and environmental impacts and the development of eco- , agro- and cultural tourism.

3. Promoting policies that allow local communities to gain optimum benefits from tourism while allowing them to determine the extent and nature of their participation.

4. Designing and implementing participatory measures to enhance employment opportunities, in particular of women, youth and persons with disabilities, including through partnerships and capacity development while conserving their natural, built and cultural heritage, especially their ecosystems and biodiversity.

5. Leveraging on the expertise of, *inter alia*, the Global Sustainable Tourism Council, UN World Tourism Organization (UNWTO) Global Observatories on Sustainable Tourism and the Global Partnership for Sustainable Tourism and other UN bodies, as well as the 10YFP on sustainable consumption and production, to provide platforms for exchanges of best practices and direct and focused support to national efforts of SIDS.

The SAMOA Declaration is a timely recognition of the linkages between climate change, international tourism, trade and innovation policy. This nexus is a particularly critical dimension of the unfolding scenario because global efforts to mitigate emissions and to promote adaptation require enhanced governance of cross-border economic activities as well as adjustments in production and consumption patterns. As such, mediation of international trade policy and boosting innovation policy are considered to be vital for implementing climate-friendly and sustainable development options.

One of the key areas associated with climate change mitigation efforts is the reduction of energy-related emissions, which are estimated to account for over 80% of global carbon dioxide (CO_2) emissions. It is estimated that international transport via air, marine vessels, road, rail and pipelines accounts for the fastest growing source of CO_2 emissions, having doubled between the period 1990 to 2010. The linkage between the aviation sector and climatic changes comes largely through carbon emissions. The major emissions of aircraft include CO_2 and water vapour (H_2O). However, aircraft also emit nitric oxide (NO), nitrogen dioxide (NO_2), sulfur oxides (SO_x) and soot (Penner *et al.*, 1999). These emissions also trigger the creation of condensation trails, which also contribute to climate change. Although there are several GHGs that contribute to the enhanced greenhouse effect and global warming, CO_2 is considered the most important GHG from human activities. For aviation alone it is estimated that by 2050, the rate of emission release by the industry would quintuple on the basis of the current rate of emissions (ICTSD, 2010).

Yet, there is potential threat from tourism source countries in terms of climate change mitigation policies such as carbon taxation schemes that will increase the cost of travel and may make long-haul travel unaffordable for a significant share of the tourist market. These policies are based on the argument that increased frequency of travel and growing numbers of long-haul tourists to SIDS will foster increases in global carbon emissions, leading to further perturbations in atmospheric systems and also leading to an enhanced greenhouse effect and global warming. However, there are concerns among tourism authorities in SIDS that as the climate change agenda becomes more accepted by the general public, this might result in many tourists choosing to remain within their own country or region during their vacation period. This shift in consumer preferences has also prompted governments and companies to adopt and promote environmentally friendly policies, for example, charges and levies, some of which will cause the cost of travel and transportation to increase. Increased travel and transportation costs will likely have adverse effects on travel and tourism to SIDS.

The intersection of these factors makes for a critical scenario for SIDS in the evolving context of climate change and trade in international services, because the air travel sector provides key services to the tourism sector in small island states, which are generally long-haul destinations from key source markets like North America and Europe.

The Aviation Sector

The aviation industry accounts for approximately 35% of international trade and the value of cargo exported by airplanes every year is US$5.3 trillion (Meltzer, 2012). By contrast, marine cargo accounts for approximately 90% of the volume of global trade and approximately 70% of its value, thereby rendering this form of transport as the primary form of transport for global trade. Although aviation transport accounts for 0.27% of global trade by volume, it contributes 14% of its value (WTO, 2009). Aviation transport is the most polluting mode of transport with a contribution of 11.2% of global emissions.

It is posited that carbon emissions from aviation have a significant impact because at flight altitude, carbon emissions have an enhanced impact on global warming due to radiative forcing. Moreover, as explained by the UNWTO (UNWTO and UNEP, 2008) in its report *Climate Change and Tourism: Responding to Global Challenges*, 'radiative forcing is thus used to calculate the entire contribution of tourist (air) travel to global warming. Radiative forcing measures the extent to which emissions of greenhouse gases raise global average temperatures now or at a specified year in the future'.[7] International aviation is presently excluded from the Kyoto Protocol because of the difficulties in allocating emissions to specific countries, but it is very much under the spotlight currently and will likely continue to be central to the post-Kyoto talks (Edwards, 2014).[8]

Notwithstanding the large volume of GHGs and carbon emitted by the aviation sector, the industry is critical to the tourism economy in SIDS. Destination tourism markets in SIDS are generally long-haul destinations from key source markets in North America and Europe. The flight distance between these long-haul destinations and the SIDS tourism markets is often prolonged due to limited capacity for direct flights. In the Caribbean, the international travel sector suffers from an absence of strategic cooperation between regional carriers (Economic Commission for Latin America and The Caribbean, 2011, p. 12). According to the Economic Commission of Latin America and the Caribbean, the Caribbean has not been able to create a competitive airline hub. As a consequence, the region's two main hubs lie in San Juan, Puerto Rico and Miami (Economic Commission for Latin America and The Caribbean, 2011, p. 12).[9]

Despite its significant contribution to GHGs, the aviation industry continues to enjoy major tax exemptions and sizeable subsidies globally. The EU has led efforts to develop strong climate policy for aviation emissions by passing the first regulation aimed at mitigating aviation emissions through the Emissions Trading Scheme (ETS) in January 2012. The ETS is essentially a cap-and-trade programme designed to reduce carbon emissions by setting a limit for the total volume of GHGs that businesses within energy-intensive industries may emit. The major emitters are obligated to obtain 'allowances' for each tonne of CO_2 or CO_2-equivalent they generate within the capped emissions limit.

When the EU initially introduced the ETS, many stakeholders in the aviation industry and government reacted with fury (Chauhan, 2011). Some governments were concerned that the ETS would lead to an unjustifiable increase in charges for air carriers outside the EU (Chauhan, 2011). The airline industry challenged the inclusion of emissions by air carriers outside the EU on the grounds, *inter alia*, that the law violated the Chicago Convention.[10] The claimants argued that the scheme constituted either an illegal charge or an illegal tax on aircraft operators, contrary to the Convention. Article 15 permits 'charges imposed for the use of airports and other facilities' but provides that 'no fees, dues or other charges shall be imposed by any contracting State in respect solely of the right of transit over or entry into or exit from its territory of any aircraft of a contracting State or persons or property thereon'.[11] The air carriers also argued that since the EU ETS would require carriers to be charged for take-offs and landings that occur outside the EU it constituted an unlawful charge under the Chicago Convention. However, in *Air Transport Association v. SS for Energy and Climate Change*, the European Court of Justice (ECJ) upheld the validity of the Directive, *inter alia*, on the grounds that 'every State has complete and exclusive sovereignty over the airspace above its territory' under international law. However, on 3 April 2014 the EU Parliament voted in favour of a deal to 'stop the clock' and exempt foreign airlines from paying for carbon emitted in EU airspace. However, German member of the European Parliament Peter Liese noted, 'if the International Civil Aviation Organization fails to deliver a global agreement [in 2016], then nobody could justify the EU maintaining such an exemption for another four years' and the system 'will then automatically snap back to full scope' (Keating, 2014). One MIT study has concluded that the EU ETS will increase return airfares from Europe to the Caribbean by 1.6% (US$19) and decrease total air passenger arrivals by 0.35%, with relatively large decreases in tourism arrivals estimated for Antigua and Barbuda and Barbados (Blanc and Winchester, 2013).

The International Air Transportation Association (IATA) has adopted a four-pronged approach to reducing emissions from the industry (Bisignani, 2007; IATA,

2007). To begin, IATA's strategy focuses upon technological advancements through research and development (R&D) in order to improve the fuel efficiency of aircraft. Second, aircraft operators are encouraged to improve their internal operations and to ascribe to high environmental standards. Further, IATA urges governments to correct infrastructural inefficiencies (particularly in airports) and to take steps to make global airspace more free for air transport carriers. Finally, economic measures are suggested as a means to boost R&D and to improve environmentally friendly practices within the industry. As such, tax credits and direct funding for ecofriendly measures, as well as an open emissions trading scheme, are asserted as possible mechanisms that can be utilized to decrease the aviation industry's contribution to climate change. As the global discourse on climate change continues to grow, more and more tourists may decide to stay at home or within their region during their vacation period since they wish to lower their own carbon footprint by travelling shorter distances. If the aforementioned measures suggested by IATA were successful in lowering emissions from the aviation industry, this might encourage more tourists to fly longer distances to SIDS, such as those in the Caribbean and the Pacific, during their vacation period.

In June 2015, President of the United States of America Barack Obama announced that his administration would take the first step toward regulating GHG emissions from airplanes (New York Times, 2015). Currently, more than one-third of global airline passengers are transported by flights in the USA, and therefore the US government's cooperation is critical to the success or failure of global policies designed to curb aircraft emissions. The cap-and-trade model has already yielded success in the USA, where it has been used to control emissions from power plants.

Marine Transport

Developed countries are estimated to account for more than 70% of global maritime emissions and more than 80% of the global shipping vessels (ICTSD, 2010). For the USA and Europe, marine transport accounts for approximately 90 and 72% of their overall trade volumes, respectively. Marine transport is currently considered to be the most environmentally efficient of all the modes of transport with a 2.7% share of total global emissions (WTO, 2009). However, according to the European Commission, emissions from shipping are likely to increase between 50% and 250% by 2050 depending on energy development and future economic growth (European Commission, 2016). Accordingly, the European Commission's 2011 White Paper on transport proposes that the EU's CO_2 emissions from maritime transportation should be reduced by no less than 40% from 2005 levels by 2050, and by 50% if feasible.

Bunker fuels[12] are one of the energy sources currently being targeted for the imposition of an environmental fuel levy in order to aid in the reduction of carbon emissions. This is only one mechanism currently being contemplated by the global community as a mechanism to assist in the reduction of global trade-related emissions. Pursuant to EU Regulations 2015/757 passed on 29 April 2015, effective 2018, large vessels traversing EU ports are required to report their verified annual emissions.

At a regional level, the EU has been vigorously proposing the inclusion of marine transport under the purview of its Emissions Trading Scheme (ETS) as well as at the International Marine Organization (IMO) negotiations (ICTSD, 2010). On the other hand, the USA is proposing, at IMO, that a trading system for energy efficient credits be created whilst at its domestic level there has been the piloting of several 'cap-and-trade' bills such as the more recent Kerry-Lieberman and Waxman-Markey Bills.

Prior to 2013, the European Union considered including shipping in its regional EU ETS. Instead, in June 2013 the EU issued a revised strategy for integrating maritime emissions into the EU's policy for reducing GHG emissions domestically. This strategy introduces a monitoring, reporting and verification (MRV) scheme for shipping (DMNVGL, 2015).[13] One major objective of the

MRV scheme is to garner more reliable data on ships' fuel consumption and energy efficiency. According to estimates by the EU the implementation of an MRV scheme is likely to reduce carbon emissions by around 2% as it will encourage ship operators to focus on their fuel consumption. Unlike maritime transport, aviation has already been included in the EU ETS which came into effect as of 2012.

An assessment of the potential impact due to the imposition of a fuel levy (i.e. CO_2 US$15–30 per tonne) on carbon emitting from maritime transport suggests that SIDS are the grouping to be most adversely affected (ICTSD, 2010). Estimates put the economic impact of increased maritime transport costs at between 0.45 and 0.89% of GDP. This is considerably higher than for Least Developed Countries, which are estimated to have a GDP impact of between 0.06 and 0.12%. The general principle is an increase in freight costs will have a greater significant impact where goods have a low value to weight ratio. The statistics indicate that the expected increase in the value of imports due to maritime transport policy would have the biggest impact on raw materials. Caribbean countries are primary exporters of raw materials. Thus the imposition of any fuel levy on maritime transport would act to reduce the competitiveness of the region's exports, particularly to the US and EU markets.

Climate Change, Travel, Tourism and SIDS

Currently 52 countries and territories are classified by the United Nations as SIDS. These SIDS are inhabited by over 50 million people, and situated across the Indian, Pacific and Atlantic Oceans. The group of SIDS is a diverse one, comprising relatively wealthy states such as Singapore and Bahamas, as well as some of the poorest countries such as Haiti. The highest concentration of SIDS is in the Caribbean and south-west Pacific. The climate change agenda is of particularly high importance since most SIDS are dependent on long-distance or long-haul source markets for tourists. It is critical, therefore, to look beyond the ecological impact of climate change to examine the economic and policy impact of the range of mitigation and adaptation policies on key economic, trade and tourism sectors.

The tourism sector is the key driver of economic growth for SIDS. Indeed, many SIDS are highly dependent upon revenue earned from tourist arrivals and through tourist-related activities like transport. Based on estimates from the UNWTO, tourism and transport receipts in SIDS are estimated at 8.2% and 1.2% of exports, respectively, which is above the world average in both instances. Among SIDS there is some variation with tourism receipts/exports being 42% in the Other SIDS category, 23.5% in the Caribbean, 16.1% in the Pacific and 4.3% in the AIMS countries (Table 7.1). Other SIDS also top the list of earners for transportation as a share of exports at 4.4%, followed by the Pacific at 3.1%, the Caribbean at 1.3% and the AIMS countries at 1.0%. Further, with regard to the Caribbean, travel and tourism is also the main contributor to GDP and employment in most SIDS. Regions like the Caribbean and the Pacific have become increasingly reliant on the sector (Reddy and Wilkes, 2012).

Data on travel to Fiji, Barbados and Mauritius from their major markets indicates that long-haul flights account for 42, 57 and 61% of all visitors to these countries, respectively, and they spend more than 4.5 h on at least one direct leg of travel before arriving at their destination (Fig. 7.1). For Mauritius it is estimated that at least 61% of its tourists spend an average of 10 h flying directly to the island, with the majority of visitors arriving from Europe, particularly France and the UK. While long-haul visitors to Barbados are estimated to be at least 57% of the island's total tourism record, typically including those from the UK, Europe and Canada, it should be mentioned that this table only includes information from two airports on the eastern seaboard of the USA, which entails a short-haul flight, rather than disaggregating by origin within the USA. As such, it is possible that this figure is underestimated. Although visitors to Fiji

Table 7.1. Travel and Tourism – Worldwide, SIDS, Caribbean, AIMS, Pacific and other SIDS, 2013 (UNWTO, 2014).

	World	SIDS[a]	Caribbean	AIMS[b]	Pacific	Other SIDS
Population (1,000)	7,162,119	67,320	39,058	10,225	10,766	7,270
Overnight visitors (1,000)	1,086,00	40,838	13,943	14,834	1,497	10,564
Overnight visitors per 100 population	15	61	36	145	14	145
Receipts in destinations (US$ million)	1,193,00	53,418	15,213	23,285	1,534	13,386
Receipts per capita	167	793	389	2,277	142	1,841
Receipts as % of exports	5.2	8.2	23.5	4.3	16.1	42.1
Int'l passenger transport (US$ million)	216,000	8,019	813	5,525	297	1,384
Transport earnings as % of exports	0.9	1.2	1.3	1.0	3.1	4.4

[a]The list of SIDS as per the definition of the UN Office of the High Representatives for the Least Developed which includes a total of 57 SIDS.
[b]The AIMS abbreviation refers to the Atlantic, Indian Ocean, Mediterranean and South China Sea, although there are currently no SIDS in the Mediterranean.

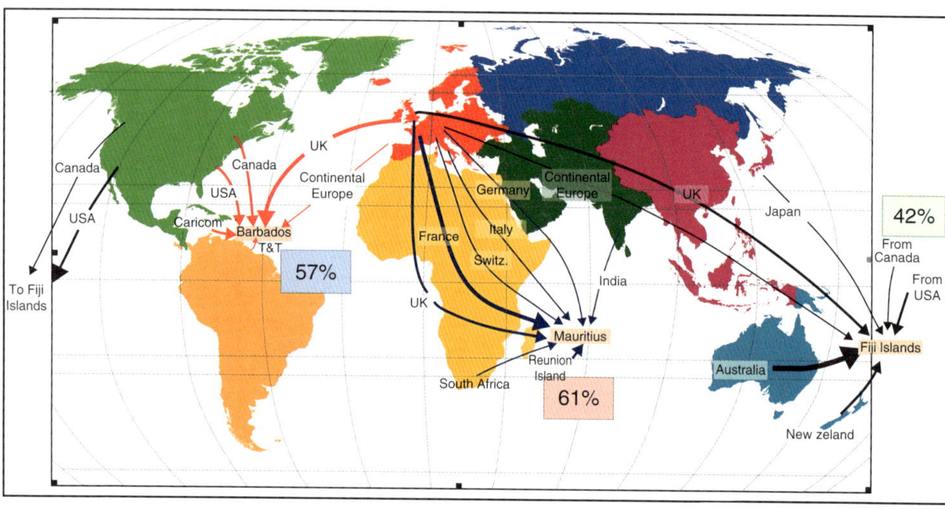

Fig. 7.1. SIDS and long-haul flights.

are generally from short-haul destinations such as Australia and New Zealand, at least 42% are long-haul visitors who come directly from Japan and the USA via Los Angeles. Furthermore, this figure also includes visitors from the UK, Canada and continental Europe, who need to take more than one flight to arrive to the islands (see Appendix 7.1 for further details).

To put this in perspective of the three case study countries of Barbados, Mauritius and Fiji Islands, Appendix 7.1 also shows estimates of carbon emissions and equivalents, which typically increase with the length of flight. Further, considering that these carbon emissions may increase by 2.7 with radiative forcing in effect,[14] and given the fact that the majority of visitors are from long-haul destinations, what this information shows is that there is indeed a strong relationship between tourism travel to SIDS and a high carbon footprint.

Certainly, the geographical, economic, physical and demographic characteristics of

SIDS create unique challenges and result in a degree of inherent vulnerability that must be due considerations as part of development strategy (Von Tigerstrom, 2005). SIDS are not major emitters of GHGs, yet they are estimated to be one of the groups of countries to be most impacted by the ecological changes associated with climate change, particularly in terms of sea level rise, coastal degradation, temperature rises, rainfall changes, coral bleaching and increased health and natural disaster risks such as hurricane damage and storm surge, particularly among the island states and low-lying coastal areas.

The predictions of scientists suggest that SIDS are highly vulnerable to the impacts of global warming, particularly sea level rise, temperature rises, rainfall changes, coral bleaching and increased storm frequency. SIDS are 'hotspots' (UNWTO and UNEP, 2008) and collectively constitute a group of countries that are estimated to be most affected by climate change. It is estimated that 72% of all SIDS are classified as either 'extremely vulnerable' or 'highly vulnerable' to climate change impact (AOSIS/UNF, 2008).

Vulnerability to climate change is determined by exposure and sensitivity to hazards as well as the capacity to adapt to those hazards. In SIDS, urban communities, cities and fragile infrastructure are often situated along the coast. Over half of the people in the Caribbean live within 1.5 km of the shoreline. Therefore rising sea levels will also impact on vital infrastructure, settlements and hotel facilities through inundation, erosion and other coastal hazards, thereby creating significant economic challenges for the tourism industry (Mimura *et al.*, 2007). Similarly, tourist climate preferences such as temperature limits for beach vacations as well as tourist perceptions of the environmental impacts of tourism-related travel and their willingness to pay to reduce this impact are likely to affect the attractiveness of SIDS for destination tourism.

Changing rainfall distribution patterns will require countries to find new and innovative solutions to establish a consistent and reliable water supply. The projected change in rainfall patterns is also likely to result in a severe decline in agricultural production for both subsistence and commercial purposes, thus threatening food security as well as agricultural export capacity. For the island and coastal regions, warming seas threaten the livelihood of commercial and artisanal fisheries and of coral reefs. This by extension has the potential to result in widespread unemployment, of fishermen and of tourism-related service providers such as scuba tourism providers or tour operators, whose income depends heavily upon the existence of healthy coral reefs. For SIDS, this predicament is exacerbated by the fact that their economies are often dependent upon mono-crop agricultural production and export, and tourism and its associated services for foreign exchange earnings, employment and contribution to GDP. To illustrate the potentially catastrophic consequences, a study of the impact of climate change on the Caribbean region paints a dismal picture (Bueno *et al.*, 2008). Based upon an analysis of hurricane damages, loss of tourism revenue, and infrastructure damage due to sea-level rise it is estimated that the cost of inaction would have a significant impact on the GDP of the countries that surround the Caribbean. The key findings of the study suggest that:

1. The costs of inaction will amount to 22% of GDP for the Caribbean as a whole by 2100.
2. The costs of inaction will reach an astonishing 75% or more of GDP by 2100 in Dominica, Grenada, Haiti, St Kitts & Nevis and Turks & Caicos.
3. The Caribbean's largest island, Cuba, faces a nearly 13% economic hit by mid-century, and a 27% loss by 2100, unless there is swift action to address climate change.
4. Losses from inaction would be less severe but still significant in Puerto Rico, reaching nearly 3% by 2050 and 6% by the end of the century.
5. The nation of Colombia, with its long Caribbean coastline, faces permanent flooding of 1900 square miles in low-lying coastal areas, affecting 1.4 million people.

The Case of the Caribbean and the UK Airline Passenger Duties

The air travel sector and the cruise ship industry provide key services to the tourism sector in small island states which are generally long-haul destinations from key source markets like North America and Europe. Notwithstanding this, the travel sector is considered a major contributor to GHGs. There is potential threat from tourism source countries through the introduction of taxation schemes and consumer movements that may deter holidaymakers from long-haul travel. The intersection of these factors makes for a critical scenario for SIDS in the evolving context of climate change and trade in international services.

An example of the challenges of such emission mitigation schemes for developing countries is the UK airline passenger duties (APD) that taxes carbon emissions on flights out of the UK. The APD was first introduced in 1994 as a flat-rate tax but as of November 2009 it operates with four distance-related bands based on the distance between London and the capital city of the destination country. In the new APD scheme, flights to the Caribbean incurred 25% and 87% more in taxes in 2009 and 2010, respectively. The Caribbean also falls into band 'C' while all of the USA, including Hawaii, is in category 'B'.

Schemes like the APD have the effect of making the cost of travel more expensive and it may also deter holidaymakers from long-haul travel thereby making the tourism product in developing country regions less competitive relative to destinations in Europe and North America. Another issue is that emission mitigation schemes may penalize airline operators from developing countries, driving them out of the market and concentrating the market on some routes such that the major carriers and their strategic partners dominate. Many developing countries may fall prey to oligopolistic practices that reduce consumer welfare.

The above analysis suggests that there is a potential threat from climate change policies to tourism source countries in terms of taxation schemes and consumer movements that may deter holidaymakers from long-haul travel. The intersection of these factors makes for a critical scenario for SIDS in the evolving context of climate change and trade in international services.

The countries of the eastern Caribbean (predominantly the OECS and Barbados) are very dependent, particularly on the UK market. Table 7.2 shows the dependence of these economies on tourism and travel as well as the importance of the UK market to other selected Caribbean countries. In 2008, Barbados topped the list with 38.7% of arrivals coming from the UK market. Antigua and Barbuda is the country with the next highest dependence on the UK market with 33.7% of total arrivals. This is of great significance given that tourism contributed as much as 47.9% and 72.4% of GDP, respectively. While other Caribbean countries are not as exposed as Barbados and Antigua & Barbuda, the level of dependence on the UK market does generate some concern, especially in the current context of austerity measures and declining consumer demand.

The region's tourism industry has experienced a drop in arrivals since the implementation of the APD but it is very difficult to untangle this from the broader impact of the global economic downturn which has been particularly challenging in the context of the UK. Table 7.3 clearly shows the drop in arrivals from the UK market registered in the selected countries. The decline was sharpest in Grenada and Antigua & Barbuda, which saw a decline of 25.9% and 25.4%, respectively. The CTO (2011) found that annual traffic data to the end of September 2010 for eight popular Caribbean destinations on Cheapflights.co.uk shows an average annual 11.6% drop in Caribbean destination searches compared to the previous year.

While adaptation is the key issue for these countries, some mitigation efforts can be achieved through pursuing green tourism to counter the shift in consumer preferences on long-haul travel. From this perspective SIDS could utilize the climate change agenda as the basis for building a low-carbon tourism strategy. This could become a driver for economic transformation considering the possibilities of restructuring current production and consumption modes in order to cope with the challenge of climate change.

Table 7.2. Importance of UK tourist market to selected Caribbean economies (CTO, 2011).

	Total stop-over arrivals 2008 (CTO data)	Stop-over arrivals from UK market 2008 (CTO data)		Travel and tourism total contribution % GDP 2008 (WTTC figures)	World Bank total GDP 2008 (US$ current)	Estimated contribution of UK stop-over arrivals 2008 (US$ current)	Estimated GDP value of a 5% reduction in UK arrivals (US$ current)
		Total	% Total arrivals				
Barbados	567,700	219,913	38.7	47.9	3,681,500,000	683,296,900	34,161845
Antigua & Barbuda	265,800	89,514	33.7	72.4	1,203,3114,815	293,395,749	14,669,787
Grenada	130,400	34,821	26.7	24.7	678,488,877	44,751,061	2,237,553
St Lucia	295,800	83,693	28.3	47.9	986,074,074	133,639,862	6,681,993
St Vincent & the Grenadines	84,100	15,442	18.4	26.9	581,722,222	28,732,637	1,436,632
Jamaica	1,767300	188,436	10.7	24.3	14,614,273,088	378,649,242	18,932,462
Dominica	78,481	7,207	9.2	26.1	374,422,222	8,974,125	448,706

Table 7.3. Total arrivals from UK market, 2008–2010 (CTO, 2011).

Country	Total Arrivals			UK market			UK % share			% Change 2010/2008	
	2008	2009	2010	2008	2009	2010	2008	2009	2010	Total	UK
Barbados	567,700	518,600	532,200	219,953	190,632	181,054	38.7	36.8	34.0	–6.3	–17.7
Antigua & Barbuda	265,800	234,400	231,300	89,514	73,251	66,747	33.7	31.3	28.9	–13.0	–25.4
Grenada	130,400	113,400	106,200	34,821	27,512	25,791	26.7	24.3	24.3	–18.6	–25.9
St Lucia	295,800	278,500	305,900	83,693	71,853	67,417	28.3	25.8	22.0	3.4	–19.4
St Vincent & the Grenadines	84,100	75,400	72,500	15,442	13,347	12,136	18.4	17.7	16.7	–13.8	–21.4
Jamaica	1,767,300	1,831,100	1,921,700	188,436	184,512	184,355	10.7	10.1	9.59	8.7	–2.2

Conclusion

SIDS are not large contributors to the problem of climate change; however, the evidence shows that these countries are vulnerable to increased ecological, health and natural disaster risks and that the travel and tourism industries will be heavily impacted. As this chapter illustrates, the challenge goes beyond ecological challenges to include issues related to the intersection between climate-change governance and international trade policy. As tourism-dependent economies, SIDS are extremely vulnerable to international climate policies within the international travel and tourism industries. The regulation of GHGs from the aviation and marine industries by the EU and North America is likely to increase the costs associated with international transportation of goods and people, with significant repercussions for international trade and trade in services like tourism.

However, there are opportunities for SIDS to utilize the climate change agenda as a driver for economic transformation considering the possibilities of restructuring current production and consumption modes in order to cope with the challenge of climate change. SIDS lack the financial resources and the technologies required to make the shift towards a low carbon economy and as such they require the transfer of efficient and clean energy technologies from developed countries. Therefore SIDS should also pursue innovation policies to facilitate more rapid adaptation responses. For example, AOSIS advocates for relevant linkages to be made between science and climate change in order to empower SIDS to design and implement policies appropriately. In 2000, AOSIS chairman observed that science provides 'an essential component in the search for feasible pathways towards the management of the environment and towards sustainable development' (Slade, 2000).

The liberalization of trade in energy-efficient goods is needed in areas where there is no competitive local production. Such trade policy measures can therefore include tax incentives or zero-tariff measures for environmentally friendly products. SIDS may also collectively forge trading agreements that are specifically tailored to address their concerns (Nurse *et al.*, 2009). These arrangements may involve formal regional blocs or organizations established by treaties or less formal bodies.

Last but not least, developing countries should support a global sectoral approach for aviation once funding is allocated for technology transfer and financing for developing country airlines. SIDS are generally policy takers in the global governance context and as such these countries tend to focus their attention on adaptation measures as well as mitigation efforts such as greening tourism and shifting to a low-carbon development strategy. This chapter suggests that SIDS also have to be vigilant of the emerging policy context and advocate for change which accords with their interest. They may also forge strategic alliances with industry stakeholders and representative international organizations as a means to bolster their leverage in international fora.

Notes

[1] Davos Declaration on Climate Change and Tourism Responding To Global Challenges at (3): 'In their choices for travel and destination, tourists should be encouraged to consider the climate, economic, societal and environmental impacts of their options before making a decision and, where possible to reduce their carbon footprint, or offset emissions that cannot be reduced directly'.

[2] In 2007 the Second International Conference on Climate Change and Tourism was convened by the World Tourism Organization (UNWTO), jointly with the United Nations Environment Programme (UNEP) and the World Meteorological Organization (WMO), supported by the World Economic Forum (WEF) and the Swiss Government in Davos, Switzerland with the primary objective of responding appropriately to climate change imperatives in the tourism sector.

[3] SIDS Accelerated Modalities of Action (SAMOA), Outcome document of the Third International Conference on Small Island Developing States (SIDS) at Art. 30.
[4] P.142 noting that 'Price-based mechanisms, such as a carbon tax on fossil fuels or a tax on energy, have been employed in several countries over the past two decades as a means of internalizing the environmental cost of greenhouse gas emissions. More recently, attention has focused on emission trading schemes. These involve fixing a cap on total emission levels, translating this into allowances to cover emissions, and creating a market to trade these allowances at a price determined by the market'.
[5] Para 24 noting that 'One study that assessed the competitiveness of islands as tourist destinations found that the cost of a holiday (price of flights and three- and four-star hotel accommodation) is a key factor determining demand for SIDS tourism'.
[6] The WTO AB concluded that the Gasoline Rule was a means of arbitrary or unjustifiable discrimination and a disguised restriction to international trade. The AB found that the USA had not taken sufficiently into account the costs that the standards imposed would incur upon importing countries. The Gasoline Rule was therefore found to be in violation of the chapeau (introductory paragraph) of Article XX of the GATT. As a result, if a country or group of countries, in their bid to combat climate change wished to take measures to conserve 'cool air' as an exhaustible natural resource, they would need to ensure that such initiatives would not constitute a means of arbitrary or unjustifiable discrimination or a barrier to international trade. Available at: www.ictsd.org/downloads/2008/08/mauritius-complete-policy-paper.pdf.
[7] UNWTO, pp. 33 (Draft).
[8] Noting that 'The parties to the Kyoto Protocol agreed that greenhouse gas emissions from international aviation should be "reduced" by developed countries (listed in an annex) through ICAO. The Protocol provides: "Parties included in Annex I shall pursue limitation or reduction of emissions of greenhouse gases ... from aviation and marine bunker fuels, working through the International Civil Aviation Organization and the International Maritime Organization, respectively." To date, all of ICAO's climate-related instruments have been resolutions or guidance to States which are not legally binding, but may form part of regional and national aviation policy'.
[9] p.12 noting that foreign carriers command the lion's share of capacity on USA- and UK-CARICOM routes, whereas local carriers provide the majority of intra-CARICOM services.
[10] Convention on International Civil Aviation, opened for signature 7 December 1944, 61 Stat. 1180, 15 U.N.T.S. 295 [hereinafter Chicago Convention]. Available at: www.icao.int/publications/Documents/7300_cons.pdf.
[11] Chicago Convention, art. 15.
[12] 'Bunker fuels' are originally the heavy, low-grade fuels used for shipping. In climate-change jargon, this is a term that covers both the maritime shipping and aviation sectors.
[13] p. 3 Noting that 'The EU-MRV regulation applies to merchant ships of 5,000 GT or above. From 1 January 2018 on, shipowners and operators (defined as "companies") will have to monitor the CO_2 emissions of their vessels per voyage, for all voyages conducted into, between and out of EU ports. Further, the monitored emissions shall be reported to the EC, which will make reported and verified emissions and related data on energy efficiency publically available for the first monitoring period on 30 June 2019.
[14] Source: Convert Units – Measurement Unit Converter, www.convertunits.com.

Bibliography

ALTA (2011) ALTA, the Latin America and Caribbean Air Transport Association, Urges Latin American Governments to Strongly Reject Unilateral EU Emissions Trading Scheme. Available at: http://goo.gl/fCj31V.

BBC News (2012) Countries rally against EU carbon tax on airlines. *BBC News*, 21 February 2012. Available at: http://www.bbc.com/news/world-europe-17114312 (accessed 9 August 2017).

Bisignani, G. (2007) Speech by Giovanni Bisignani former Director General of the International Air Transport Association delivered at World Air Transport Forum, Cannes, 17–19 October 2007. Available at: http://www.iata.org/pressroom/speeches/Pages/18-10-2007-01.aspx (accessed 9 August 2017).

Blanc, É. and Winchester, N. (2013) The impact of the EU emissions trading system on air passenger arrivals in the Caribbean. *Journal of Travel Research* 52(3), 353–363.

Brandi, C. (2010) Climate and Trade Governance from a Normative Perspective – The Case of Carbon Border Adjustments. Available at: http://goo.gl/lijcwh.

Bueno, R., Herzfeld, C., Stanton, E.A. and Ackerman, F. (2008) *The Caribbean and Climate Change: The Costs of Inaction*. Tufts University, Somerville, MA. Available at: http://ase.tufts.edu/gdae/Pubs/rp/Caribbean-full-Eng.pdf.

Carbon Dioxide Information Analysis Center (CDIAC) (2011) Ranking of the World's Countries by 2008 per Capita Fossil-fuel CO_2 Emission Rates. Available at: http://cdiac.ornl.gov/trends/emis/top2008.cap (accessed 9 August 2017).

Caribbean Tourism Organization (2011) Caribbean Response to the UK Government's Consultation on Reform of Air Passenger Duty. Available at: http://goo.gl/Nq52Um (accessed 9 August 2017).

CARICOM (2011) Aviation Tax: A Levy on Regional Development – CARICOM SG. Available at: http://goo.gl/mDvzlZ.

Centre for International Sustainable Development Law (CISDL) (2012) Legal Analysis on the Inclusion of Civil Aviation in the European Union Emissions Trading System (Verki Michael Tuntang ed., 2012). Available at: http://goo.gl/vDmwR0 (accessed 9 August 2017).

Chauhan, C. (2011) India threatens to move WTO against EU's carbon tax. *Hindustan Times* 28 July 2011. Available at: www.hindustantimes.com/india-news/newdelhi/india-threatens-to-move-wto-against-eu-s-carbon-tax/article1-726730.aspx.

Chidiak, M. and Tirpak, D. (2008) Mitigation Technology Challenges: Considerations for National Policy Makers to Address Climate Change. Available at: http://goo.gl/fODHIk.

Congressional Research Service (2011) Summary of H.R. 2454: American Clean Energy and Security Act of 2009. Available at: http://goo.gl/OJenqO (accessed 9 August 2017).

Cosbey, A. (2010) Competitiveness Impacts of Climate Change on LDC's Export Trade. Available at: www.iadb.org/intal/intalcdi/PE/2011/08296.pdf (accessed 9 August 2017).

Curran, L. (2009) Carbon Taxing Imports – Can the North Reduce Global Warming While Avoiding Negative Economic Implications for the South?. Available at: http://goo.gl/7j3FE3 (accessed 9 August 2017).

De la Torre, A., Fajnzylber, P. and Nash, J. (2008) Low Carbon, High Growth: Latin American Responses to Climate Change. Available at: http://goo.gl/m0wyXw (accessed 9 August 2017).

Department of Energy and Climate Change (UK) (2011) The Aviation Greenhouse Gas Emissions Trading Scheme (Amendment) Regulations 2011. Available at: http://goo.gl/r7LXEj.

DMN-VGL (2015) EU-MRV REGULATION MARITIME SAFER, SMARTER, GREENER. European Union's regulation on monitoring, reporting and verification of CO_2 emissions from maritime transport (EU-regulation 2015/757).

Economic Commission for Latin America and The Caribbean (2011) An Assessment of The Economic Impact of Climate Change on the Transportation Sector in Barbados.

Edwards, D. (2014) Comparative climate change law & policy: Challenges for regulating aircraft emissions in the United States and the European Union. *Issues in Aviation Law and Policy* Vol. 14 No. 1, pp. 103–127, Autumn 2014. De Paul University College of Law International Aviation Law Institute.

European Commission (2016) Reducing Emissions from the Shipping Sector. Available at: http://ec.europa.eu/clima/policies/transport/shipping/index_en.htm (accessed 10 August 2017).

Faber, J. and Brinke, L. (2011) The Inclusion of Aviation in the EU Emissions Trading System: An Economic and Environmental Assessment. Available at: http://goo.gl/BjzrDF.

Grubb, M. (2004) Technology innovation and climate change policy: an overview of issues and options. *Keio Economic Studies* 41(2), 103–132.

Holmes, P. (2011) Border Carbon Adjustment Measures and Small States. Available at: http://goo.gl/HXlyle.

International Air Transport Association (IATA) (2007) Press Release No. 37 dated 18 October 2007. Available at: www.iata.org/pressroom/pr/Pages/18-10-2007-01.aspx (accessed 10 August 2017).

International Centre for Trade and Sustainable Development (2008) *Climate Change and Trade on the Road to Copenhagen: A Policy Discussion Paper*. International Centre for Trade and Sustainable Development, Geneva, Switzerland.

International Centre for Trade and Sustainable Development (2009) *Scoping Paper on Climate Change Mitigation Policies and Measures in Select OECD Countries: Trade and Development Implications in Developing Countries*. International Centre for Trade and Sustainable Development, Geneva, Switzerland.

International Centre for Trade and Sustainable Development (ICTSD) (2010) International Transport, Trade and Climate Change, Information Note Number 17, October 2010. Available at: http://ictsd.org (accessed 20 April 2012).

Jessop, D. (2011a) What Future for Tourism? Available at: www.bbc.co.uk/caribbean/news/story/2011/03/printable/110314_jessop_tourism_apd.shtml (accessed 10 August 2017).

Jessop, D. (2011b) Push-back Grows against Europe's Emissions Tax Scheme. Available at: http://jamaica-gleaner.com/gleaner/20110612/business/business7.html (accessed 10 August 2017).

Keating, D. (2014) MEPs back deal to change ETS rules for aviation. *European Voice* 3 April 2014. Available at http://goo.gl/34DIHI (accessed 10 August 2017).

Lamy, P. (2009) Global problems do not respond to unilateral fixes: Copenhagen must be our focus. *WTO News*, 26 June 2009.

Melendez-Ortiz, R., Biswas, T. and Jegou, I. (2010) Anticipating Pressing Issues in Integration: The Trade, Climate and Sustainable Development Nexus.

Memorandum of Understanding between Guyana and Norway. Available at: www.lcds.gov.gy/images/stories/Documents/MOU.pdf.

Mimura, N., Nurse, L., McLean, R.F., Agard, J., Briguglio, L. *et al.* (2007) Small islands. Climate change 2007: impacts, adaptation and vulnerability. In: Parry, M.L., Canziani, O.F., Palutikof, J.P., van der Linden, P.J. and Hanson, C.E. (eds) *Contribution of Working Group II to the Fourth Assessment Report of the Intergovernmental Panel on Climate Change*. Cambridge University Press, Cambridge.

New York Times (2015) EPA Takes Step to Cut Emissions From Planes. Available at: http://goo.gl/T5MRg7 (accessed 10 August 2017).

Nurse, K. (2008) Climate change, tourism and services in small island developing economies. *Biores* 2(4). Available at: http://goo.gl/nl5Fim.

Nurse, K. (2010) International transport, tourism and services. In: Gueye, K., Sell, M. and Strachan, J. (eds) *Trade, Climate Change and Sustainable Development: Key Issues for Small States, Least Developing Countries and Vulnerable Economies*. Commonwealth Secretariat, London, pp. 49–58.

Nurse, K. (2015) Coping with Climate Change in the Caribbean: Challenges, Opportunities and Strategies for Tackling Climate Change.

Nurse, K., Niles, K. and Dookie, D. (2009) Climate change policies and tourism competitiveness in small island developing states. Paper presented at *NCCR Swiss Climate Research Conference on the International Dimensions of Climate Policies*, 21–23 January 2009. University of Bern, Bern.

Official Journal of the European Union (2009) COMMISSION DECISION of 8 June 2009 on the detailed interpretation of the aviation activities listed in Annex I to Directive 2003/87/EC of the European Parliament and of the Council. Available at: www.lne.be/en/ets-aviation/scope-of-the-eu-ets-directive-for-aircraft-operators/090608-guidelines-on-the-detailed-interpretation-of-aviation-activities-pb-l-eng.pdf.

Reddy, M.V. and Wilkes, K. (2012) *Tourism, Climate Change and Sustainability*. Routledge.

Slade, T.N. (2000) Overview Address By H.E. Ambassador Tuiloma Neroni Slade, Permanent Representative of Samoa to the United Nations, Chairman of the Alliance of Small Island States (AOSIS). Statement at the Pacific Islands Climate Change Conference, Rarotonga, Cook Islands, 3–7 April 2000, entitled 'Linking Science And Climate Change Policy.' Available at: www.sidsnet.org/aosis/statements/08.html (accessed 19 June 2008).

United Nations Conference on Trade and Development (UNCTAD) (2014) *Small Island Developing States: Challenges In Transport And Trade Logistics. D/B/C.I/MEM.7/8.*

United Nations Environmental Programme (UNEP) (2009) Climate and Trade Policies in a Post-2012 World. Available at: http://goo.gl/krTjpl.

UNWTO (2007) Davos Declaration on climate change and tourism responding to global challenges. *Second International Conference on Climate Change and Tourism*, Davos, Switzerland, 3 October 2007.

UNWTO (2014) *Tourism in Small Island Developing States (SIDS): Building a More Sustainable Future for People of the Islands*. World Tourism Organization, Madrid.

UNWTO and UNEP (United Nations World Tourism Organization and United Nations Environment Programme) (2008) *Climate Change and Tourism – Responding to Global Challenges*.

Vergara, W., Kondo, H., Pérez, E., Pérez, J.M., Rueda, V.M. *et al.* (2007) Visualizing future climate in Latin America: Results from the application of the Earth Simulator. *Latin America and Caribbean Region Sustainable Development Working Paper*, 30.

Von Tigerstrom, B. (2005) Small island developing states and international trade: special challenges in the global partnership for development. *Melbourne Journal of International Law* 6(2), 402–436.

WTO (1996) *United States – Standards for Reformulated and Conventional Gasoline*. 6.37, WTO Doc. WT/DS2/R (Panel Report) (29 January 1996).

WTO (World Trade Organization) (1998) *United States Import Prohibition of Certain Shrimp and Shrimp Products*. WTO Doc. No. WT/DS58/AB/R (12 October 1998), reprinted in 41 I.L.M. 149 (2002).

WTO (2009) Trade and Climate Change: A WTO-UNEP Report by Ludivine Tamiotti, Robert Teh, Vesile Kulaçoğlu, Anne Olhoff, Benjamin Simmons and Hussein Abaza. Available at: www.wto.org/english/res_e/booksp_e/trade_climate_change_e.pdf (accessed 10 August 2017).

Appendix 7.1. Long-haul flights to Barbados, Mauritius and Fiji islands (www.convertunits.com; Barbados Statistical Service, www.visitbarbados.org; Mauritius Digest of International Travel and Tourism Statistics 2004, www.gov.mu.org; Fiji Islands Bureau of Statistics, www.statsfiji.gov.fj).

	% of visitors (average 1999–2003)	Airport (direct flights only)	Distance	Flight duration	Carbon emissions
Barbados – Grantley Adams International Airport (BGI) to:					
United Kingdom	40.1%	London mean (LON)[a]	4207 miles 6771 km	8 h, 25 min	1576 lbs CO_2 715 kg CO_2e; 0.72 t CO_2e
United States of America	22.2%	Miami International (MIA)[b]	1599 miles 2573 km	3 h, 13 min	740 lbs CO_2 336 kg CO_2e; 0.34 t CO_2e
		New York – John F. Kennedy International (JFK)[b]	2079 miles 3346 km	4 h, 10 min	962 lbs CO_2 436 kg CO_2e; 0.44 t CO_2e
CARICOM (excl. T&T)	12.2%				
Canada	10.2%	Toronto – Pearson International (YYZ)[a]	2424 miles 3901 km	4 h, 51 min	908 lbs CO_2 412 kg CO_2e; 0.41 t CO_2e
Europe (excluding UK)	6.1%				
Trinidad and Tobago	4.7%	Piarco International (POS)[b]	215 miles 346 km	25 min	100 lbs CO_2 45 kg CO_2e; 0.05 t CO_2e
Mauritius – Sir Seewoosagur Ramgoolam International (MRU) to:					
France	29.8%	Paris – Charles de Gaulle International (CDG)[a]	5857 miles 9426 km	11 h, 43 min	2194 lbs CO_2 995 kg CO_2e; 1.00 t CO_2e
Reunion Island	13.9%	Roland Garros (RUN)[a]	143 miles 230 km	17 min	66 lbs CO_2 30 kg CO_2e; 0.03 t CO_2e
United Kingdom	11.6%	London mean (LON)[a]	6056 miles 9746 km	12 h, 7 min	2269 lbs CO_2 1029 kg CO_2e; 1.03 t CO_2e
Germany	7.8%	Frankfurt International (FRA)[a]	5704 miles 9179 km	11 h, 24 min	2137 lbs CO_2 969 kg CO_2e; 0.97 t CO_2e;
South Africa	7.1%	Johannesburg – OR Tambo International (JNB)[b]	1908 miles 3070 km	3 h, 49 min	883 lbs CO_2 400 kg CO_2e; 0.40 t CO_2e
		Cape Town International (CPT)[a]	2561 miles 4122 km	5 h, 7 min	960 lbs CO_2 435 kg CO_2e; 0.44 t CO_2e
Italy	5.8%	Milan – Malpensa International (MXP)[a]	5485 miles 8828 km	10 h, 58 min	2055 lbs CO_2 932 kg CO_2e; 0.93 t CO_2e
India	2.9%	Mumbai – Chatrapati Shivaji International (BOM)[a]	2904 miles 4674 km	5 h, 49 min	1088 lbs CO_2 494 kg CO_2e; 0.49 t CO_2e
Switzerland	2.8%	Zurich International (ZRH)[a]	5579 miles 8978 km	11 h, 9 min	2090 lbs CO_2 948 kg CO_2e; 0.95 t CO_2e

Continued

Appendix 7.1. Continued.

	% of visitors (average 1999–2003)	Airport (direct flights only)	Distance	Flight duration	Carbon emissions
Fiji – Nadi International Airport (NAN) to:					
Australia	29.4%	Sydney – Kingsford Smith International (SYD)[b]	1970 miles 3170 km	3 h, 56 min	911 lbs CO_2 413 kg CO_2e; 0.41 t CO_2e
		Melbourne – Tullamarine International (MEL)[a]	2403 miles 3868 km	4 h, 48 min	900 lbs CO_2 408 kg CO_2e; 0.41 t CO_2e
New Zealand	17.6%	Auckland International (AKL)[b]	1336 miles 2150 km	2 h, 40 min	618 lbs CO_2 280 kg CO_2e; 0.28 t CO_2e
USA	15.6%	Los Angeles International (LAX)[a]	5518 miles 8881 km	11 h, 2 min	2068 lbs CO_2 938 kg CO_2e; 0.94 t CO_2e
UK	10.2%	(no direct flights – stopovers in LAX)			
Japan	6.8%	Tokyo – Narita Godzilla International (NRT)[a]	4409 miles 7095 km	8 h, 49 min	1652 lbs CO_2 749 kg CO_2e; 0.75 t CO_2e
Continental Europe	6.2%				
Canada	3.0%	(no direct flights – stopovers in LAX or AKL)			

Notes and Assumptions on Visitor Information: Visitor arrivals by country of residence averaged over the period 1999–2003.
Notes and Assumptions on Flight Information: Formulas are based on the UK DEFRA specifications for domestic, short-haul and long-haul international flights. Exact times may vary with wind speeds. Does not account for radiative forcing which may increase carbon dioxide emissions by 2.7 times.
CO_2e: Carbon dioxide equivalents. Assumes an average flight speed for a commercial airliner of 500 mph, which is equivalent to 805 km/h or 434 knots.
[a]Long haul international flight (calculation includes an assumption of a 79.7% load factor with 346 seats on the B747-400 and 261 seats on the B767-300ER)
[b]Short haul international flight (calculation includes an assumption of a 65% load factor with 139 seats)

8 Case Study Ireland: Coastal Tourism and Climate Change in Ireland

J.A.G. Cooper* and Stephen W. Boyd
Ulster University, Northern, Ireland

Introduction

The study of the relationship between coastal regions and tourism is a relatively recent one compared to other landscapes (Agarwal and Shaw, 2007). An interest in the sea and sea-bathing culture can be traced back to Britain in the 1750s which would quickly spread to the northern coastline of France (1780s), Germany and the north-east coastline of America (1790s) and the northern coastline of Spain (1830s) (Towner, 1996). However, it would not be until the Edwardian and Victorian era that Britain's desire to 'go to the seaside' was ushered in and with it, the major transformation of that geographic setting. Coldwater bathing environments would not be eclipsed by warmer seas until the 1960s when the north side of the Mediterranean quickly replaced colder waters of northern Europe, ushering in the start of what is recognized as the era of modern tourism with a major appeal of sun, sand and sea destinations. Gormsen (1981; cited in Agarwal and Shaw, 2007) illustrates that the appeal of coastal areas and the development of coastal resorts would quickly diffuse in a series of waves to Turkey and the northern African coastline of the Mediterranean, followed by the Caribbean, Indian Ocean, South-east Asia and South America and Oceania regions. This chapter focuses on the transformation of Ireland's coastline from a tourism perspective, addressing first the context in which tourism occurs, the nature of visitor flows and change, the mix of visitor attraction and the position coastal tourism attractions and beaches play in overall visitor choice, and the economic benefits of tourism. Coastal regions as tourist space are discussed along with how specific focus has been directed toward purposeful development of the majority of Ireland's coastline in terms of both product development and visitor experience. Present climate and future changes are discussed as are the effects such changes bring to coastal regions and their tourism product mix.

Tourism in the Republic of Ireland and Northern Ireland

The history against which tourism has developed in the Republic of Ireland and Northern Ireland is one of contrast. The former has always enjoyed the required antecedents

*Corresponding author, e-mail: jag.cooper@ulster.ac.uk

for tourism to develop (an attractive, safe and accessible environment, with an identifiable brand product and associated experience), whereas the latter has suffered from a negative image of a dangerous environment where 30 years of civil unrest has made it difficult to perceive the destination as having potential for out-of-state visitors (Boyd, 2013a). Since 2000 (2 years after the signing of the Good Friday Agreement which saw devolved power return to Northern Ireland with the setting up of shared governance that embraced all sectors of Northern Ireland society), visitor numbers have been on an upward trajectory and now double those compared to the late 1960s just prior to the start of the violence era (Boyd, 2013a). Much of this appeal is linked to heritage of which the coastal environment is an integral part of the natural, and to a lesser extent, built, heritage capital that appeals to visitors and is part of the region's unique selling point and competitive advantage to similar peripheral regions (Boyd, 2016).

In 2014, 8 million visitors came to the Republic of Ireland (this does not take into account the 7.3 million classed as domestic visitors), compared to 2.2 million to Northern Ireland (an almost equal number, 2.3 million, were domestic visitors), where the industry is worth €6.6 billion and £751 million, respectively (NISRA, 2015; Fáilte Ireland, 2016). When direct and indirect spending is accounted for, tourism in 2014 directly accounted for 58,000 jobs in Northern Ireland and 205,000 jobs in the Republic of Ireland (NISRA, 2015; Fáilte Ireland, 2016). In terms of gross national product (GNP), tourism in the Republic of Ireland is worth 2.5% of export earnings; when domestic tourism expenditure is added in revenue terms tourism overall accounts for 4.2% of GNP (Fáilte Ireland, 2016). In contrast, in Northern Ireland tourism accounts for 2.5% of gross value added (GVA), the equivalent of £3.3 billion (NISRA, 2015).

National statistics do not provide a detailed breakdown of the value of the tourism industry based around coastal settings and attractions, involving also marine activities. What, however, recent (2014) overseas holiday-maker surveys reveal is the importance of beautiful scenery (90%), offering a good range of natural attractions (87%) and enjoying natural unspoilt environments (85%); all aspects that the whole island of Ireland can provide (Fáilte Ireland, 2016). We will return later in the chapter to examine the profile of those that choose Ireland (including Northern Ireland) as a holiday destination, but it is necessary first to set out the wider context of visitor flows and the principal attractions that draw visitors to Ireland, both north and south.

Visitor Flows, Participation and Principal Attractions

Tourism statistics in 2014 for both regions of the island reveal a distinctive pattern of purpose of visit. The majority of visitors to the Republic of Ireland come for a holiday (48%), followed by visiting friends and relatives (VFR) (32%), with business accounting for 16% of all visits (CSO/NISRA, 2016). In contrast, the majority of visits to Northern Ireland are to visit friends and relatives (51%), followed by holiday (31%; that figure rises to 45% for foreign visitors) and then business (26%; but only 8% for foreign visitors) (CSO/NISRA, 2016). Northern Ireland has always had a higher proportion of VFR visitors where 'holiday/leisure' was not the main purpose, though this position is starting to be reversed as it is now a very clear second main reason, with business-related visits a distant third reason for visiting. When statistics for overnight visits to Northern Ireland are examined, 'holiday' is ranked as the most popular reason (45%), compared to VFR (42%), and in the trend data between 2011 and 2014, the largest increase in reason for visit were holidaymakers (a +21% increase between 2013 and 2014) as compared to VFR (a +13% rise) (NISRA, 2015).

Only the Republic of Ireland statistics for 2014 record the main activities that visitors engage in. Hiking and walking remain the most popular activities by visitors (63% or 1,193,000 visitors), followed by cycling (15% or 286,000 visitors), golf

(9% or 172,000 visitors), angling (8.3% or 157,000 visitors) and equestrian (4% or 79,000 visitors) (Fáilte Ireland, 2016). There have been some notable changes compared to the statistics presented in 2008 which recorded visitor activity profiles on surveys reported by Fáilte Ireland. The rise in cycling in the 2014 figures could be accounted for by the fact that Ireland (both north and south) hosted the Grande Departe (opening stages) of the Giro D'Italia in May of that year.

The above activities align nicely with Fáilte Ireland's development agenda that is aimed at specific target markets by identifying what are Ireland's best prospects for growing tourism in its main overseas markets. Three priority segments have emerged, including 'social energisers' (in search of 'wow' experiences, seeking fun and excitement and immersing themselves in destinations), 'culturally curious' (out to broaden their minds and expand their experiences by exploring new landscapes, history and culture) and 'great escapers' (visitors who are adventurous and committed to spending quality time in a breath-taking place) (Fáilte Ireland, 2016). It is this last segment that best characterizes visitors keen to take part in activities and get physical with nature, albeit the smallest market segment of the three (ranging from 9% to as little as 3% of the key overseas markets of America, Britain, Germany and France) (Fáilte Ireland, 2016). However, as will be addressed later in the chapter, tourism product development that has been directly linked with Ireland's coastline, in the form of a number of driving routes, taps into the other two holidaymaker segments that emerge as being more relevant for Ireland's major overseas visitors. In particular, 45% and 34% of American and German visitors relate to being 'social energisers'.

Less detail on tourist activities is provided by Tourism Northern Ireland (the national tourism body for Northern Ireland). In 2008 the focus was on identifying key markets through what they then had termed 'winning themes'. The activity tourism market, previously discussed, was one of those winning themes, along with city breaks, culture and heritage, business and excellent events. Currently, under a larger banner of business support, major events and business have been given their own focus with culture, heritage and activity tourism being grouped as one. In this latter grouping, a number of subcategories stand out, including golf tourism (valued at £33 million to the Northern Ireland economy; with 139,300 golf visitors in 2014) and activity tourism (valued at £100 million to the Northern Ireland economy), involving over 200 private sector operators providing over 40 different activities. According to a 2010 outdoor activities survey (NITB, 2011), some differences emerged over participation between overseas and domestic visitors. Walking was ranked the highest by overseas and domestic visitors (63% and 32%, respectively), whereas domestic participation was greater for canoeing and watersports (18% compared to 8% by overseas visitors), cycling (14% compared to 6%) and fishing (12% compared to 4%). Only golf saw greater overseas participation (14% compared to 12%). The survey also recorded the number of provider sites across Northern Ireland for activities. Most sites (424) were associated with angling opportunities across the six counties' lakes, rivers and coast line. Other sites included cycling (136 sites), golf (128 sites), watersports (89 sites) and adventure activities (72 sites), with the least number of sites (65) identified for equestrian opportunities. Other sub-themes draw on visitor interest in gardens and with sites associated with the HBO series *Game of Thrones* which is filmed in Northern Ireland. Both these sub-themes will be returned to later in the chapter as there is a close association between the opportunity spaces for these experiences and Northern Ireland's coastal environment.

A cursory glance at what attractions receive the most visitors (Table 8.1) reveals that the top attractions are mostly found in coastal regions, particularly in Dublin, Counties Clare and Cork in the case of the Republic of Ireland, and Belfast, the North and South-east Coasts for Northern Ireland (Fig. 8.1). It is not surprising that Dublin and Belfast receive the most visitors to Ireland overall, recording 5.4 million (of which 4.1 million are overseas visitors) and

Table 8.1. Top ten visitor attractions in Northern Ireland and the Republic of Ireland, 2014 (NISRA, 2015; Fáilte Ireland, 2016).

Northern Ireland	Visitor numbers (2014)	Republic of Ireland	Visitor numbers (2014)
Giants Causeway Visitor Centre (north coast)	788,000	Guinness Storehouse (Dublin)	1,269,371
Titanic Belfast	634,000	Cliffs of Moher Visitor Experience (Clare)	1,080,501
Ulster Museum (f) (Belfast)	466,000	Dublin Zoo	1,076,876
Historic Walls of Derry (f)	370,000	National Aquatic Centre (Dublin)	931,074
W5 museum (Belfast)	324,000	Book of Kells (Dublin)	650,476
Carrick-a-Rede Rope Bridge (f) (north coast)	323,000	National Gallery of Ireland (f) (Dublin)	593,183
The Guildhall (f)(Derry)	299,000	National Botanic Gardens (f) (Dublin)	541,946
Belfast Zoo	253,000	Doneraile Wildlife Park (f) (Cork)	460,000
Pickie Fun Park (f) (Bangor; south-east coast)	243,000	St Patrick's Cathedral (Dublin)	457,277
Murlough Reserve (f) (south-east coast)	241,000	Tayto Park (Meath)	450,000

(f), free attractions

2.9 million (of which 1.75 million are overnight visits), respectively (Fáilte Ireland, 2016; www.belfastcity.gov.uk as accessed on 15 March 2016).

Table 8.1 also reveals that while many of the top visitor attractions are found in coastal cities (Belfast and Dublin), few are natural heritage attractions located in coastal regions. Instead, the focus is with heritage and cultural visitor centres and recreation/entertainment spaces. Those attractions that are directly located on the coastline are the Giant's Causeway, County Antrim (a World Heritage Site inscribed in 1986), Carrick-a-Rede Rope Bridge, County Antrim and the Cliffs of Moher (recently declared as a GeoPark), County Clare. Statistics are not collected on visitor numbers to beaches in the Republic of Ireland and visitor numbers exist only for those beaches in Northern Ireland that are under National Trust ownership. Of these, Portstewart Strand beach was last ranked in the top ten visitor attractions in 2000 (ranked as 8th) with over 140,000 recorded visitors. No urban beach today is highly ranked; Crawfordsburn County Park near Belfast prior to 2009 recorded approximately 600,000 visitors annually to its beaches, most of whom are local residents walking dogs and exercising. There are no figures post-2009.

Statistics also fail to link activities that visitors engage in with where these engagements occur. It is reasonable to assume that a considerable proportion of hiking and walking and golf takes place along the coastlines of both the Republic of Ireland and Northern Ireland. The main activities listed by Nairn (2005) on the Irish coast were surfing, sailing, windsurfing, scuba diving, sea kayaking, sea and shore angling, birdwatching, whale and dolphin watching, seal watching, coastal walks and rock climbing.

As mentioned earlier in the chapter, the product portfolio for the Republic of Ireland focuses on scenery, culture, heritage and friendly and hospitable people, and sells the Irish experience by encouraging visitors to explore an 'Island of unique character and characters'. This is supported by the 2014 overseas holidaymakers survey which reiterates Ireland's strong performance as a destination that receives very high satisfaction scores for friendly and hospitable people (97%), beautiful scenery (95%), as a safe and secure destination (95%), having interesting history and culture (93%), a good range of natural attractions (91%) and a natural unspoilt environment (90%). Attractive cities and towns, as well as offering an easy, relaxed pace of life also scored well

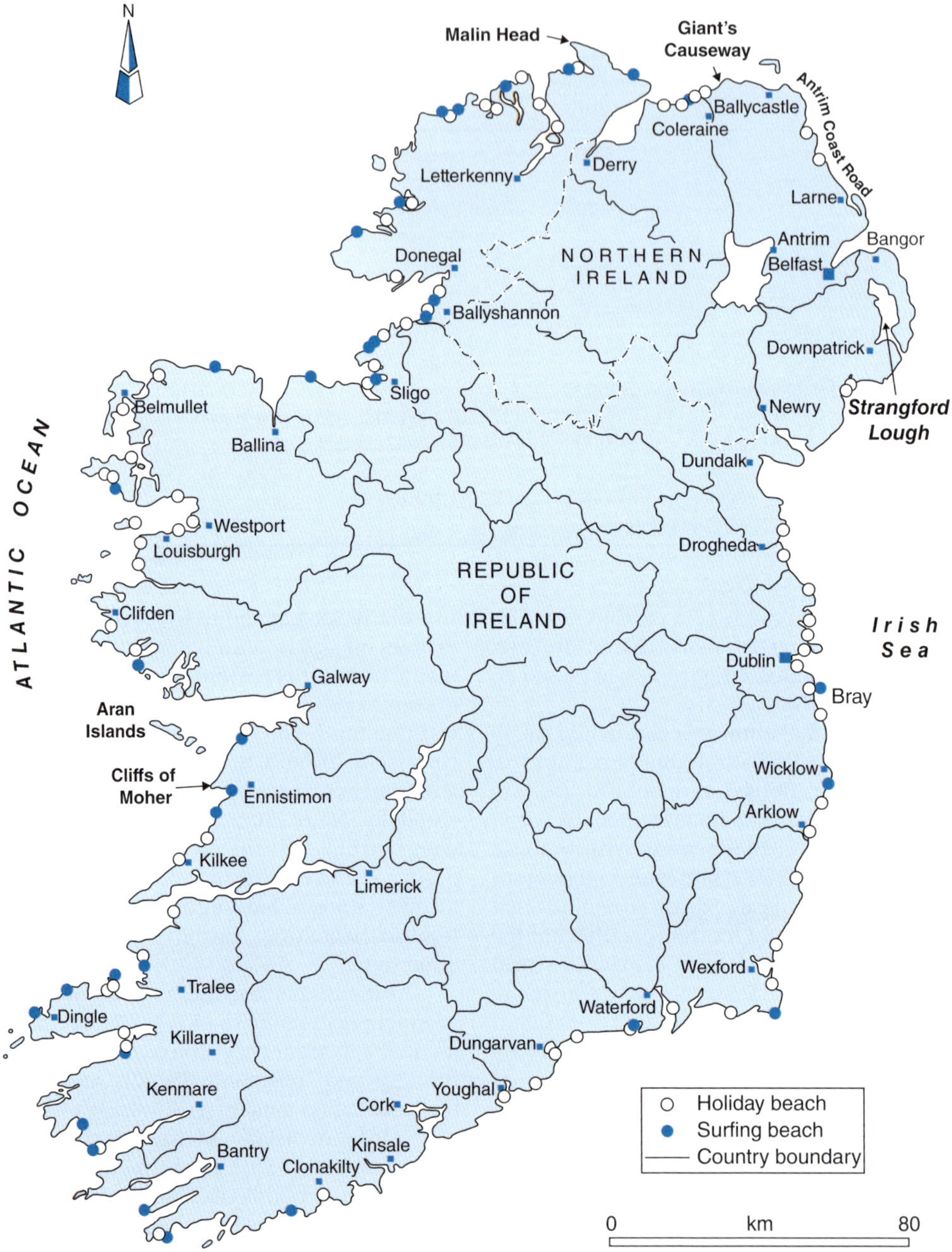

Fig. 8.1. Map of Ireland showing main surfing and holiday beaches and localities referred to in the text (after Nairn, 2005).

with 90% and 89% satisfaction, respectively (Fáilte Ireland, 2016). Combined, these satisfaction measures provide the Republic of Ireland with a competitive advantage over other similar northern European peripheral destinations. As for Northern Ireland, visitors are encouraged to visit a destination that is undergoing a post transformation out of conflict and experience 'our awakening', immerse themselves in a friendly and hospitable culture, visit international stand-out signature attractions, taste excellent locally produced food and drink, and a destination with vibrant city cultures and host of major events (www.nitb.com as accessed on 14 March 2016).

Coastal Regions as Tourist Spaces

The coastline and the coastal regions around the island have played an integral role in meeting visitor expectation of beautiful scenery, of experiencing 'wow' destinations as well as being given opportunities to engage in a wide variety of sports tourism and water-based activities. Both governments over the past decade have deliberately developed the coastline as part of a wider product offer, where coastal road networks were transformed from organic transport corridors to purposive driving tourism routes (Timothy and Boyd, 2015). In 2007, the then Northern Ireland Tourist Board (now Tourism Northern Ireland) marked out the first section of the island's coastline with a purposive driving route called the 'Causeway Coastal Route' (CCR); an 80 miles (130 km) long essential Irish journey that took in windswept cliffs, spectacular scenery, unspoilt beaches, historic castles and forts. It would later provide visitors with a purposive driving route along the north-east and north coastline of Northern Ireland, which would connect a number of signature projects including the improved realm of the Derry Walls, the building of the Titanic Belfast visitor attraction in the Titanic Quarter region of Belfast, with a midpoint signature attraction of a new visitor center at the Giant's Causeway World Heritage Site (Boyd, 2013b). The route was described by Jacobs Creek in 2008 as one of the top touring routes in the world. A 2009 experience survey of 1500 visitors to the Giant's Causeway, Antrim and Causeway Coast revealed this route as a 'must see' attraction to uncover stories and enjoy authentic, unique, precious and iconic stops (Boyd, 2013b). In terms of visitor experience the top five were noted as memorable (75%), natural (75%), unique (52%), historical (49%) and inspiring (47%) (NITB, 2012).

Part of the visitor experience along the CCR is gardens and sites associated with the HBO series *Game of Thrones*. Regarding the former, three of the nine stops of the County Antrim Garden Trail are in coastal settings along the CCR. As for *Game of Thrones* film settings, of the 14 locations across Northern Ireland, eight are coastal along the north/north-east and south-east coastlines, to include stops such as Ballintoy harbour and Downhill beach (www.nitb.com as accessed on 14 March 2016).

The east and south-east coastline of Northern Ireland saw the marking out of the Mourne Coastal Route in 2013. This is an equally spectacular journey from the coastal community of Bangor (in close proximity to Belfast), taking visitors down the Ards peninsula through an array of coastal communities that are popular domestic holiday destinations, to the tip of the peninsula at Portaferry. Visitors cross over the mouth of Strangford Lough in a ferry to the community of Strangford before continuing along the south-east coastline through small fishing communities to the coastal resort of Newcastle and Annalong before ending at the mouth of Carlingford Lough and driving along the Narrow water to the city of Newry, where on the outskirts of the city, the Mourne Coastal Route ends. There are fewer tourist stops along this route, but the journey is equally impressive (www.nitb.com as accessed on 14 March 2016).

Recently, in 2013 Fáilte Ireland marked out the Wild Atlantic Way (WAW) as a coastal route along the west coast of Ireland. It is the world's longest defined coastal touring route, approximately 2500 km in length with 142 designated discovery points and 15 signature discovery points along its route. The route stretches from Muff, County Donegal in the north, to Kinsale, County Cork in the south. The WAW is divided up into six geographic regions: Northern Headlands, The Surf Coast, the Bay Coast, the Cliff Coast, Southern Peninsulas and the Haven Coast. There is yet to be undertaken any extensive study of the economic impact of the WAW and the extent to which visitor numbers have increased in the west of Ireland. One deliberate strategy by Fáilte Ireland in marking out the WAW was to draw visitors out from the core and hub region of Dublin to peripheral regions and to enable the tourism industry in peripheral coastal regions to benefit from a greater proportion of visitors (www.Fáilteireland.gov.ie). At the time of writing a number of research clusters were in the process of being set up to study the impact of the WAW, profiling the visitor and their experience, assessing the opportunity for sustainable community tourism development, and the contribution of food turnover to hotel revenue along the WAW (WAW Scoping Report, 2016).

These three driving tourism purposive routes enable visitors (international and domestic) to engage directly with Ireland's coastline, where much of the visitor experience is associated with coastal resorts, key coastal stops and the coastline itself. There are ongoing discussions to develop another route around the east and south-east coastline of the Republic of Ireland thereby creating a tourism circuit coastal product and experience.

As for signature natural environment stops along its 7500 km coast, Ireland has some stunningly beautiful beaches, both urban and rural (Fig. 8.1). It is, however, important to make the distinction between urban and rural beaches. The former (Fig. 8.2) are linked to seaside resorts that have a long history of resort development stemming back to the late 1800s/early 1900s (e.g. Bundoran, County Donegal; Portrush, County Antrim; Newcastle, County Down; and Bray, County Wicklow), where the beach surroundings have been modified with urban sea walls, promenades running parallel to the coastline, and visitor accommodation has been built in close proximity to the beach or overlooking it. In contrast, the vast majority of beaches around Ireland are rural in character, with little modification and the virtual absence of a built environment designed to accommodate visitors (Fig. 8.3).

Fig. 8.2. Facilities at Portrush are typical of urban seaside resorts.

Fig. 8.3. Kinnego Bay, County Donegal, is typical of the scenic attraction of Ireland's many undeveloped rural beaches.

In recent years, authorities have sought to achieve certain award status for beaches (e.g. Blue Flag) and stretches of coastline (e.g. Green Coast). Blue Flag status is awarded to beaches for their cleanliness, water quality, safety and level of environmental protection (www.blueflag.org). In contrast, the more recent Green Coast award is made to more rural beaches that are recognized for their natural unspoilt beauty, water quality and cleanliness (www.tidynorthernireland.org). In 2013, 113 awards were given out in the Republic of Ireland, comprising swimming spots, urban beaches, world-renowned surfing spots and first-class marinas; in particular, 74 Blue Flags (includes four blue marinas) and 45 National Green Coast awards (Table 8.2). In 2015 Northern Ireland had ten Blue Flag beaches, four Green Coast Awards and ten Seaside Award beaches (Table 8.3). The Green Coast awards in association with Coca-Cola recognize beaches and surrounding communities managed for their natural quality and environmental value, whereas the Seaside awards rate resorts on the criteria of very good water quality, and being excellent places to visit and enjoy.

While there are no formal statistics recorded on the levels of domestic visitor numbers to Ireland's coastline, domestic visitors are as important as out-of-state visitors to the economic life of Ireland's resort communities. Over the past decade, there has been considerable change to the mix of accommodation available with the decline of guest houses and bed and breakfast establishments and the rise of second homes along the coastline (Fig. 8.4) and within the resort communities of the north coast to the extent that in some resorts second home ownership comprises between 33% and 65% of all housing stock (Boyd, 2006). This has had many unforeseen social implications (Cooper and McKenna, 2009a). Many resorts

Table 8.2. Blue Flag beaches by county in the Republic of Ireland in 2013 (www.antaisce.ie).

Region	Number in 2013
Louth	3
Dublin	2
Wicklow	3
Waterford	2
Wexford	3
Cork	5
Kerry	12
Clare	6
Galway	7
Mayo	12
Sligo	2
Donegal	13

throughout Ireland have recently undergone redevelopment in an attempt to move away from stagnation and avoid the decline stage of the Butler resort model (Butler, 1980). Initial plans in 2007 to rejuvenate many of Northern Ireland's coastal communities have not materialized as a result of a combination of the financial global crises of 2007–2009 and the consequent reduced block grant that the Westminster government provides the Northern Ireland Assembly. Instead, small incremental ad hoc changes have occurred in a number of coastal resorts in Northern Ireland with funding provided by the Department of Social Development (DSD) through its funding programme for regeneration. The resort community of Newcastle (south-east coastline) has seen a coastal promenade and revived main street in terms of public realm and the resorts of Portrush and Portstewart (north coastline) saw public realm improvements and cleaned frontage ahead of the Irish Golf Tournament hosted in the region in 2012 and one of the stages of the Giro D'Italia in 2014 taking in the north and north-east coastline of Northern Ireland. Apart from these exceptions of urban resort change, Northern Ireland's coastal communities lack extensive tourism infrastructure, in particular indoor recreation facilities and quality hotels. This lack of development is important given the climate of Ireland and its relation to tourism and the extent to which a lack of year-round attractions and product development brings a high degree of seasonality to how visitors engage with Ireland's coastal environment, an area this chapter now turns to address.

Table 8.3. Blue Flag, Green Coast and Seaside awards for Northern Ireland in 2015 (www.tidynorthernireland.org).

(a) Blue Flag beaches

Name of beach	County	Beach characteristics
Magilligan, Benone Strand	Londonderry	sandy
Cranfield Bay	Down	sandy
Portstewart Strand	Londonderry	sandy, rocky
Murlough	Down	sandy
Portrush West Strand	Antrim	sandy, rocky
Whiterocks	Antrim	coastal
Tyrella	Down	sandy
Downhill	Londonderry	sandy
Crawfordsburn	Down	sandy
Castlerock	Londonderry	sandy

(b) Green Coasts

Name of beach	County
Cranfield Bay	Down
Murlough	Down
Tyrella	Down
Helen's Bay	Down

(c) Seaside Award beaches

Name of beach	County
Downhill (R)	Londonderry
Portrush West (R)	Antrim
Portrush East (R)	Antrim
Portstewart Strand (R)	Antrim
Carnlough (r)	Antrim
Ballygally (r)	Antrim
Millisle (r)	Down
Cranfield Bay (R)	Down
Murlough (R)	Down
Tyrella (R)	Down

R, Resort; r, rural

The Climate of Ireland in Relation to Tourism

A number of studies have shown a strong link between tourism and climate (Mieczkowski, 1985; Lise and Tol, 2002). Tourism

Fig. 8.4. Second home development at Culdaff typified the development boom of the early 2000s on the Irish coast. During this phase many traditional homes, hotels and businesses were converted to apartments and additional new builds proliferated at the coast.

is, however, influenced by many other aspects of the destination such as cultural, landscape and historical attributes, recreational opportunities and family ties. Factors such as cost, proximity, ease of access and language also influence choice of destination.

Ireland's climate is influenced mainly by its position as an island on the margin of the Atlantic Ocean (Rohan, 1986). It is warmed by the north-flowing North Atlantic Drift and its maritime climate is characterized by high levels of rainfall, strong winds and low hours of sunlight. The west coast has high wave energy that diminishes rapidly into the Irish Sea around both north and south coasts. Off the west coast the average wave height is 2.5–3 m whereas on the east coast it is 0.5 m or less.

The following characteristics of Ireland's climate derive from the Irish Meteorological Service (www.met.ie/climate/climate-of-ireland.asp). Average annual temperature is about 9°C. In the east, summer mean daily maximum is about 19°C and winter mean daily minimum is about 2.5°C. The mean daily maximum temperature at coastal stations of Valentia (west coast) and Dublin (east coast), in the warmest month (July), is between 12 and 19°C. Air temperature falls below zero on only about 10 days per year at the coast, compared to about 40 days per year inland. Ireland normally experiences only between 1400 and 1700 h of sunshine annually and on about 50% of days there is 100% cloud cover.

The west and north-west coasts of Ireland are renowned as the windiest places in Europe. At Malin Head, gales are recorded on more than 50 days each year on average. Only between 0.8% and 3.3% of the time are calm conditions recorded at the coast compared to up to 9% inland. Mean annual wind speed ranges from 7 m/s in the north-west to 4 m/s in the east midlands.

Rainfall is normally associated with Atlantic low pressure systems and there is consequently a strong rainfall gradient (around 1000–1250 mm/year in the west to 750–1000 mm/year in the east). The south-east coast has only around 150 days per year of rain (>1 mm), compared to 225 in the west. Although rain is common, it is not particularly intense; average hourly rainfall amounts in Ireland usually range from 1 to 2 mm. High rainfall intensity (15 to 20 mm/h) has a recurrence interval of about 5 years. Rainfall is highest in winter and lowest in early summer; April is the driest month in most of Ireland, except in the south, where June holds that distinction.

Ireland's seawater is from 7°C to 8°C warmer than the average global sea temperature at such latitudes. Surface water temperatures at Malin Head in 2008 ranged from a low of 7.9°C in February to a high of 15.1°C in August. Average February sea temperatures vary from 10°C in the south-west to <7°C in the north-east while August temperatures range from 16°C in the south to 13°C in the north-east.

In terms of its role in tourism, climate contributes largely to the landscape character of Ireland and also its resulting culture. It is also an important factor in limiting some outdoor recreational activities (e.g. sea bathing) and promoting others (e.g. surfing). The weather on any given day plays an important role in tourist behaviour and selection of visitor sites (Fahy *et al.*, 2005). It is, however, probably true to say that Irish tourism exists in spite of its climate and weather conditions.

Climate Change and the Irish Coast

Several historical records and other indicators provide historical evidence of climate change in Ireland (Carter, 1988). These records are of variable length and are unevenly scattered geographically; however, collectively they provide an impression of the main changes that have been occurring in Ireland historically. Since the 1980s there has been a general warming trend of 0.3–0.7°C per decade in Irish waters (Dunne *et al.*, 2009). On land, Sweeney *et al.* (2003) noted a mean annual temperature increase of 0.5°C during the 20th century, with a more rapid rise in the 1990s. Significant decreases in frost frequency and increases in 'hot day' frequencies also occurred during this time and substantial increases in winter precipitation occurred in the north-west with a decrease in summer rainfall in the south-east (Sweeney *et al.*, 2003).

Climate change in Ireland has been the subject of several recent reports (Sweeney *et al.*, 2002, 2003; Dunne *et al.*, 2009) and its implications for the marine environment specifically were considered by Boelens *et al.* (2005), who noted that:

- Land temperatures will increase by up to 1.5°C in winter and 2.0°C in summer by 2055;
- Average winter rainfall will increase by 10% by 2055 while summer rainfall will fall by between 10% and 40% in summer;
- An increase in sea levels similar to the global mean prediction of 0.49±0.08 m by the end of the century; and
- An increase in the number of extreme events (e.g. more intense storms, hotter summers).

Dunne *et al.* (2009) presented the following scenarios on the basis of a suite of models of global atmospheric circulation, ocean surge and wave climate:

- Temperature increases of 3°C to 4°C towards the end of the century;
- Increase in autumn and winter rainfall by 15–25% towards the end of the century;
- Summers to become drier by 10–18%;
- No change in mean wind speeds over the coming decades, but an overall reduction by 4–5% towards the end of the century;
- Increased frequency of very intense cyclones;
- Seawater warming of 0.3–0.4°C per decade in the Atlantic and 0.6–0.7°C per decade in the Irish Sea;

- Increased frequency of storm surge events; and
- Extreme wave heights to increase in most regions.

The effect of global sea level rise at the coast is expressed as relative sea level change (the relationship between land and sea) and this is markedly variable around Ireland because of the underlying effect of land uplift. In the north, relative sea levels have changed little in the past 50 years (Orford *et al.*, 2006) because land uplift has approximately equalled sea level rise, whereas in the south they have increased at rates similar to the global average (Devoy, 2008). Recent projections (Orford *et al.*, 2007) suggest sea levels will rise around the entire island during the next few decades as global sea level rise begins to exceed rates of land uplift.

Climate Change and Irish Coastal Tourism

Climate change not only affects human activity directly through temperature, rainfall and wind, but also indirectly through changes in the natural landscape, its morphology, habitats and species driven mainly by changes in sea level and wave conditions. These indirect impacts arise through coastal erosion and accretion and flooding. The response by humans to perceived impacts can also have profound implications for the environment and human activities (Cooper, 2009).

Increased air and water temperatures, reduced rainfall and wind are likely to improve the conditions for outdoor activities in Ireland. Using the Tourism Climatic Index (TCI; Mieczkowski, 1985) Amelung and Viner (2006) suggest that north-west Europe may develop a more favourable climate than Mediterranean destinations leading to increases in home holidays. Ireland could attain from two to four 'very good months' by the end of the century (it currently has none) (Nicholls and Amelung, 2008). Since visitor numbers are strongly correlated with daily temperature (Viner and Agnew, 1999) this will likely increase visitor numbers to the coast. While this might be regarded as good for the tourism economy, it creates some issues. Since most visitors use a car to travel to the coast, the current carrying capacity as determined by car parking availability and road traffic volumes might be exceeded. Traffic is currently an issue at some rural Irish beaches and resorts in summer. Rossnowlagh, County Donegal, for example, has an emergency traffic management scheme for such occasions (Cooper and McKenna, 2009b). Large numbers of pedestrians can also cause serious erosion problems on dunes as documented at Brittas Bay, County Wicklow by Quinn (1977), and this can require control measures to be put in place. Increases in numbers can also increase pressure for good quality environments and enhanced facilities (Phillips and Jones, 2006).

Coasts respond to increased wave height and sea level rise by changes in shape and position. The most common manifestations of this change are coastal erosion and flooding in low-lying areas. In Ireland, this will simply exacerbate trends that are already evident in the south, but lead to a complete change in the north where sea levels have been stable historically and coastal erosion has not been a major issue. Structures on or near the shore will be at risk from such coastal changes but the geological record shows that in general terms, coastal depositional systems (beaches, tidal flats, marshes) continue to exist but tend to move upwards and landwards as sea level rises.

With accelerated sea level rise cliffs may erode at faster rates and shore platforms will become more active than during the recent past. Increasing contrasts between dry and wet periods and increased storm intensity are likely to contribute to enhanced cliff instability. At the Giant's Causeway, the rise in sea level may mean that access to parts of the site becomes more difficult as extreme water levels occur more frequently. Cliff paths may become more difficult to maintain in the face of more instability (Orford *et al.*, 2007).

Many of the most iconic cultural sites on the Irish coast are historical defensive sites such as Dun Aengus, County Galway or Dunluce Castle, County Antrim. Built on cliff-top locations, these sites are prone to periodic erosion. The increased frequency of cliff erosion will likely make this more of a management issue in the future. An indication of the importance of such sites is that in County Londonderry the Mussenden Temple, an iconic 18th-century building on the cliff top, was protected by strengthening of the surrounding cliff edge.

When cliff recession threatens infrastructure, management problems can arise. The main access road to the Dingle peninsula is heavily utilized by tourist traffic during the peak season from June to September. Erosion of sea cliffs caused part of the road to collapse in April 2007. This threatened the entire Dingle tourism industry and a rapid intervention by Kerry County Council (at a cost of €4 million) was necessary to stabilize the cliff and rebuild the road (O'Connor *et al.*, 2009). At a smaller scale, access to several (mainly holiday) homes was cut by coastal erosion of an access road built on a soft dune cliff at Kilmichael, County Wexford in December 2007 (Cooper and McKenna, 2008). Popular tourist drives, such as the Antrim Coast Road, are likely to require more frequent maintenance and even capital upgrades in the face of increased erosion threats from sea level rise. The same issues pertain to many access roads and railway lines in Ireland which follow the coastline. The east coast railway line has had to be defended against erosion because of its coastal location. Occasional blockage of roads and railway lines by rock falls and landslides may increase in frequency with the projected changes in precipitation and greater contrasts between wet and dry periods.

Beaches respond much more rapidly to changed environmental conditions than rock coasts. Ireland's beaches are overwhelmingly rural in nature and, left to themselves, they can adjust readily to changes in sea level by changing their morphology. There is, however, an apparent aversion in Ireland to permitting such entirely natural behaviour and many beaches and dunes have been damaged by rock armour and sea walls. In Northern Ireland about 25% of sandy beaches are backed by some kind of defence (Cooper *et al.*, 2016). Indicative of the public perception is a statement by an Irish government minister (Browne, 2008) that:

> Ireland must continue to protect our coastline, sections of which are constantly under threat from erosion. Being an island community, our coastline is one of our most important natural resources and is of particular importance to our tourism industry. The coastal protection programme aims to construct works that slow or even halt erosion in places where the coastline is at its most vulnerable.

Similarly an MEP stated: 'The ongoing erosion at Lisfannon threatens the future of Inishowen's most popular beaches, as the Council's recent efforts have failed to stop erosion at the beach it is time to look at the possibility of creating a permanent erosion prevention structure'. These sentiments illustrate one of the major misconceptions highlighted by Cooper and McKenna (2008) regarding the meaning of coastal protection. In these statements, 'protection' means only protection of land and infrastructure. In many cases this unfortunately also means degradation or destruction of the natural coastal system, particularly beaches. Clearly, degradation is not beneficial to many aspects of tourism which data show to be reliant on a high quality environment and scenic beauty.

Links-type golf courses are major offenders here (Fig. 8.5), but caravan parks (Fig. 8.6) and, worse still, houses built in ill-advised locations (where erosion is active or highly likely, Fig. 8.7) have all been defended even when the erosion is purely a seasonal or episodic response to storms and when the infrastructure is poorly sited. Under a rising sea level coastal defences cut off the sediment supply to the beach causing it to narrow and even disappear altogether. Similar effects can be produced by efforts to halt erosion of cliffs, which cuts off the supply of sediment to some beaches and leads to their demise.

It is a particular challenge in Ireland to maintain the coastal scenery and maintain the coastal resilience by not armouring and finding alternative approaches such as the proactive one of not permitting building in high risk zones and the responsive one of

Fig. 8.5. Rock armouring of a dune to protect a golf course at Portrush.

Fig. 8.6. Informal armouring of the coast at a caravan site creates an unsightly landscape.

Fig. 8.7. Walls built to defend private homes degrade the beach environment. Millisle, County Down.

relocating threatened infrastructure. The human response to coastal erosion is therefore critical in influencing its effect on future tourism.

On Ireland's resort beaches the choices are more difficult, because many beaches have been backed by seawalls and promenades for more than a century (Fig. 8.8) and this infrastructure, plus the buildings and infrastructure behind it, have become an integral part of the tourism landscape. The 'bucket and spade, ice cream and arcade entertainment spaces' experience still holds resonance with a large sector of the population; any physical changes that climate change may bring to these environments will not lessen the demand for this type of product. The narrowing and loss of beaches in front of such sea walls (Taylor *et al.*, 2004), however, leads to loss of recreational space accompanied by loss of environmental quality. There has been significant historical loss of beach area at Bray, County Wicklow and Portrush, County Antrim, both of which have sea walls and promenades at the rear of the beach. To maintain the resource in such instances will precipitate several choices, all involving expenditure. With increased sea level and storminess, sea walls themselves will come under increasing wave attack, requiring maintenance and even enlargement. To maintain the beach requires either that the sea wall be shifted landward or removed (a difficult choice given the infrastructure that is typically located behind such structures), or that the beach be artificially nourished with sand brought in from an external source. This is common practice in other parts of the world, but in Ireland has been limited to only a few undertakings at Rosslare, Courtown and Cahore (all in County Wexford). Beach nourishment is an expensive activity that replaces sand or gravel being removed by wave action. Nourishment requires an open-ended commitment if the beach is to

Fig. 8.8. A seawall and promenade at Portrush prevents adjustment of the beach and has degraded the natural beach resource. It provides amenities for the traditional Irish seaside resort experience.

be maintained and is usually confined to resorts whose economic situation depends on maintenance of the beach. In many cases it is a response to human-induced erosion. Worryingly, nourishment was also proposed proactively as a remedial action to counter the effects of erosion predicted to be caused by the construction of a marina at Greystones, County Wicklow.

Sea level rise in estuaries in particular has implications for the habitats that occur there. Some of the most important birdwatching sites in Ireland are associated with tidal flats and salt marshes in estuaries (e.g. the RSPB bird reserve in Belfast, North Bull Island, Dublin, Shannon Estuary, Strangford Lough). Left to themselves, these environments tend to evolve under rising sea levels by extending landward. Unfortunately, the shorelines of many estuaries are now armoured. Most of central Dublin and central Belfast is built on formerly intertidal land that was progressively reclaimed (Fig. 8.9). This land thus needs to be defended against flooding. With rising sea levels, the defences themselves face increased risk of overtopping, which means that they are likely to require capital upgrading or enhanced maintenance in the future. Many of Ireland's coastal cities (Cork, Dublin, Belfast) experience episodic problems with flooding. A storm surge in the Irish Sea in February 2002 led to the highest water levels recorded in Dublin since 1923. The water level was about 50 cm above the highest water level experienced each year in the Port. That storm caused damage throughout the Irish Sea with flooding from Dublin to Belfast and the collapse of the sea wall in front of the Slieve Donard Hotel in Newcastle, County Down. Insurance losses in Ireland amounted to €37 million. A rise in sea level of 50 cm would mean that the 2002 extreme water levels would be reached almost every year (Cooper, 2009).

Fig. 8.9. Central Dublin and its port is built largely on reclaimed intertidal land that is at increased risk of flooding under rising sea levels.

The scale of losses sustained in 2002 could clearly not be withstood at this frequency and the most probable response in urban areas is likely to be improved defences. In some cases, such as Belfast, tidal barrages built to reduce flood risk have become tourism attractions in themselves. The Belfast system is interpreted in the Lagan Lookout Centre. In less-developed areas, other options may prove both economically and environmentally more appropriate.

In highly developed urban settings the only short-term option for coping with sea level rise is likely to be to continue defending the shorelines. This means that tidal flats and salt marshes will reduce in area or even be squeezed out altogether. At sites with less development, other options exist that might permit intertidal habitats to evolve. A number of sites have already experienced collapse of sea defences as it has become uneconomic to maintain them. Near Rathmullan, County Donegal, for example, a formerly reclaimed agricultural area has been flooded by the sea since the earth embankments were breached and not repaired. The fields are now reverting to salt marsh.

In Strangford Lough, a mainly rural area in County Down, a 1 m sea level rise could cause the loss of 50% of the intertidal area (Orford *et al.*, 2007). At this site up to 60,000 wildfowl and wading birds (including 75% of the world population of Light-bellied Brent Geese) congregate in winter where they feed on the resources provided by the tidal flats. Management options include accepting the loss, creating compensatory habitat elsewhere, removing the sea defences and permitting the habitats to migrate. All such options have important consequences for the natural habitats and associated tourism activity. There are potential additional linked impacts of habitat loss for tourism activities. Loss of habitat affects the entire ecosystem and has implications for angling species as well as bait organisms (much bait digging is done on intertidal flats).

The retreat or coastal realignment option has the potential to enable coastal systems to operate naturally and maintain high quality coastal landscapes. The landward removal of tourism infrastructure and historic buildings has also been considered elsewhere. Such options, however, are not currently supported by current practice or public opinion in Ireland. The fact that Ireland has had one of the highest rates of coastal development of any EU country between 1990 and 2000 (EEA, 2006) is inevitably going to lead to increased future calls for 'protection' of property and consequent environmental degradation under existing practices. The long-term implications of this approach for tourism in Ireland (outside major cities and a few resorts) are likely to be highly damaging.

References

Agarwal, S. and Shaw, G. (eds) (2007) *Managing Coastal Tourism Resorts: A Global Perspective*. Channel View Publications, Clevedon, UK.

Amelung, B. and Viner, D. (2006) The sustainability of tourism in the Mediterranean: exploring the future with the tourism climatic index. *Journal of Sustainable Tourism* 14, 349–366.

Boelens, R., Minchin, D. and O'Sullivan, G. (2005) Climate change: Implications for Ireland's marine environment and resources. *Marine Foresight Series* 2. Marine Institute, Dublin.

Boyd, S.W. (2006) Second homes and tourism development on Northern Ireland's north coast: Social benefit, social cost. *Paper Presented at the IGU Regional Conference*, Queensland University of Technology, Brisbane, 3–7 July.

Boyd, S.W. (2013a) Tourism in Northern Ireland: before violence, during and post violence. In: Butler, R.W. and Suntikul, W. (eds) *Tourism and War*. Routledge, London, pp. 176–192.

Boyd, S.W. (2013b) The Causeway Coastal Route and Saint Patrick's Trail: Heritage Tourism Route Development. In: Garrod, B. and Fyall, A. (eds) *Contemporary Cases in Heritage*. Goodfellow Publishers, Oxford, pp. 204–228.

Boyd, S.W. (2016) Heritage as the USP for tourism in Northern Ireland: attraction mix, effective storytelling and selling of a dark past. In: Hooper, G. (ed.) *Tourism and Heritage in Britain and Ireland*. Palgrave, London.

Browne, J. (2008) Announcement by John Browne T.D., Minister of State at the Department of Agriculture, Fisheries and Food, of the Coastal Protection Programme, 2008. Department of Agriculture, Food and the Marine. Available at: www.agriculture.gov.ie/press/pressreleases/2008/april/title,13657,en.html (accessed March 2016).

Butler, R.W. (1980) The concept of a tourist area cycle of evolution: implications for management of resources. *Canadian Geographer* 24(1), 5–12.

Carter, R.W.G. (1988) *Coastal Environments*. Academic Press, Cambridge, Massachusetts.

Central Statistics Office (CSO)/Northern Ireland Statistics and Research Agency (NISRA) (2016) *Visitors to Ireland and Northern Ireland 2014: A Statistical Profile of Tourism*. Cork/Belfast.

Cooper, J.A.G. (2009) Coastal economies and people. In: Baxter, J.M., Buckley, P.J. and Frost, M.T. (eds) *Marine Climate Change Ecosystem Linkages Report Card 2009*. Online science reviews, p. 18. Available at: www.mccip.org.uk/elr/coasts (accessed 12 July 2009).

Cooper, J.A.G. and McKenna, J. (2008) Working with natural processes: the challenge for Coastal Protection Strategies. *Geographical Journal* 174, 315–331.

Cooper, J.A.G. and McKenna, J. (2009a) Boom and bust: the influence of macroscale economics on the world's coast. *Journal of Coastal Research* 25, 533–538.

Cooper, J.A.G. and McKenna, J. (2009b) Managing cars on beaches: a case study from Ireland. In: Williams, A.T. and Micallef, A. (eds) *Beach Management*. Earthscan, pp. 136–146.

Cooper, J.A.G., O'Connor, M.C. and McIvor, S. (2016) Coastal defences versus coastal ecosystems: a regional appraisal. *Marine Policy*. DOI: 10.1016/j.marpol.2016.02.021.

Devoy, R. (2008) Coastal vulnerability and the implications of sea-level rise for Ireland. *Journal of Coastal Research* 2, 325–341.

Dunne, S., Hanafin, J., Lynch, P., McGrath, R., Nishimura, E. *et al.* (2009) *Ireland in a Warmer World – Scientific Predictions of the Irish Climate in the Twenty-First Century (2001-CD-C4-M2) STRIVE Report.* Environmental Protection Agency, Dublin.

EEA (European Environment Agency) (2006) The changing faces of Europe's coastal areas. *EEA Report* 6/2006. EEA, Copenhagen.

Fahy, G., Griffin, J., Henry, D., Jones, M., Kelly, D. *et al.* (2005) *Climate Change and the Visitor Economy of the Sefton Coast. School of Planning and Landscape.* University of Manchester, Manchester, UK.

Fáilte Ireland (2016) *Tourism Facts 2014*. Fáilte Ireland.

Lise, W. and Tol, R.S.G. (2002) Impact of climate on tourist demand. *Climatic Change* 55, 429–449.

Mieczkowski, Z. (1985) The tourism climatic index: a method of evaluating world climates for tourism. *Canadian Geographer* 29, 220–233.

Nairn, R. (2005) *Ireland's Coastline*. The Collins Press, Dublin.

Nicholls, S. and Amelung, B. (2008) Climate change and tourism in north-western Europe. *Tourism Analysis* 13, 21–31.

NITB (2011) *Activity Tourism: Sharing Success Report*. Northern Ireland Tourist Board, Department of Enterprise, Trade and Investment, London.

NITB (2012) *Signature Project Experience: Giants Causeway, Antrim and Causeway Coast Experience Report.* Northern Ireland Tourist Board, Belfast.

Northern Ireland Statistics and Research Agency (NISRA) (2015) *Northern Ireland Annual Tourism Statistics 2014*. Northern Ireland Tourist Board.

O'Connor, M.C., Lymbery, G., Cooper, J.A.G., Gault, J. and McKenna, J. (2009) Practice versus policy-led coastal defence management. *Marine Policy* 33, 923–929.

Orford, J.D., Murdy, J. and Freel, R. (2006) Developing constraints on the relative sea-level curve for the northeast of Ireland from the mid-Holocene to the present day. *Philosophical Transactions of the Royal Society A* 364, 857–866.

Orford, J.D., Betts, N., Cooper, J.A.G. and Smith, B. (2007) *Future Coastal Scenarios for Northern Ireland*. Report to National Trust, Northern Ireland.

Phillips, M.R. and Jones, A.L. (2006) Erosion and tourism infrastructure in the coastal zone: problems, consequences and management. *Tourism Management* 27, 517–524.

Quinn, A.C.M. (1977) *Sand Dunes: Formation, Erosion and Management*. An Foras Forbartha, Dublin.

Rohan, P.K. (1986) *The Climate of Ireland*. The Stationery Office, Dublin.

Sweeney, J., Donnelly, A., McElwain, L. and Jones, M. (2002) *Climate Change: Indicators for Ireland (2000-LS-5.2.2-M1)*. Environmental Protection Agency, 54 pp.

Sweeney, J., Brereton, T., Byrne, C., Charlton, R., Emblow, C. *et al.* (2003) *Climate Change: scenarios and impacts for Ireland (2000-LS-5.2.1-M1)*. Environmental Protection Agency, 229 pp.

Taylor, J.A., Murdock, A.P. and Pontee, N.I. (2004) A macroscale analysis of coastal steepening around the coast of England and Wales. *The Geographical Journal* 170, 179–188.

Timothy, D.J. and Boyd, S.W. (2015) *Tourism and Trails: Cultural, Ecological and Management Issues*. Channel View Publications, Bristol, UK.

Towner, J. (1996) *An Historical Geography of Recreation and Tourism in the Western World 1540–1940*. John Wiley and Sons, London.

Viner, D. and Agnew, M. (1999) Climate change and its impact on tourism. Report for WWF-UK. Climatic Research Group, University of East Anglia, Norwich, UK.

Wild Atlantic Way (WAW) (2016) *Scoping Report*. Fáilte Ireland.

Websites

www.antaisce.ie An Taisce announce 74 International Blue Flag and 45 National Green Coast Awards announced for Ireland in 2013 with stricter bathing water quality standards, accessed 10/03/2016.

www.belfastcity.gov.uk Belfast Tourism Monitor Report 2012 accessed 15/03/2016.

www.blueflag.org Blue Flag History, accessed 10/03/2016.

www.Fáilteireland.gov.ie accessed 10/03/2016.

www.met.ie/climate/climate-of-ireland.asp MET Eireann, Climate of Ireland, accessed 14 March 2016.

www.nitb.com Business support: culture, heritage and activity tourism, accessed 14/03/2016.

www.tidynorthernireland.org 10 NI beaches awarded Blue Flag status, accessed 10/03/2016.

9 Case Study Italy: The Tourism Management of Climate Change in the Mediterranean Region: Adaptation Strategies in Sardinia and Sicily

Rita Cannas*

Department of Economics & Business, University of Cagliari, Sardinia (Italy)

Introduction

The Mediterranean (Fig. 9.1) is the world's leading tourism destination attracting almost 200 million tourism arrivals which represent 18.9% of total international flows, and it is forecast that tourism arrivals will reach 264 million in 2030 (UNWTO, 2015). As a consequence of such predictions, Europe, which currently attracts more than a half of the total international tourism arrivals, will lose 10% of its total share in 2030, and 4% on the Mediterranean (UNWTO, 2015), likely due to the rise of Asian and Pacific tourism flows. Nevertheless, such numbers forecast that the Mediterranean will maintain its popular and successful power of attraction among international visitors. In parallel, particularly in the latest decades, there have been rising concerns both in literature (Agnew and Viner, 2001; Maddison, 2001; Amelung and Viner, 2006; Perry, 2006) and media (as critically underlined by Rutty and Scott, 2010) regarding dramatic climate change effects and the scenario on the Mediterranean, by which it will become less attractive in the summer, being too hot for tourists. The same increasing temperatures that would make the Mediterranean too hot would make conditions in Northern Europe more suitable for tourism. As a consequence of such a phenomenon, Northern European tourists would spend their holidays in higher latitudes instead of travelling to the Mediterranean (Hamilton *et al.*, 2005; Amelung and Viner, 2006; Amelung *et al.*, 2007).

Indisputably, it has been largely demonstrated that global climate is changing; in addition, the Mediterranean is sensitive and tightly related to climate and its changes, as climate is one of the most important tourism resources. In fact, the Mediterranean is considered as a tourism destination highly vulnerable to climate change effects by relevant bodies such as the Intergovernmental Panel on Climate Change (IPPC) and the United Nation World Tourism Organization (UNWTO), as well as by scholars (Alcamo *et al.*, 2007; Moreno, 2010; Rutty and Scott, 2010): the ascertained increase of temperature is one of the major threats to Mediterranean tourism. Nevertheless, it is less deterministic and difficult to predict in a long time scale, for the next 30–50 years, what would be the responses of tourists, policymakers, businesses, local communities, etc., to climate

*E-mail: rita.cannas@unica.it

Fig. 9.1. Map of the Mediterranean Sea (author's elaboration from NASA map).

change effects. For instance, Rutty and Scott (2010) point out that tourism flows may not necessarily decline as a result of climate change; they suggest that changes caused by climate change will depend on the abilities of destinations to adapt or take advantage of the shoulder season climate. Thus, adaptation responses by tailored strategies are crucial for the Mediterranean.

Moving from this scenario, the study focuses on the issue of climate change adaptation in Mediterranean coastal destinations, attempting to answer these research questions: how are the tourism firms changing/adapting their decisions and strategies to respond to the climate change challenges, particularly in the Mediterranean coastal destinations? Are they fully aware about the effects of climate change? In the attempt to answer that, the study analyses the case studies of the biggest islands of the Mediterranean, Sardinia and Sicily (Italy), through secondary data, and by conducting interviews with key stakeholders.

Overview on Climate Change Adaptation

The tourism industry, particularly tourism based on nature activities, is sensitive to forecasted climate change impacts (Saarinen and Tervo, 2006; Scott, 2006; Scott *et al.*, 2007), which also makes tourism-dependent communities vulnerable to global climate change and uncertain of their local outcomes. Altogether, the estimated impacts of climate change and the vulnerability of the tourism industry and related social systems call for the research and development of adaptation strategies and policies. While the need to adapt to climate change is absolutely crucial, this need is still not fully understood in the tourism sector (Bramwell and Lane, 2008; Weaver, 2011; Kaján and Saarinen, 2013).

Nevertheless, adaptation, which is considered as the appropriate response to the measures taken to reduce climate change impacts on the tourism industry (Patterson *et al.*, 2006; Becken and Hay, 2007), is a key theme within the realm of tourism climate change studies. Becken (2013) states that a half of published papers focus on climate change impacts and adaptation, particularly on winter tourism. In terms of number of studies, less attention has been devoted to coastal tourism (Lohmann, 2001; Moreno, 2010), while island tourism and disappearing destinations are recently attracting scholars' attention (Becken *et al.*, 2011; Jones and Phillips, 2011).

Adaptation refers to both the process of adapting and the condition of being adapted (Smit *et al.*, 1999). Particularly, while contextualizing adaptation for natural and human systems, these scholars stress the attention on three points: (i) adaptation to what; (ii) who or what adapts; and (iii) how does the adaptation occur. The first point is largely discussed in climate change literature, e.g. variability to temperature and precipitation, or extreme weather conditions. Regarding who or what adapts in the adaptation process, there emerge four general thematic groups: businesses, tourists, destinations, and policies and frameworks. Looking into tourism businesses, the most adaptation responses are focused on *diversification*, which can be referred to products, as well as to demands (Kaján and Saarinen, 2013).

In coastal destinations, diversification of the products can deal with additional services around the seaside attractions, e.g. cultural tours, new sport activities, or alternative products. Not always diversifying into alternative products or increasing attractions in new seasons may bring significant benefits into seaside tourism destinations; for instance, branching out into winter tourism may be difficult because of competition with more winter-oriented destinations. Conversely, developing new products in the absence or scarcity of snow in winter-sport destinations may not guarantee alternative tourism flows. This suggests that climate change adaptation responses should be 'adapted' to the specific environmental and socio-economic characteristics of destinations. Quoting Kaján and Saarinen (2013, p. 188), 'standardized adaptation strategies are impossible to implement due to each community's unique characteristics and adaptation capabilities'. Moreover, these scholars underline the need to combine market-led and business-driven research on climate change adaptation to a more holistic and community-centred analysis.

Diversification of demands can be seen as the other side of the coin. It may imply the attraction of new segments of tourists due to marketing strategies, and re-branding initiatives undertaken by both businesses and public organizations. Market diversification is strictly linked with tourist perceptions. Plausibly, tourist perceptions cannot be encapsulated into linear thinking models. For instance, Rutty and Scott (2010) state that travellers to the Mediterranean have different perceptions of unacceptable warm temperatures, depending on the type of their destination: they express more adaptive behaviour toward a coastal beach resort, and less to an urban tour.

Gössling *et al.* (2012) analyse the response of tourists to the climate change impacts particularly by framing their perceptions to adaptation and mitigation actions. They point out that tourist perceptions based on various aspects such as perceived authenticity, tourist resources, e.g. climate, travel time and travel cost, perceived safety and security, existing facilities and host hospitality, can have direct influence on the destination attractiveness. Therefore, understanding tourist perceptions of the impact of climate change is useful to anticipating seasonal shift in tourism demand, as well as the decline or increase of specific tourism segments.

Tourist adaptation responses to climate change are also strictly related to processes of representation of places (Buzinde *et al.*, 2010). Destinations affected by climate change impacts, such as beach erosion, may undertake representational strategies while restoring their landscapes, in order to influence tourists. The mentioned scholars propose two main strategies: the first, static, is addressed to maintaining the outdated image of the destination by hiding or ignoring the beach erosion. The second implies a dynamic representation of the destination through an active relationship among tourism promoters and tourists, e.g. by promoting among tourists understanding of the causes of beach erosion, involving them in new activities on the beach, etc. Indeed, representational strategies are crucial for re-branding destinations, but changing a destination image requires efforts not only from local business but also from the host community (Kaján and Saarinen, 2013).

Exploring Adaptation Responses in Sicily and Sardinia

Tourism features

Sicily and Sardinia are the largest islands of the Mediterranean Sea. From an administrative point of view, they both belong to Italy and due to their geographical insularity, and specific historical and cultural characteristics, have been recognized as autonomous regions of Italy.

Sicily has a land area of 26,600 km^2, a coast line of 2100 km^2 and a resident population of more than 5 million. In Sicily, there are impressive archaeological sites hosting extraordinary cultural evidence from the past, e.g. the Valley of Temples in Agrigento, the Baroque monuments in Noto, the Norman cathedrals in Palermo and so on, all treasures that have been included in the World Heritage List (WHL) of UNESCO. In fact, Sicily has the highest number of recognized UNESCO heritage sites, both cultural and natural, and intangible heritage (Figs 9.2–9.4) (OTIE, 2015).

In Sicily, the historical roots of tourism are strictly linked with the Grand Tour phenomenon. In fact, Sicily was a location for educational tours of young aristocrats coming from northern European kingdoms during the 17th to 19th centuries. Nowadays, the most attractive areas for tourism in Sicily are Messina, Palermo and Trapani, situated in the northern coastal zone of the island. The sun, sea and sand holiday represents the largest share of the Sicilian tourism market; consequently, seaside tourism is also spatially concentrated on coastal areas, and temporally concentrated in summer (as shown on Figs 9.5–9.7). However, particularly in recent years, cultural tourism is an increasing segment of the tourism demand. This fact is crucial for reshaping the seasonal patterns which characterized Sicilian tourism, especially by considering that cultural tourism can reduce seasonality (Cuccia and Rizzo, 2011).

Looking into the main tourism features, the Sicilian tourism sector holds more than 5400 accommodation units and around 203,000 beds (Osservatorio sul Turismo, 2014).

Fig. 9.2. The Selinunte Temple (source: Rita Cannas).

Fig. 9.3. The San Vito Lo Capo Beach (source: Maurizio Giannone).

Fig. 9.4. The Sicilian Puppets (source: Maurizio Giannone).

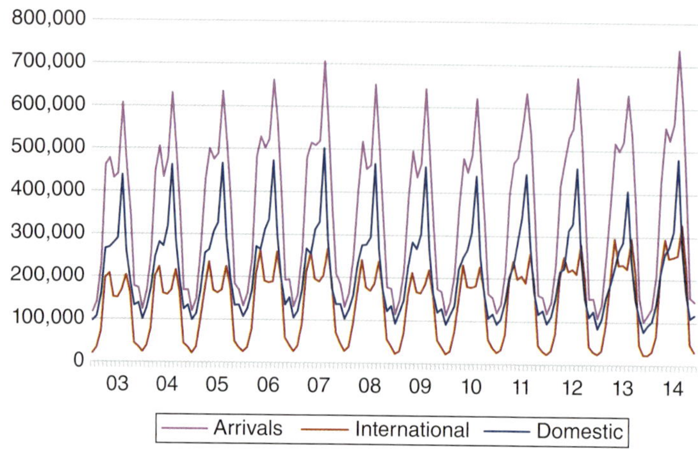

Fig. 9.5. Monthly tourist arrivals in Sicily 2003–2014 (based on ISTAT, 2015).

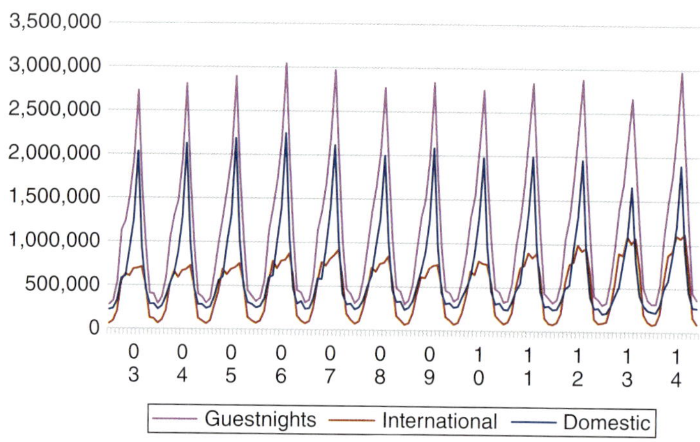

Fig. 9.6. Monthly tourist guestnights in Sicily 2003–2014 (based on ISTAT, 2015).

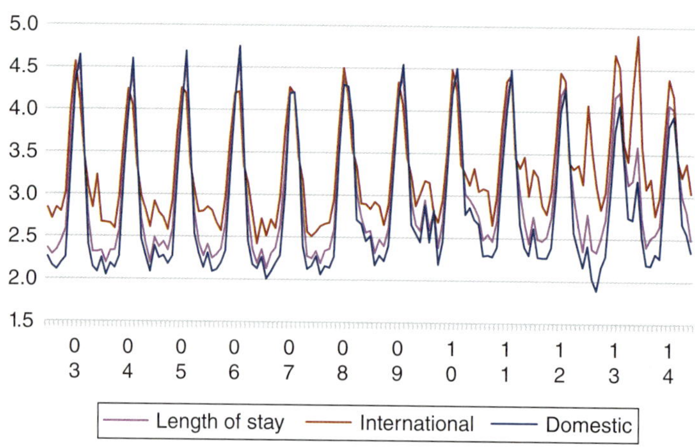

Fig. 9.7. Monthly tourist length of stay in Sicily 2003–2014 (based on ISTAT, 2015).

Among such tourism accommodation capacity, the hotels represent one-third of the units and two-thirds of the Sicilian beds. Regarding tourist flows, in 2014 Sicily received more than 4.5 million tourists and registered almost 15 million guest nights. Domestic tourists are still predominant in terms both of total arrivals (55.1%) and guest nights (50.1%) (Osservatorio sul Turismo, 2014). The main international markets come from France, Germany, the UK, the USA and Russia. The average length of stays is 3.2 nights whereas the international length of stays is slightly higher (3.5).

According to Volo (2010) the international visitor market to Sicily is much less seasonally peaked than the Italian market; in fact, the international tourist arrivals data show two peaks, one in May and the other in September. Moreover, Volo (2010) notices a difference in arrivals and overnight stays between Italians and international tourists, where international visitors spend more nights in Sicily than do Italians in the months of May and October.

Sardinia has a land area of 24,000 km^2, a coast line of almost 1900 km and a resident population of 1.6 million, resulting in a low population density of 68 inhabitants per km^2. Sardinian tourism development started in Costa Smeralda (or Emerald Coast) located in the north-east of the island, which is one of Italy's best known destinations. It was created as a destination in the 1960s by the Prince Aga Khan, contributing extensively to the development of a territory whose economy was based primarily on agriculture and farming (Del Chiappa, 2012). For several decades, the luxury pole of Costa Smeralda has been perceived by the Sardinians, particularly from those not directly involved in the tourism business, as an exogenous and colonial tourism model totally shaped by external investors (Cannas and Giudici, 2015). However, within tourism businesses, policymakers and stakeholders today there is a large consensus on considering Costa Smeralda as the main economic driving force for tourism both to north-east Sardinia and to the whole island.

Like many other tourist islands, Sardinia depends heavily on the sun, sea and sand holiday, but it has an enormous potential for sustainable tourism, including nature watching, heritage and archaeology, food and wine tourism, trekking and other outdoor activities (Pulina *et al.*, 2013). Sardinia has a unique history, such as the Nuraghic civilization developed around 3500 years ago, and the most important Nuraghic site of Barumini has been listed as a UNESCO World Heritage Site in 1997 (Figs 9.8–9.10). Sardinia also has its own distinctive culture and identity diffused on many traditions, typical foods and wineries, which boast an ancient origin.

Looking into the main Sardinian tourism features, the accommodation facilities consist of more than 4500 units and the bed capacity is around 207,000 (ISTAT, 2015). Although there are only 935 hotels, they hold one-half of the Sardinian bed place capacity. The northern areas of Olbia-Tempio (where the Costa Smeralda is located) and Sassari are the most accommodation facility-equipped areas, as well as the most tourist-attractive areas. Another important tourism pole is located on the southern coastal areas around Cagliari.

With regard to the tourist flows, the official statistics of the Italian Statistics Office (ISTAT, 2015) show that there are almost 2.4 million tourist arrivals and around 11.3 million overnights. In particular, the Italian market share is predominant, representing 54.2% arrivals and 53.3% overnights. The average length of stay is 4.8 nights and there is no significant difference between the Italian segment and the international tourist demand. In recent years, there has been an increase in the share of foreign markets coming predominantly from Germany, France and the UK, probably due to the rise of low-cost flights.

The Sardinian tourism industry lacks strong products and effective branding strategies (Cannas, 2017). In particular, tourism public policy framework seems lacking on a strategic vision, and not integrated with the private businesses' views and strategies. Tourism represents less than 9% of the Sardinian GDP, the seaside market is still predominant and concentrated mainly in coastal areas and in summertime. The extreme

Fig. 9.8. The Nuraghe of Ponte (source: Paolo Lombardi).

seasonality of the Sardinian tourism market is clearly represented in Figs 9.11–9.13; in 2014, between June and September, there were 72% of the arrivals and 84% of the overnights (ISTAT, 2015).

Comparing tourism seasonality between Sicily and Sardinia, it clearly emerges the highest stage of seasonality in the latter. Cultural tourism seems to mark the difference in Sicily, as well as the whole of international flows that reach peaks in shoulder seasons. This confirms what Rutty and Scott (2010, p. 278) point out: 'demand may not necessarily decline as a result of climate change, but rather contribute to a shift in the timing that tourists visit in the [Mediterranean] region by mid to late century'. However, the aim of this study is not to compare the two cases to each other; rather it focuses on the exploration of adaptation responses to climate change given particularly by tourism businesses of Sicily and Sardinia.

Fig. 9.9. Southern west coast view of Sardinia (source: Rita Cannas).

Fig. 9.10. Sardinian folklore: celebration of Saint Efis (source: Rita Cannas).

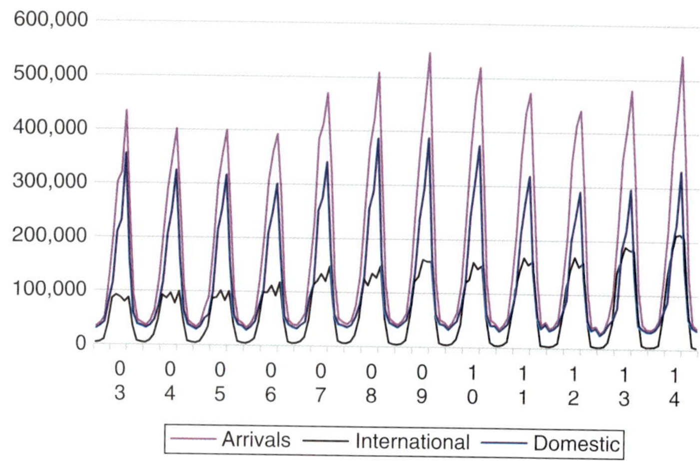

Fig. 9.11. Monthly tourist arrivals in Sardinia 2003–2014 (based on ISTAT, 2015).

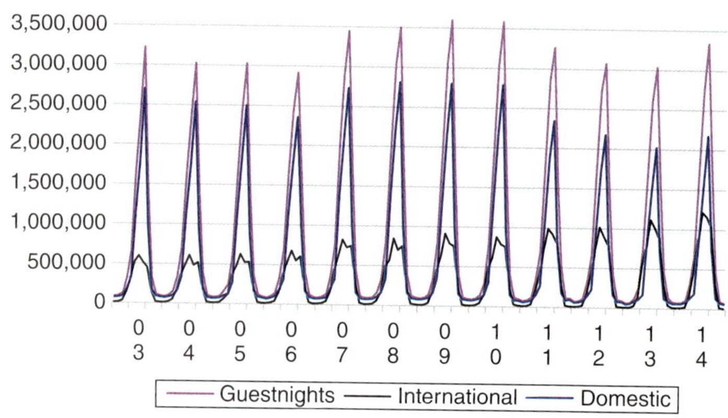

Fig. 9.12. Monthly tourist guestnights in Sardinia 2003–2014 (based on ISTAT, 2015).

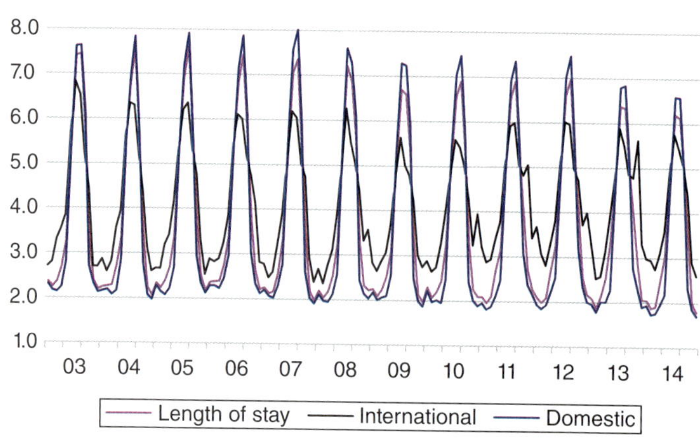

Fig. 9.13. Monthly tourist length of stay in Sardinia 2003–2014 (based on ISTAT, 2015).

Climate change adaptation views of tourism stakeholders

The exploration of adaptation responses to climate change in Sicily and Sardinia has been realized by interviewing a selected panel of eight tourism businesses and representatives of the public sector in the two chosen cases. In particular, leaders of the main accommodation sector associations, as well as key representatives of public regional authorities, were interviewed between January and February 2016.

Unanimously, the whole tourism stakeholders interviewed for the purpose of this study underline as a strong weakness the lack of a public policy framework on tourism both in Sicily and Sardinia, mainly due to the wide undervaluation, within national and local governors, of the tourism sector as a strategic economic driving force for local development. Thus, the research confirms that although Sicily and Sardinia are not independent states, 'the general lack of consideration of tourism as an economic sector in national policies can hinder the development of climate change-related policies in the sector' (Becken and Clapcott, in Kaján and Saarinen, 2013, p. 182). In fact, both the Sicilian and Sardinian interviewees stress the lack of climate change tourism policies, as well as of business strategies that take into consideration climate change impacts: in the agenda of policymakers and tourism businesses climate change issues seem to be missing. The main reason tourism stakeholders appear quite unprepared to face climate change is the lack of awareness about its impacts in the tourism sector.

Adopting the categories of anticipatory/reactive adaptation (Gössling *et al.*, 2012), there emerges that instead of carrying out *anticipatory* strategies to minimizing forthcoming climate change impacts, tourism businesses undertake *reactive adaptation* actions toward impacts already occurred, most of the time unconsciously. For instance, implementations of green labels/schemes into hotel management are seen as marketing tools for attracting green-sensitive demands, or/and means to reducing operational costs, while contributions given to reduce climate change impacts do not emerge likewise. Also, interlinked with unaware adaptation is the uncertain geopolitical scenario in which not only climate change but also wars and terrorist attacks in and near the Mediterranean destinations are impacting on tourism features; the thought of some interviewees both of Sicily and Sardinia who said we are selling security without any direct merit' is quite emblematic. This means that seasonal shifts or change in tourism flows may be caused by external factors rather than being outcomes of specific strategies in which, again, climate change issues do not seem contemplated. The claim expressed by Jones and Phillips (2011) that potential threats from climate change are likely to create considerable adverse impacts unless managed effectively, seems totally unknown.

Diversification of products seems a popular aim with the public and private tourism operators, probably because it is directly related to tourism seasonality issues for which policymakers' sensitivity appears high. For instance, in the Tourism Development Plan of Sicily 2015–2017 (Regione Siciliana, 2014), extending seasonality is clearly mentioned as the fourth strategic goal, but is not linked with the climate change phenomenon. From businesses' views, product diversification strategies aim to satisfy the need to modify seasonality, in order to counteract the extreme effects of such a phenomenon, particularly in Sardinia. Some businesses' activities deal with weddings on the beach, diving sport activities throughout the year, and link between the opera season (e.g. in Cagliari) and seaside tourists, etc. Tourism businesses state that opportunities which might pertain to the wide spectrum of de-seasonality actions, e.g. sport tourism such as sailing, cycling, diving, hiking, golf, etc. are not fully recognized. They also note that opportunities deriving from favourable temperatures in October and November may be captured by equipping accommodation facilities with heating systems, or the simplest – by keeping the hotels open.

Diversification of demands toward gastronomic niches and cultural tourists are undertaken particularly by Sicilian operators. Again, the reasons for attracting new demands or enhancing types of clients already existing seem not linked with climate change issues. In fact, looking at the series of statistics (Figs 9.5–9.7 and 9.11–9.13) it does not seem that domestic or international tourists of both islands have been influenced by the rise of temperatures or the beach erosion phenomena related to climate change. Unfortunately, in Sardinia, irresponsible tourists are a direct cause of beach erosion, as ascertained by the fact that every summer thousands of kilogrammes of stolen sand are confiscated from tourists by the airport police. Nevertheless, beyond such deplorable behaviours, climate change effects may become very critical in the next decades to destination management, if not seriously taken into account by tourism stakeholders.

With regard to representational strategies, interviewees do not highlight significant issues. Climate change effects seem invisible to them at the moment. However, there emerge some cases of hoteliers that have changed their communication strategies from an emphasis on accommodation features to an emphasis on landscapes and natural environment, such as marine protected areas, in which hotels are located. Finally, tourism interviewees stress the attention upon the lack of engagement procedures and practices among politicians and tourism businesses (e.g. as suggested by Becken, 2013), which appear as major obstacles on facing the climate change challenge.

Conclusions

Although climate change is globally becoming more and more an urgent and essential issue, in Mediterranean coastal destinations like Sicily and Sardinia, characterized by strong tourism seasonality features, it is still far to be embedded into their public policies and businesses strategies. The lack of awareness on climate change impacts, particularly in the tourism management realm, shows how long and hard is the path to face the climate change in destinations, as well as how unavoidable is the need to share knowledge on that topic among tourism stakeholders. In fact, climate change adaptation looks like a passive and an unaware process, probably because climate change impacts are still not fully perceived by tourism stakeholders. Thus, the study confirms such warning by scholars according to which, although the need to adapt to climate change is imperative, this is not understood in the tourism sector. Diversification of products and demands seem to be the businesses' responses to the desire of extending the summer season, rather than being planned strategies aimed to prevent climate change impacts on management destinations.

Although the study has an exploratory purpose, and greater depth is needed through further investigations, it provides some recommendations in order to recognize and manage climate change challenges, as well as to remove some critical obstacles that affect the two Mediterranean destinations. Briefly, the study recommends the following: recognizing *tourism as a key sector* both in Sicily and Sardinia; satisfying the need for *public policy frameworks* on tourism and climate change under a holistic view and by the *direct engagement* of tourism businesses into the public policy creation process; promoting *actions* to enhance knowledge and provide practices on climate change impacts and scenarios within both public and private stakeholders; catching up opportunities deriving from climate change through *proactive strategies*. Finally, following the rising interest in Sicily in the form of cultural tourism, which positively impacts on the seasonal shift due to climate change, such interest should be enhanced, particularly in Sardinia, as well as in all the Mediterranean coastal destinations.

References

Agnew, M.D. and Viner, D. (2001) Potential impacts of climate change on international tourism. *Tourism and Hospitality Research* 3(1), 37–60.

Alcamo, J., Flörke, M. and Märker, M. (2007) Future long-term changes in global water resources driven by socio-economic and climatic changes. *Hydrological Sciences Journal* 52(2), 247–275.

Amelung, B. and Viner, D. (2006) Mediterranean tourism: exploring the future with the tourism climate index. *Journal of Sustainable Tourism* 14(4), 349–366.

Amelung, B., Nicholls, S. and Viner, D. (2007) Implications of global climate change for tourism flows and seasonality. *Journal of Travel Research* 45, 285–296.

Becken, S. (2013) A review of tourism and climate change as an evolving knowledge domain. *Tourism Management Perspectives* 6, 53–62.

Becken, S. and Clapcott, R. (2011) Developing public policy for climate change in the tourism sector. *Journal of Policy Research in Tourism, Leisure and Events* 3(1), 1–17.

Becken, S. and Hay, J. (2007) *Climate Change and Tourism: From Policy to Practice*. Routledge, London.

Becken, S., Hay, J. and Espiner, S. (2011) The risk of climate change for tourism in the Maldives. In: Carlsen, J. and Butler, R. (eds) *Island Tourism: Sustainable Perspectives*. CAB International, Wallingford, UK, pp. 72–84.

Bramwell, B. and Lane, B. (2008) Priorities in sustainable tourism research. *Journal of Sustainable Tourism* 16, 1–4.

Buzinde, C.N., Navarrete, D.M., Kerstetter, D. and Redclift, M. (2010) Representations and adaptation to climate change. *Annals of Tourism Research* 37(3), 581–603.

Cannas, R. (2017) Embedding social values in tourism management: community currencies as laboratories of social entrepreneurship? In: Dredge, D. and Gymothy, S. (eds) *Tourism and the Collaborative Economy: Perspectives, Politics, Policies and Prospects*. Springer, Tourism on the Verge Series Cham, CH, pp. 223–240.

Cannas, R. and Giudici, E. (2015) Tourism relationships between Sardinia and its Islands: collaborative or conflicting? In: Baldacchino, G. (ed.) *Archipelago Tourism: Policies and Practices*. Ashgate, Farnham, UK, pp. 67–81.

Cuccia, T. and Rizzo, I. (2011) Tourism seasonality in cultural destinations: empirical evidence from Sicily. *Tourism Management* 32, 589–595.

Del Chiappa, G. (2012) Community integration: a case study of Costa Smeralda, Italy. In: Fayos-Sola, E., Silva, J. and Jafari J. (eds) *Tourism Strategies: Development, Governance, Climate Change*. Emerald, Bingley, UK, pp. 243–263.

Gössling, S., Scott, D., Hall, M.C., Ceron, J.P. and Dubois, G. (2012) Consumer behaviour and demand response of tourists to climate change. *Annals of Tourism Research* 39(1), 36–58.

Hamilton, J.M., Maddison, D.J. and Tol, R.S.J. (2005) Climate change and international tourism: a simulation study. *Global Environmental Change* 15, 253–266.

ISTAT (National Statistics Office) (2015) *Movimento dei clienti negli esercizi ricettivi [Clients flow in the accommodation facilities]*, years 2003–2014.

Jones, A. and Phillips, M. (2011) Introduction. In: Jones, A. and Phillips, M. (eds) *Disappearing Destinations: Climate Change and the Future Challenges for Coastal Tourism*. CAB International, Wallingford, UK.

Kaján, E. and Saarinen, J. (2013) Tourism, climate change and adaptation: a review. *Current Issues in Tourism* 16(2), 167–195.

Lohmann, M. (2001) Coastal resorts and climate change. In: Lockwood, A. and Medlik, R. (eds) *Tourism and Hospitality in the Twenty First Century*. Butterworth-Heinemann, Oxford, pp. 284–295.

Maddison, D. (2001) In search of warmer climates? The impact of climate change on flows of British tourists. *Climatic Change* 49(1/2), 193–208.

Moreno, A. (2010) Mediterranean tourism and climate (change): a survey-based study. *Tourism and Hospitality Planning & Development* 7(3), 253–265.

Osservatorio sul Turismo (2014) Il Turismo in Sicilia. Rapporto 2012–2013. *Regione Siciliana*. Available at: http://pti.regione.sicilia.it/portal/page/portal/PIR_PORTALE/PIR_LaStrutturaRegionale/PIR_TurismoSportSpettacolo/PIR_Turismo/PIR_6584964.046479105/report_2012-13.pdf (accessed 25 February 2016).

OTIE (Observatory on Tourism in the European Islands) (2015) Economia del Turismo in Sicilia. Filiera 5: Il Turismo nei Siti UNESCO. Available at: www.otie.org/uploads/9/8/9/6/9896763/filiera_5_il_turismo_nei_siti_unesco.pdf (accessed 29 January 2016).

Patterson, T., Bastianoni, S. and Simpson, M. (2006) Tourism and climate change: two-way street, or vicious/virtuous circle? *Journal of Sustainable Tourism* 14, 339–348.

Perry, A. (2006) Will predicted climate change compromise the sustainability of Mediterranean tourism? *Journal of Sustainable Tourism* 14(4), 367–375.

Pulina, M., Meleddu, M. and Del Chiappa, G. (2013) Residents' choice probability and tourism development. *Tourism Management Perspectives* 5, 57–67.

Regione Siciliana (2014) *Piano triennale di sviluppo turistico regionale 2014–2020 Programma Triennale 2015–2017*. Available at: http://pti.regione.sicilia.it/portal/page/portal/PIR_PORTALE/PIR_LaStrutturaRegionale/PIR_TurismoSportSpettacolo/PIR_Turismo/PIR_Osservatorio_Turistico/PIR_Pianotrennale20142016 (accessed 27 February 2016).

Rutty, M. and Scott, D. (2010) Will the Mediterranean become 'Too Hot' for tourism? A reassessment. *Tourism and Hospitality Planning & Development* 7(3), 267–281.

Saarinen, J. and Tervo, K. (2006) Perceptions and adaptation strategies of the tourism industry to climate change: the case of Finnish nature-based tourism entrepreneurs. *International Journal of Innovation and Sustainable Development* 1, 214–228.

Scott, D. (2006) Global environmental change and mountain tourism. In: Gössling, S. and Hall, C.M. (eds) *Tourism and Global Environmental Change*. Routledge, London, pp. 54–75.

Scott, D., Jones, B. and Konopek, J. (2007) Implications of climate and environmental change for nature-based tourism in the Canadian Rocky Mountains: a case study of Waterton Lakes National Park. *Tourism Management* 28(2), 570–579.

Smit, B., Burton, I., Klein, R.J.T. and Street, R. (1999) The science of adaptation: a framework for assessment. *Mitigation and Adaptation Strategies for Global Change* 4, 199–213.

UNWTO (2015) Tourism Highlights 2015. Available at: www.e-unwto.org/doi/pdf/10.18111/9789284416899 (accessed 2 February 2016).

Volo, S. (2010) Research Note: Seasonality in Sicilian tourism demand – an exploratory study. *Tourism Economics* 16(4), 1073–1080.

Weaver, D. (2011) Can sustainable tourism survive climate change? *Journal of Sustainable Tourism* 19(1), 5–15.

10 Case Study Portugal: Addressing Tourism development and Climate Change in Small Atlantic Islands: the Case of the Azores

Helena Calado,*[1] Paulo Borges,[2] Kiat Ng[3] and Marta Vergílio[4]

[1]*MARE – Marine and Environmental Sciences Centre, and FCT, University of the Azores, Portugal;* [2]*Department of Geosciences, University of the Azores, Ponta Delgada, Portugal;* [3]*Faculty of Sciences, University of Lisbon, Portugal;* [4]*CIBIO – Research Center in Biodiversity and Genetic Resources, University of the Azores, Ponta Delgada, Portugal*

Introduction

In the North Atlantic, the Macaronesian Biogeographic Region encompasses four archipelagos: Cape Verde Republic; the Spanish Autonomous Region of the Canary Islands; and the Portuguese Autonomous Regions of Madeira and the Azores. Sharing some features, namely in biodiversity, these archipelagos present a series of economic, political and biophysical differences that largely influence their development. At different stages of the tourism life cycle, the four archipelagos have one of the major contributions for growth and development in this activity. Being remote, isolated and Atlantic, they all face challenges due to climate change impacts. With large differences between the dry and more southern archipelagos (Cape Verde and the Canaries) on the climatic influence of the coast of Africa, and the rainy subtropical archipelagos of Madeira and the Azores, the four archipelagos aim at the same goal: maintaining or expanding their touristic market.

The Azores, consisting of nine volcanic islands with stunning landscapes, coast, and year-round mild weather, are one of the nine Outermost Regions (ORs) in the European Union (EU). The EU recognized that ORs possess specific problems pertaining to small islands such as remoteness, insularity, small size, difficult topography and climate, and economic dependence on a few products (European Commission, 2014). However, ORs also have major assets and potential to contribute to their own development and to Europe as a whole acting, for example, as laboratories for studying and combating effects of climate change (European Commission, 2014).

Accessibility is vital to the current and future Azorean regional development, both due to the distance that separates the archipelago from the Portuguese mainland (around 1500 km) and other countries, and

*Corresponding author, e-mail: helena.mg.calado@uac.pt

the inter-island discontinuity of nine small islands spread over 600 km of ocean (Fig. 10.1). This issue became even more prominent with the new strategies for the tourism sector that became one of the main pillars of the regional economy.

Tourism in the Azores only started growing in the mid-1990s despite its obvious potential. Since then, tourism has grown rapidly over the last decade and has gained considerable interest and growing visitation figures. To illustrate its tourism quality, the Azores was ranked several times in top place: it was second among 111 world island destinations for sustainable tourism by National Geographic Traveler in 2008, it was ranked as one of the top ten best trips of summer 2011 by the editors of National Geographic Traveler, and the National Geographic Traveler of the Netherlands has selected the Azores as first in their selection of the 20 most beautiful places in the world to visit in 2016. In addition to the economic benefits of tourism to the region, it is also socially beneficial in terms of demographics as it may contribute to the settlement of young and innovative people in the smaller and less developed islands through new opportunities of services for tourists.

Concurrently, climate change concerns in the Azores are gaining interest due to increasing global warming awareness, as well as EU and global climate change policies. Research on climate change and its associated impacts and climate change policies on the Azores are already ongoing. Tourism has been increasingly established as one of the most important economic sectors for the Azores with the coasts as one of the main tourist attractions. Therefore, it is necessary to deepen knowledge and understanding of the impacts of climate change and especially of how current EU and regional policies could impact on coastal tourism. This chapter thus addresses the impact of climate change on coastal tourism in the Azores and, in particular, discusses whether the EU and regional climate change policies are reasonable and feasible at a local scale for the Azores. The chapter will present a

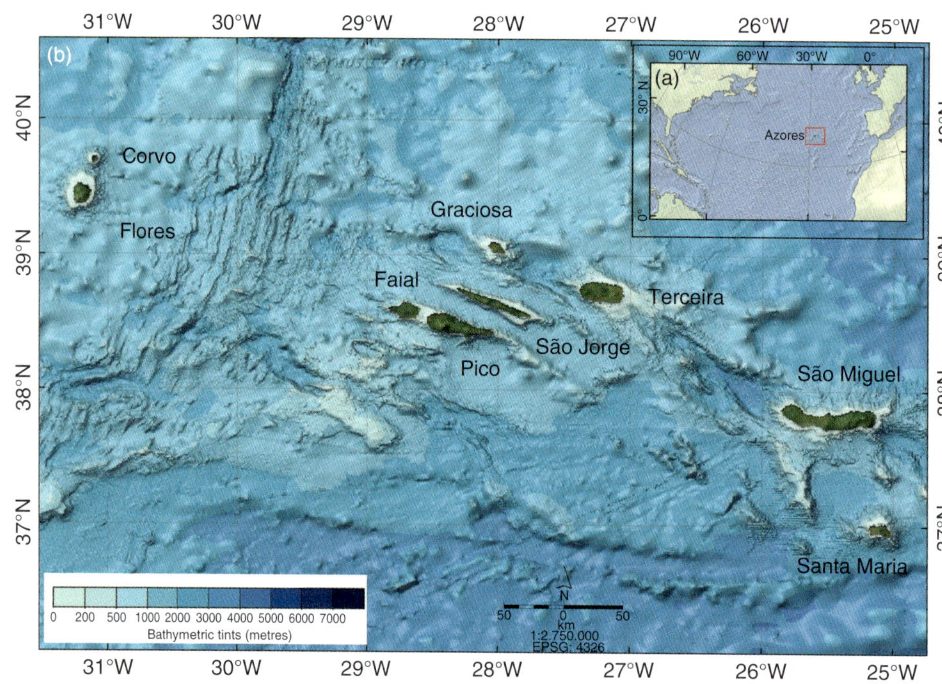

Fig. 10.1. Study area: (a) Azores islands in the North Atlantic Ocean; (b) bathymetric map of the Azores (imagery reproduced from the GEBCO_2014 Grid, version 20150318, www.gebco.net).

background study of the Azores, an overview of coastal tourism in the Azores, climate change and its associated effects in the Azores, climate change policies, impacts and current effort of the Azores in combating climate change, and finally conclusions and recommendations.

This chapter is based on the initial findings of Calado *et al.* (2011a). The conceptual framework was restructured according to new global developments and data and complementary information was updated and the discussion/conclusions revised.

Background on the Study Area

The Azores archipelago, a Portuguese Autonomous Region, is located in the North Atlantic between 36°55′ and 39°43′ North and 24°45′ and 31°17′ West, approximately 1500 km from Europe and is formed by nine volcanic islands and a few islets scattered along 600 km and roughly aligned WNW–ESE and rising from the Azores Plateau, defined by the 2000 m bathymetric contour (Fig. 10.1).

The islands of the Azores archipelago are geologically young and are located in a tectonically and volcanically active region near the triple junction of the American (west), Eurasian (north-east) and African (south) plates.

The Azores occupies an area of 2333 km^2 and has a population of roughly 240,000 inhabitants. Despite its small land area, the Azores archipelago encompasses an Exclusive Economic Zone (EEZ) of approximately 940,000 km^2, claiming an extension of the continental shelf of more than 2 million km^2 and has a coastline of circa 850 km.

The Azores islands littoral, although it is a monotonous region from the point of view of the materials that support the shore, has coastal forms as diverse as low rocky coasts to bluffs, plunging cliffs, pocket beaches, dunes and lagoons. Borges (2003) classified the Azorean coasts into two major categories: ‘primary coasts’ and ‘secondary coasts’, these latter being more mature and shaped essentially by marine processes (cf. Shepard, 1973).

The littoral of the different islands of the archipelago is essentially constituted by ‘secondary coasts’, with São Jorge Island being that with the most extensive shoreline of that category (86.7%), in contrast with Pico Island, which has the shortest extension (63%; Borges, 2003). In the sub-class ‘erosion forms’, varied units may be recognized, namely mixed, lava-flow, tephra, landslide, composite, alluvial fan and hyaloclastite coasts (Borges, 2003).

The coastal sectors corresponding to lava-flow, alluvial fan or landslide processes make contact with the sea via small cliffs (modally less than 10 m high) while the remnant types of coasts develop plunging cliffs of considerable height (exceeding 200 m) or are bordered by gravel beaches (Borges, 2003).

According to Borges (2003), the ‘secondary coasts’, sub-class ‘construction forms’ are in the minority. In fact the units of sandy beaches cannot be observed on Flores, Corvo, São Jorge and Pico islands. Sandy beaches develop in pocket beaches, mostly aimed south, preferably developing according to the W–E direction with average of over 200 m. In general its cross-section, variable between 10–50 m, has one or more berms and the beach face has a modal inclination of 6°. Typically, they are associated, directly or indirectly, with fluvial outlets. Most of these features are permanent throughout the year, but correspond to thin veneers of sand deposited over rocky or shingle surfaces (Borges, 2003). Black, clean sand, with low contents of carbonates, medium to coarse and moderately to moderately well sorted sand of volcanic nature forms the most common fill of these beaches (Borges, 2003).

Units of sandy beach with dunes are very rare and can be found only on São Miguel, Faial and Terceira Islands. The sand from the dunes is medium sized and well sorted to very well sorted (Borges, 2003).

Fajã da Caldeira and Fajã dos Cubres at São Jorge Island are the only units of lagoon with gravel barrier (Borges, 2003).

The littoral corresponding to primary coasts, sub-class constriction, is less representative; however, several units can be identified such as landslide, alluvial fan and hyaloclastite coasts and the more frequent lava-flow coasts (Borges, 2003).

Coastal Tourism in the Azores

Coastal tourism emerged as a new sector in the mid-1990s. Until then, there were very few hotels and direct air connections. The adoption of tourism growth-enhancing policies by the regional government led to the provision of air connections and promotion of brand awareness spawning a boom in hotel construction with the total number of hotel beds growing from 5098 in 2001 to 9976 in 2011 (SREA, 2011). Tourist nights spent in touristic accommodation increased from 864,766 in 2001 to over 1,150,449 in 2011 (SREA, 2011). Economically, the total tourism revenue increased from about €32 million in 2001 to over €47 million in 2011. With regard to air connections, currently there are several air companies operating in the regional space and offering several direct connections and tour packages to, among others, the Nordic countries, Germany, the UK, Spain, the Netherlands, the USA and Canada. As a means to alleviate regional economic asymmetries, the Azores Regional Government (2008) promotes inter-island hopping to spread the economic benefits from tourism across the islands.

De Menezes *et al.* (2008) stated that nature, landscape, remoteness and weather are parameters that will promote longer expected stays in the Azores. Coastal recreational activities are certainly major tourist attractions. Some of the main activities include sailing, yachting, boat tours, cruise tourism, scuba diving, whale and dolphin watching, swimming with dolphins and sport fishing.

During 2015, the Regional Secretariat for Tourism and Land Transportation is working on a new 'Strategic and Marketing Plan for the Azores Tourism – Horizon 2020'. This plan aims to 'increase the development of anchor products for Azorean islands and at the same time create measures for their qualification and development' as well as 'the expansion of a culture of strategy for tourism, shared by the whole society that will raise the ongoing development of hospitality'. It also focuses on the construction of touristic destinations built on two main vectors: 'structuration, qualification, valorization of offers' and 'promotion'.

Climate Change Policies

Increasing awareness on climate change and its impacts has been one of the supports for raising the political effort globally. The recently agreed targets reached at the Climate Action & UNEP, COP21-Sustainability, Climate and Latest Green Technologies Forum (2015, Paris), present ambitious mitigation aims of sustaining emissions within limits compatible to an increase of 2°C global maximum temperature. This global effort will be transferred to nations' internal mitigation policies on Energy, Transportation, Cities, Industry and specific strategies for adaptation. In the EU, several sectored policies coordinate Member States into specific actions, one of the most important being the Climate & Energy Package 2020 (European Commission, 2016) targeted towards reducing emissions from energy production by 20% and a 20% saving from efficient use. In direct consequence, Member States develop strategies for alternative energy production, mainly through renewable sources. In parallel to this effort, considerable investment in adaptation is foreseen through the EU Strategy on Adaptation to Climate Change (COM, 2013) transposed to National Adaptation Plans.

Current Effort in the Azores

In the Azores, since 2011, the Regional Strategy for Climate Change (Government Council Resolution No. 123/2011; Azores Regional Government, 2011) has established several actions and among the most important is the development of the Regional Operational Plan for Climate Change, initiated in November 2015. The plan will: downscale climatic scenarios for the Azores until 2030; determine the regional contribution to global emissions; establish mitigation measures; present a set of adaptation measures and an according implementation/ financial programme. However, measures for diversifying energy production through renewable energy (especially geothermal), reducing emissions, reducing consumption and increasing efficiency are already taking place in line with the Regional Strategic Plan for Energy in the Azores.

The Azorean Government is also developing efforts in order to comply with the European Floods Directive (Directive 2007/60/ EC; EU, 2007). This directive establishes 'a framework for the assessment and management of flood risks, aiming at the reduction of the adverse consequences for human health, the environment, cultural heritage and economic activity associated with floods in the Community'. The EU Member States shall elaborate different instruments, such as flood hazard maps, flood risk maps and flood risk management plans, which are already being developed in the Azores.

The Azores Region is an insignificant emissions producer of greenhouse effect gases (GEG), less than 2.5% of country total (APA, 2014). Nevertheless, the Azores initiated a series of projects on efficient energy production and use. These projects range from the regional energy production company to the regional transport, housing and services. Research on compensation-accounting mechanisms such as carbon storage and sequestration by endemic vegetation and forest systems was initiated (M. Vergílio *et al.*, unpublished results) as an additional effort for mitigation.

In another initiative, the Municipality of Vila Franca do Campo, on São Miguel Island, is one of the partners on the ClimaAdaPT.local Project (ClimAdaPT.Local, 2015). It promotes the establishment of Municipal Climate Change Adaptation Strategies throughout the country and includes 24 municipalities, universities and research centres. In this project a first attempt to establish some downscaled trends on future climate in the Azores were established, and are presented in the next sections as a basis for discussion. The municipality has drawn up its Adaptation Strategy and Tourism, especially coastal and maritime-related tourism activities, which is one of the activities seen as a potential opportunity of climate change (with trends for drier hot summers).

In 2011, one of the main concerns for tourism development in the Azores was the imposition of an allowance amount on the airline companies by the Emission Trading Scheme (Directive 2008/101/EC; EU, 2008). The system, which entered into force in 2012, should have aggravated the price of transportation into the Azores. However, the liberalization of the air flight transportation market into São Miguel Island and the incoming low-cost companies have counterbalanced, annulated and even cheapened the price of flight tickets and increased tourism statistics (SREA, 2011).

Climate Change in the Azores and its Associated Effects on the Coasts

The Azores climate is influenced by its geographic location in an open oceanic basin, open to the North Pole and the tropics (Ferreira, 1980). Marine air masses interact with cold or temperate air masses from the pole. In addition, local factors such as distance from the coast, altitude and the exposed island slopes are major influences (Agostinho, 1941). The Azores climate can be considered as marine temperate, which is reflected by the low thermal amplitude, high precipitation, high air humidity and persistent wind. The altitude and the exposure of the terrain to the dominant winds are important variables that regulate the spatial distribution of precipitation in the archipelago (Agostinho, 1941; Bettencourt, 1979). Naturally, the large-scale atmospheric circulation also plays an important role in the precipitation distribution among the different groups of islands in the Azores (Andrade *et al.*, 2008; Marques *et al.*, 2008).

As discussed by Calado *et al.* (2011b), seven coastal environmental hazards have been identified as posing significant threats to the Azores coasts. They are mainly: (i) sea-level rise; (ii) coastal storm; (iii) stream flooding; (iv) coastal erosion; (v) tsunami; (vi) landslide; and (vii) seismicity and volcanism. Although not listed specifically as a coastal environmental hazard, Calado *et al.* (2011b) stated that human activity contributes significantly to human vulnerability along the coast.

For the Azores islands the main forcing factors are storminess (heavy rain, coastal storms and hurricanes), increase in the wave climate in the Atlantic and sea level rise.

Climate change in the Azores archipelago is producing changes in the precipitation pattern, such as the annual precipitation distribution, with an increase in winter precipitation and decreases in other seasons (Santos *et al.*, 2004; Miranda *et al.*, 2006), with an increase in heavy rain episodes also estimated (Miranda *et al.*, 2006; Tomé, 2013). Therefore the increased winter precipitation will intensify the occurrence of landslides, some of them at sea cliffs, resulting in coastal retreat, but will also escalate stream flooding in streams that are usually ephemeral and torrential. Likewise it will trigger flash floods in the streams with a perennial regime, which typically have their flood plains occupied by settlements because these are among the few attractive coastal fringes of the islands. Additionally, those areas frequently experience simultaneous stream and coastal flooding due to the heavy rain events which occur, commonly associated with coastal storms (cf. Raposo, 1998; Borges, 2003).

The long fetch that characterizes the Azores results in a high-energy wave climate where both sea and swell are relevant sources of energy to the coast, with the northern to north-western facing shores of each island being more exposed in general (Borges *et al.*, 2002; Calado *et al.*, 2015). Additionally, the wave climate of the North Atlantic is increasing according to Bacon and Carter (1991), Bouws *et al.* (1996), Young *et al.* (2011) and Bertin *et al.* (2013). In line with this observed phenomenon, the storminess of the Azores increased slightly over the last 150 years (Borges, 2003), which seems to be consistent with the significant increase in Atlantic storm activity stated by Vecchi and Knutson (2008) for a similar period of time. The storminess in the Azores is characterized by its elevated inter-annual and inter-decadal variation. Andrade *et al.* (2008) indicate that the average storm lasts for 2.3 days and the average storm frequency is 3.1 storms/year. According to Borges *et al.* (2011), lower intensity storms occur four times every 5 years, while an extreme storm (e.g. hurricane, tropical storm) occurs on average once every 7 years and these extreme events present a tendency towards being shorter, but more frequent and more intense (Borges, 2003), which seems to be consistent with global observations made by Young *et al.* (2011). The extreme storm usually occurs with south-westerly waves of maximum significant wave height of 11.7 m and highest wave height of 22.2 m (Borges, 2003), resulting in coastal flooding and coastal erosion (land and sediment loss). There is a scenario of increasing trend in the wave climate of the North Atlantic (Bertin *et al.*, 2013), as well as storminess (Vecchi and Knutson, 2008) and a slight increase of hurricanes in the same basin (Webster *et al.*, 2005; Holland and Webster, 2007), although Murakami *et al.* (2014) state that the change in the frequency is not significant. This means that the Azores will most certainly be affected more regularly by hurricanes (Bengtsson *et al.*, 2007) and the eastward and pole-ward migration of genesis (Zhao and Held, 2012), along with the increase in the storminess of the Azores archipelago (Borges, 2003). It is expected that these factors will, in the near future, strongly contribute to amplifying coastal erosion rates on the Azorean shores and will escalate the events of coastal flooding and coastal landslides. Upon comparison of the Azorean recession rates (e.g. 0.8 m/year on bluffs; Borges, 2003) with other places in the world (Sunamura, 1992) and the materials eroded from the scarps and cliffs, the result suggests that the mesoscale values of the Azores recession rates are in general less significant (Borges, 2003). However, it is important to note that in relation to the relatively small size of the Azores, the mesoscale coastal erosion rates might represent a significant reduction of the already limited land availability there.

Sea-level rise figures prominently among the consequences of climate change, and global projections for the end of the 21st century indicate an increase ranging from 0.26 m to 0.82 m (Church *et al.*, 2013), which may reach 1.10 m according to Jevrejeva *et al.* (2012). Ng *et al.* (2014) analysed data from the tidal gauge station of Ponta Delgada (São Miguel Island) and the sea level analyses between 1978 and 2007 indicated a statistically significant rising trend (2.5±0.4 mm/year; p=0.000), while between 1996 and 2007 it was 3.3±1.5 mm/year (p=0.025), agreeing with other global sea level studies (Ng *et al.*, 2014). Kopp *et al.* (2014) project an increase for the Azores region of 0.8 m by the end of this century. In the Azores archipelago, as elsewhere, the local sea-level rise will cause saltwater intrusion into aquifers and surface waters and will dramatically reshape the erosion of beaches and bluffs, coastal flooding and storm damage, boosting the effect of wave climate and storminess. With the actual rising trend of 2.5 mm/year, presented by Ng *et al.* (2014), applying the 'Bruun model' (Bruun, 1962, 1983), the sandy or gravel beaches of the Azores have an erosion rate of 0.25 m/year, which is consistent with field data obtained by Borges (2003).

Impact of Climate Change on Coastal Tourism in the Azores

Climate change in the Azores might affect some factors with impact on tourism with diverse intensities, such as weather, freshwater supply, accommodation, recreational activities and safety.

The slight temperature increase, with probably warmer winters (Santos *et al.*, 2004; Miranda *et al.*, 2006), might benefit tourism in the Azores. More tourists would visit the archipelago if winter becomes milder as they will be able to enjoy more outdoor recreational activities in pleasant weather. However, the precipitation pattern is also showing a slight difference with drier summers and more intense torrential rainfall in winter (ClimAdaPT.Local, 2015).

Unlike weather, water resources in the Azores might be reduced and freshwater supply may become one of the limiting issues for tourism growth in the Azores. With a reduction in precipitation in summer, the risk of saltwater intrusion in groundwater and an increased demand by growing tourism in the summer in the Azores, there is a likelihood of a freshwater shortage in the future.

With increasing threat of sea-level rise, extreme events, coastal erosion and landslides, there will be a restriction on the number of hotels and commercial buildings that can be constructed along the coasts. The reduction in ocean-view hotels could have a minor impact on tourism.

Although outdoor recreational activities might benefit from an increase in temperature and a decrease in precipitation, there might be negative impacts concerning coastal recreational activities. Scenarios concerning sandy beaches are uncertain: they could be reduced due to coastal erosion or benefit from the more sediments arising from the probable increase of cliff and bluff erosion, coastal landslides and stream flooding, the main sources of sediment for Azorean sandy beaches, according to Borges (2003). Nevertheless, with limited beaches in the Azores, mitigation measures are essential in curbing further beach erosion. On the other hand, without negative impacts, marine species of interest for tourism are more commonly spotted in the Azores waters over the last few years. This could be due to more interest and observers of marine species now than before, or it could also be due to change in migratory patterns as a result of a possible change in water conditions (Bentz *et al.*, 2013).

Regarding safety, climate change could increase coastal storms, coastal erosion and landslides. However, with appropriate mitigation measures in place, such as coastal protection and setback rules on construction, tourist vulnerability along the coasts might not be significantly increased. The main doubt remaining is the capacity of the Regional Government to implement effectively some of the unpopular measures like setback or reallocation in a region with a traditional distinctive pattern of coastline settlement.

Torrential rainfall episodes, landslides and coastal flooding have been largely reported by the media. Detailed news on losses and destruction create an image of insecurity that can have negative marketing outcomes for the regional tourism strategy.

Coastal Vulnerability Assessment: a Tool for Atlantic Island Management

Managing climate risk contributes to climate change adaptation. Well-informed adaptation requires understanding of good practices and application of sustainable interventions which increase resilience and reduce vulnerability (European Commission, 2014).

Vulnerability might be defined as the propensity or predisposition to be negatively affected (Adger, 2006; Oppenheimer *et al.*, 2014). It is a central concept in climate change research, even if it is conceptualized in different ways depending on the research communities using it, such as communities dealing with natural hazards and disaster management, ecology, sustainability science and land change (Füssel, 2007). However, it is consistent throughout the literature that the vulnerability of any system is a function of the exposure and sensitivity of that system to perturbations or external stresses and the (lack of) capacity of the system to deal with the effects of those conditions (Adger, 2006; Smit and Wandel, 2006; Oppenheimer *et al.*, 2014).

Exposure and sensitivity are intrinsic properties of the system and depend on the characteristics of the system and on the characteristics of the stimulus, such as climate change (Smit and Wandel, 2006). Oppenheimer *et al.* (2014) refer to exposure as 'the presence of people, livelihoods, species or ecosystems, environmental functions, services, and resources, infrastructure, or economic, social, or cultural assets in places and settings that could be adversely affected'. Adger (2006) refers to sensitivity as 'the degree to which a system is modified or affected by perturbations'.

Vulnerability to climate change is thus a characteristic of the system and a function of exposure, sensitivity and adaptive capacity, resulting from dynamic biophysical and social processes that shape local conditions and have to be considered when measuring vulnerability (Adger, 2006).

Coastal areas in the Azores comprise both natural areas with high value for conservation and strongly humanized areas where main urban areas are located. Thus, when measuring coastal vulnerability, exposure indicators must include both natural and social areas. Natural areas are mainly represented by classified areas included in the Island Natural Parks (Regional Legislative Decree No. 15/2012/A; Azores Regional Government, 2012), such as nature reserves, natural monuments, protected areas for habitat and species management, protected landscapes and protected areas for resource management. Social areas are mainly represented by: (i) people and groups, such as sensitive groups (e.g. the elderly and children), housing and social equipment; (ii) heritage, namely architectural heritage and classified buildings; and (iii) the economic system, such as tourist facilities, industry, forests and agriculture.

Sensitivity of coastal areas in the Azores is mainly dependent on: (i) coastal inundation; (ii) type of coast; (iii) wave regime and storminess; and (iv) existence of coastal defence constructions. Coastal inundation is a determinant in measuring sensitivity in the Azores considering that main urban areas are located on the coastal fringe and will certainly be affected by coastal hazards deriving from climate change. The type of coast is directly linked to coastal erosion. Wave climate and storminess determine the sensitivity to coastal events, such as coastal erosion and coastal floods. The existence of coastal defences properly applied might decrease damage and impacts on coastal populations.

Small islands are vulnerable to the effects of climate change, sea-level rise and extreme events (Mimura *et al.*, 2007). In light of global sea-level rise and storm trends (i.e. frequency, intensity and track), there is a need to understand coastal vulnerability to these potential impacts based on available, accessible and best known information. Therefore Borges *et al.* (2014) tested a coastal vulnerability assessment that employs a modified Coastal Vulnerability Index (CVI) for Pico Island, which has been successfully undertaken in other locations (Palmer *et al.*, 2011; Davies, 2012). Borges *et al.* (2014) used four indicators for this vulnerability assessment: (i) type of coastal cliffs (Fig. 10.2a); (ii) coastal exposure to swell and storm waves; (iii) outcrop flooded; and (iv) coastal defences. The above-mentioned researchers obtained a five-grade CVI for Pico Island ranging from extremely low to very high (Fig. 10.2b) and their work suggested that coastal vulnerability in Pico is relatively uniform and practically the entire coast corresponds to class 3 – moderate (Fig. 10.2b). A few months after the presentation of the CVI for Pico Island, a coastal landslide and associated sea cliff retreat occurred at one of the segments with moderate CVI ranking (São Miguel Arcanjo settlement;

Fig. 10.2. Coastal vulnerability assessment for Pico: (a) type of coasts (SM – soft material, HM – hard material; modified from Borges, 2003); (b) simplified CVI grades (modified from Borges *et al.*, 2014); (c) scar of the June 2014 coastal landside at a segment of the coast of São Miguel Arcanjo settlement (from: Photograph taken by Ana Malheiro, senior researcher of the Regional Civil Engineering Laboratory).

Fig. 10.2c). Although this event occurred on a coastal segment with moderate vulnerability it served somehow to validate the CVI and contributed to the refinement of the preliminary coastal vulnerability assessment that is being undertaken for the entire archipelago.

Conclusions and Recommendations

Climate change research in the Azores needs an increase in development. To date, research carried out on the impact of climate change on the Azores has not yet produced consistent methods to support decision-making. Based on preliminary findings as discussed above, the most significant effects of climate change for coastal tourism in the Azores will be the reduction in coastal developments of resorts due to coastal erosion, flooding, storm surge and landslide hazards. However, this would have to be translated into regional spatial planning and tourism strategies.

Without doubt, the policies of climate change of the EU could be beneficial to combat global warming. However, being a remote, isolated and small island with great dependence on air transportation for its tourism, the resulting higher cost of airfares could have a significant impact on the Azores and all the Macaronesian islands. The tourism growth strategy for this emerging sector greatly depends upon frequent air transportation with competitive prices. Accessibility is thus an indispensable element to maintain the attraction and competitiveness of the destination. The role of low-cost airlines in this case is of utmost importance as it directly contradicts the taxation of transport as responsible for heavy emissions. In this sense, there is no evidence that actual or near future policies and strategies for climate change will have a major impact on the islands' tourism. In the case of the Azores, as well as the other Macaronesian islands, tourism growth seems to be more sensitive to external economical fluctuations and alternative markets/destinations oscillations than the consequences of restrictive or proactive policy.

The modified Coastal Vulnerability Index (CVI), developed for Pico Island, has been validated and represents a powerful tool to identify areas that are more vulnerable to coastal hazards. Based on this information, spatial planning instruments might establish use conditions in more vulnerable coastal areas. In addition, tourist equipment and activities can be better (re)allocated, avoiding direct negative prejudice and also indirect consequences on tourism marketing related to safety issues.

The Azores is not a disappearing destination based on the current preliminary assessment of climate change impact on coastal tourism. However, it is important to note that the main convenient mode of transportation to this remote archipelago is via air travel. In consequence, the EU climate change policies cannot result in higher air travel cost. These policies would largely impede the growth of tourism in the Azores unless they were adapted for the Azores by taking into consideration the following five main factors: (i) current and projected climate change impact on the Azores; (ii) the contribution of the Azores to the global warming-based GHE emission; (iii) the greater effort than other developed Member States that the Azores has made in combating climate change with respect to being self-sufficient in energy; (iv) the accessibility issue, which explains the dependence on air travel; and (v) the lack of competitiveness and economic repercussions for the Azores, which the EU is trying to improve. This is also applicable to all the other archipelagos in Macaronesia.

More research needs to be carried out to understand better the predicted impact and corresponding mitigation strategies and implementations to combat climate change in the Azores. However, the EU climate change policies also need to be tailored to a local scale level in order for the Azores, and all the Macaronesian archipelagos, to remain competitive. Considering the difficult mobility and accessibility conditions of the islands, which have no alternative to air transportation, there needs to be a special local-specific applicable system for the region.

References

Adger, W.N. (2006) Vulnerability. *Global Environmental Change* 16, 268–281. DOI: 10.1016/j.gloenvcha.2006.02.006.

Agostinho, J. (1941) Clima dos Açores. *Açoreana* 2(4), 224–267 (in Portuguese).

Andrade, C., Trigo, R.M., Freitas, M.C., Gallego, M.C., Borges, P. *et al.* (2008) Comparing historic records of storm frequency and the North Atlantic Oscillation (NAO) chronology for the Azores region. *The Holocene* 18(5), 745–754. DOI: 10.1177/0959683608091794.

APA (2014) Relatório do Estado do Ambiente 2014. Agência Portuguesa do Ambiente. (Report in Portuguese). ISBN 978-972-8577-69-8. Available at: http://sniamb.apambiente.pt/infos/geoportaldocs/REA/rea2014.pdf (accessed December 2015; in Portuguese).

Azores Regional Government (2008) *An assessment of 'Strategy for the Outermost Regions: Achievements and Future Prospects'.* Regional Government, The Autonomous Region of the Azores (in Portuguese).

Azores Regional Government (2011) Government Council Resolution No. 123/2011 of 19 October. Approves the Regional Strategy for Climate Change of the Azores. Available at: http://servicos-sraa.azores.gov.pt/grastore/SRAM/Resolu%C3%A7ao%20-%20estrat%C3%A9gia%20para%20as%20altera%C3%A7%C3%B5es%20clim%C3%A1ticas.pdf (accessed December 2015; in Portuguese).

Azores Regional Government (2012) Regional Legislative Decree No. 15/2012/A of 2 April. Establishes the Legal Regime of Nature Conservation and Biodiversity Protection. Available at: http://dre.pt/pdf1sdip/2012/04/06600/0162501713.pdf (accessed November 2015).

Bacon, S. and Carter, D. (1991) Wave climate changes in the North Atlantic and the North Sea. *International Journal of Climatology* 11, 545–558. DOI: 10.1002/joc.3370110507.

Bengtsson, L., Hodges, K.I., Esch, M., Keenlyside, N., Kornblue, L. *et al.* (2007) How may tropical cyclones change in a warmer climate. *Tellus* 59A, 539–561. DOI: 10.1111/j.1600-0870.2007.00251.x.

Bentz, J., Dearden, P. and Calado, H. (2013) Strategies for marine wildlife tourism in small islands – the case of the Azores. In: Conley, D.C., Masselink, G., Russell, P.E. and O'Hare, T.J. (eds) *Proceedings 12th International Coastal Symposium. Journal of Coastal Research* Special Issue 65, 874–879.ISSN 0749-0208.

Bertin, X., Prouteau, E. and Letetrel, C. (2013) A significant increase in wave height in the North Atlantic Ocean over the 20th century. *Global and Planetary Change* 106, 77–83. DOI: 10.1016/j.gloplacha.2013.03.009.

Bettencourt, M.L. (1979) *O Clima de Portugal.* Instituto Nacional de Meteorologia e Geofísica, Portugal, XVIII, 103 pp. (in Portuguese).

Borges, P. (2003) Ambientes litorais nos grupos Central e Oriental do arquipélago dos Açores, conteúdos e dinâmica de microescala. PhD thesis, University of Azores, 413 pp. (in Portuguese).

Borges, P., Andrade, C. and Freitas, C. (2002) Dune, bluff and beach erosion due to exhaustive sand mining – the case of Santa Bárbara, S. Miguel (Azores, Portugal). In: Cooper, J.A.G. and Jackson, D.W.T. (eds) *Proceedings 7th International Coastal Symposium. Journal of Coastal Research,* Templepatrick, Northern Ireland, Special Issue 36, 89–95.

Borges, P., Ng, K. and Calado, H. (2011) Coastal hazards in the Azores archipelago – coastal storms and flooding. In: Micallef, A. (ed.) *MCRR3-2010 Conference Proceedings. Journal of Coastal Research,* Grosseto, Tuscany, Italy, Special Issue 61, p. 474. ISSN 0749-0208.

Borges, P., Phillips, M.R., Ng, K., Medeiros, A. and Calado, H. (2014) Preliminary coastal vulnerability assessment for Pico Island (Azores). In: Green, A.N. and Cooper, J.A.G. (eds) *Proceedings 13th International Coastal Symposium. Journal of Coastal Research,* Durban, South Africa, Special Issue 70, 385–388. ISSN 0749-0208.

Bouws, E., Jannink, D. and Komen, G. (1996) The increasing wave height in the North Atlantic. *Bulletin of the American Meteorological Society* 77(10), 2275–2277.

Bruun, P. (1962) Sea-level rise as a cause of shore erosion. *Journal of the Waterways and Harbors Division* 88, 117–130.

Bruun, P. (1983) Review of conditions for uses of the Bruun Rule of erosion. *Coastal Engineering* 7, 77–89.

Calado, H., Ng, K., Borges, P., Alves, F. and Sousa, L. (2011a) Climate change and coastal tourism in the Azores Archipelago. In: Jones, A. and Philips, M. (eds) *Disappearing Destinations, Climate Change and Future Challenges for Coastal Tourism.* CAB International, Wallingford, UK, pp. 111–122. DOI: 10.1079/9781845935481.0111.

Calado, H., Borges, P., Phillips, M., Ng, K. and Alves, F. (2011b) The Azores archipelago, Portugal: improved understanding of small island coastal hazards and mitigation measures. *Natural Hazards* 58, 427–444. DOI: 10.1007/s11069-010-9676-5.

Calado, H., Botelho, A.Z., Costa, A.C., Michael, J., Medeiros, A. *et al.* (2015) LocAqua – Modelo de determinação de locais com potencial para a instalação de unidade de Aquicultura na Região Hidrográfica dos Açores (RH9). Unpublished Technical Report, CIBIO Universidade dos Açores/Fundação Gaspar Frutuoso. Project supported by Regional Department of Sea Science and Technology (in Portuguese).

Church, J.A., Clark, P.U., Cazenave, A., Gregory, J.M., Jevrejeva, S. *et al.* (2013) Sea Level Change. In: *Climate Change 2013: The Physical Science Basis*. Contribution of Working Group I to the Fifth Assessment Report of the Intergovernmental Panel on Climate Change (Stocker, T.F., Qin, D., Plattner, G.-K., Tignor, M. Allen, S.K. *et al.* (eds)). Cambridge University Press, Cambridge, pp. 1137–1216.

ClimAdaPT.Local (2015) Municipalities and Climate Change - EMAAC - Municipal Strategies for Adaptation to Climate Change. FCUL (Coord.). Financed by EEA Grants/Agência Portuguesa do Ambiente (APA). Available at: http://climadapt-local.pt/en/project (accessed January 2016; in Portuguese).

COM (2013) 216 Final. An EU Strategy on Adaptation to Climate Change. Communication from the Commission to the European Parliament, the Council, the European Economic and Social Committee and the Committee of the Regions, COM(2013) 216 Final, Brussels, 11 pp. Available at: http://ec.europa.eu/transparency/regdoc/rep/1/2013/EN/1-2013-216-EN-F1-1.pdf (accessed December 2015).

Davies, W.T.R. (2012) Applying a Coastal Vulnerability Index (CVI) to the Westfjords, Iceland: a preliminary assessment. Master's thesis, University of Akureyri, Ísafjörður, Iceland, 109 pp.

De Menezes, A., Moniz, A. and Vieira, J. (2008) The determinants of length of stay of tourists in the Azores. *Tourism Economics* 14, 205–222.

European Commission (2014) *The Economic Impact of Climate Change and Adaptation in the Outermost Regions - Final Report*. Publications Office of the European Union, Luxembourg. DOI: 10.2776/83383. Available at: http://ec.europa.eu/regional_policy/sources/activity/outermost/doc/impact_climate_change_en.pdf (accessed November 2015).

European Commission (2016) 2020 Climate & Energy Package. European Commission. Available at: http://ec.europa.eu/clima/policies/strategies/2020/index_en.htm (accessed February 2016).

European Union (2007) Directive 2007/60/EC of the European Parliament and of the Council of 23 October 2007 on the assessment and management of flood risks (Text with EEA relevance). *Official Journal of European Union*. Available at: http://eur-lex.europa.eu/legal-content/EN/TXT/PDF/?uri=CELEX:32007L0060&from=EN (accessed November 2015).

European Union (2008) Directive 2008/101/EC of the European Parliament and of the Council of 19 November 2008 amending Directive 2003/87/EC so as to include aviation activities in the scheme for greenhouse gas emission allowance trading within the Community. *Official Journal of European Union*. Available at: http://eur-lex.europa.eu/LexUriServ/LexUriServ.do?uri=OJ:L:2009:008:0003:0021:en:PDF (accessed November 2015).

Ferreira, D.B. (1980) *Contribution á l'Étude des Ventes et de l'Humidité dans les Îles Centrales de l'Archipel des Açores*. Centro Estudos Geográficos, Lisbon (in French).

Füssel, H. (2007) Vulnerability: a generally applicable conceptual framework for climate change research. *Global Environmental Change* 17, 155–167. DOI: 10.1016/j.gloenvcha.2006.05.002.

GEBCO_2014 Grid (2014) The GEBCO_2014 Grid, version 20141103. Available at: www.gebco.net (accessed 31 October 2016).

Holland, G.J. and Webster, P.J. (2007) Heightened tropical cyclone activity in the North Atlantic: natural variability or climate trend? *Philosophical Transactions of the Royal Society A* 365, 2695–2716. DOI: 10.1098/rsta.2007.2083.

Jevrejeva, S., Moore, J.C. and Grinsted, A. (2012) Sea level projections to AD2500 with a new generation of climate change scenarios. *Global and Planetary Change* 80–81, 14–27. DOI: 10.1016/j.gloplacha.2011.09.006.

Kopp, R.E., Horton, R.M., Little, C.M., Mitrovica, J.X., Oppenheimer, M. *et al.* (2014) Probabilistic 21st and 22nd century sea-level projections at a global network of tide-gauge sites. *Earth's Future* 2, 383–406. DOI: 10.1002/2014EF000239.

Marques, R., Zêzere, J., Trigo, R., Gaspar, J. and Trigo, I. (2008) Rainfall patterns and critical values associated with landslides in Povoação County (São Miguel Island, Azores): relationships with the North Atlantic Oscillation. *Hydrological Processes* 22, 478–494. DOI: 10.1002/hyp.6879.

Mimura, N., Nurse, L., McLean, R.F., Agard, J., Briguglio, L. *et al.* (2007) *Small islands – Climate Change 2007: Impacts, Adaptation and Vulnerability*. Contribution of Working Group II to the Fourth Assessment Report of the Intergovernmental Panel on Climate Change (Parry, M.L., Canziani, O.F., Palutikof, J.P., van der Linden, P.J. and Hanson, C.E. (eds)). Cambridge University Press, Cambridge, pp. 687–716.

Miranda, P.M.A., Valente, M.A., Tomé, A.R., Trigo, R., Coelho, M.E. *et al.* (2006) O clima de Portugal nos séculos XX e XXI. In: Santos, F.D. and Miranda, P.M.A. (eds) *Alterações Climáticas em Portugal: Cenários, Impactos e Medidas de Adaptação (Projecto SIAM II)*. Gradiva, Lisbon, pp. 47–113 (in Portuguese).

Murakami, H., Hsu, P.-C., Arakawa, O. and Li, T. (2014) Influence of model biases on projected future changes in tropical cyclone frequency of occurrence. *Journal of Climate* 27, 2159–2181. Available at: http://dx.doi.org/10.1175/JCLI-D-13-00436.1 (accessed 11 August 2017).

Ng, K., Phillips, M.R., Borges, P., Thomas, T., August, P. *et al.* (2014) Maintaining a way of life for São Miguel Island (the Azores archipelago, Portugal): an assessment of coastal processes and protection. *Science of the Total Environment* 481, 142–156. Available at: http://dx.doi.org/10.1016/j.scitotenv.2014.01.067 (accessed 11 August 2017).

Oppenheimer, M., Campos, M., Warren, R., Birkmann, J., Luber, G. *et al.* (2014) Emergent risks and key vulnerabilities. In: Field, C.B., Barros, V.R., Dokken, D.J., Mach, K.J., Mastrandrea, M.D. *et al.* (eds) *Climate Change 2014: Impacts, Adaptation, and Vulnerability. Part A: Global and Sectoral Aspects*. Contribution of Working Group II to the Fifth Assessment Report of the Intergovernmental Panel on Climate Change. Cambridge University Press, Cambridge, pp. 1039–1099.

Palmer, B.J., van der Elst, R., Mackay, F., Mather, A.A., Smith, A.M. *et al.* (2011) Preliminary coastal vulnerability assessment for KwaZulu-Natal, South Africa. *Proceedings 11th International Coastal Symposium. Journal of Coastal Research*, Poland, Special Issue 64, 1390–1395. ISSN 0749-0208.

Raposo, A.G.B. (1998) Os desabamentos na bacia hidrográfica da Ribeira Quente e sua freguesia, na madrugada de 31 de Outubro de 1997. *Açoreana* 8(4), 571–590 (in Portuguese).

Santos, F.D., Valente, M.A., Miranda, P.M.A., Aguiar, A., Azevedo, E.B. *et al.* (2004) Climate change scenarios in the Azores and Madeira Islands. *World Resource Review* 16(4), 473–491.

Shepard, F. (1973) *Submarine Geology*, 3rd edn. Harper & Row, New York, 517 pp.

Smit, B. and Wandel, J. (2006) Adaptation, adaptive capacity and vulnerability. *Global Environmental Change* 16, 282–292. DOI: 10.1016/j.gloenvcha.2006.03.008.

SREA (2011) Statistic Series 2001–2011, Chapter 9 – Tourism. Available at: http://estatistica.azores.gov.pt/upl/%7Bc65c1938-9640-43f7-92b8-594f387d1d9c%7D.pdf (accessed October 2015).

Sunamura, T. (1992) *Geomorphology of Rocky Coasts*. Wiley, Chichester, UK, 302 pp.

Tomé, R. (2013) Mudanças climáticas nas regiões insulares. PhD thesis, University of Azores, 116 pp. (in Portuguese).

Vecchi, G.A. and Knutson, T.R. (2008) On estimates of historical North Atlantic tropical cyclone activity. *Journal of Climate* 21, 3580–3600. DOI: 10.1175/2008JCLI2178.1.

Webster, P.J., Holland, G.J., Curry, J.A. and Chang, H.-R. (2005) Changes in tropical cyclone number, duration, and intensity in a warming environment. *Science* 309, 1844–1846. DOI: 10.1126/science.1116448.

Young, I.R., Zieger, S. and Babanin, A.V. (2011) Global trends in wind speed and wave height. *Science* 332, 451–455. DOI: 10.1126/science.1197219.

Zhao, M. and Held, I.M. (2012) TC-Permitting CGM Simulations of Hurricane Frequency Response to Sea Surface Temperature Anomalies Projected for the Late-Twenty-First Century. *Journal of Climate* 25, 2995–3009. DOI: http://dx.doi.org/10.1175/JCLI-D-11-00313.1.

11 Case Study Malta: Climate Change and Tourism: Risks, Hazards and Resilience – An Island Perspective

A. Jones*
Institute for Tourism Travel and Culture, The University of Malta; and Island & Small States Institute, The University of Malta

Introduction

This chapter focuses on current strategic management issues that are critical to the growing complexity of relationships between global tourism, predicted climate change and policies for tourism destination management using Malta as a case example. The Maltese government's own assessment (MTA, 2012) has highlighted significant threats posed to the tourism industry from climate change. Such recognition stems from earlier assessments particularly from the Maltese Ministry for Resources and the Environment (2004) which, at that time, highlighted potential threats from adverse climate change predictions. These included, for example, the deterioration of potable water supplies and quality, more frequent extreme weather events, soil degradation, erosion and an accentuated desertification process, threats to public health, changes in seawater mass characteristics and effects on fish stocks, coastal erosion and inundation together with biodiversity reduction. The resultant possible impact on the tourist economy of the islands was specifically highlighted. In turn, the Malta Tourism Authority's (MTA, 2012) Strategy for Tourism 2012–2016 and the Ministry for Tourism (2015) draft National Tourism Policy 2015–2020, albeit rather ambiguous on the subject, both to some extent highlight the need to promote better sustainable approaches for future tourism development across the Maltese islands. Little advice, however, on the tangible ways in which this can be achieved are effectively offered.

Malta and the Mediterranean – Tourism Growth and Climate Change

Key to these debates are, of course, the impact that current and predicted environmental and climate change threats are currently having on tourism within the Maltese islands and over the broader Mediterranean basin. From 2010 to 2015, tourism attained record yearly performances mainly attributed to increased air routes and more effective marketing initiatives placing Malta as a year-round destination. Statistics for 2013–2014 show that Maltese tourism grew by an average 9.3% per annum and despite a slight divergence to niche markets traditional forms of tourism based upon sun and sea still remained the predominant market share. MTA statistics illustrated arrivals exceeded 1.8 million visitors in 2015 (MTA, 2015).

*E-mail: andrew.jones@um.edu.mt

According to studies done by the MTA's Research Unit (MTA, 2015), tourism has seen a steady growth year after year and it now accounts for 29% of the GDP and the largest contributor to the market services sector. Tourism now accounts for 22% of government income, 11% of imports and outflows and 17% of fulltime equivalent employment. This continued growth in tourism numbers, particularly in the peak summer period, is already creating environmental strains which are now leading to carrying capacity issues, resource, waste and pollution impacts. Such issues have been explored by authors such as Dodds (2007), Austin (2012) and Anon (2014). That said, the tourism policy for the Maltese Islands 2012–2016 (MTA, 2012) and the more recent Ministry for Tourism's 2015–2020 strategy (Ministry for Tourism, 2015) primarily sets out to ensure that tourism works towards the sustainable development of Malta's tourism industry. In turn it encourages the tourism industry to adapt to tourism trends as they evolve. The Maltese Government's policy directions for future tourism growth are thus fairly clear. However, impact forecasts for climate change particularly for Malta and its tourism industry and for the wider region as a whole remain less clear and less well-defined. A report from the *Malta Independent* (2009) did attempt to start a serious debate on the consequences of climate change for tourism on the islands. This highlighted predicted threats from severe climate events, flooding, infrastructure damage and adverse ecological change. In a similar vein the Maltese Government also established a Climate Change Committee on Adaptation, which reported on a national climate change adaptation strategy (CCCA, 2010), which again contributed to the ongoing debate and stressed the tourism industry as a specifically vulnerable sector. However, it is the IPPC (2007) fourth assessment report that set the framework for understanding the nature of climate change across the Mediterranean with predictions of increased fires, heatwaves, droughts and floods, more pest outbreaks, long-term sea-level change, increased sea temperatures and large-scale disease-related mortality events.

Such reports have provided a framework for the contemporary understanding of threats in Malta. More recent research by the European Union (EU) does provide a number of additional indicators and more tangible forecasts. For example, the Commission and other international bodies have gone some way to address current predicted forecasts for environmental change for the Mediterranean region. A joint report by the IUCN, MedPlan and WWF (2012) clarifies the predicted changes to the Mediterranean marine environment illustrating considerable increases already to sea temperature and salinity over the last 40-year period. In a similar vein a report by NASA (2013) maps global temperature increases between 2008 and 2012, showing an average temperature rise across the Mediterranean of over 3°C during this period. The European Environmental Agency's (EEA, 2012) report on climate change vulnerability in Europe adds some quite stark predictions for environmental change across the Mediterranean region with forecasts measuring significant increases in temperatures, proliferation of more simultaneous hot days and nights, intensification of drought, the rise of solar radiation and surges in insect infestation, adverse ecological change together with significant decreases in water availability. The report particularly highlights the vulnerability of the Southern Mediterranean regions, pinpointing coastal environments, areas of high population and high dependency on summer tourism at the forefront of current risks. These are very much key characteristics of the tourist economy of the Maltese islands and provide some profound warnings in this respect. The report concludes that 'the suitability of Southern Europe for tourism would decline markedly during key summer months' (EEA, 2012, p. 209).

Current Risks and Hazards

Apart from such data which provide some substantiated research on current changes to Mediterranean environments, there remains little first-hand empirical data supporting evidence of climate change and impacts on tourism. This is particularly true when data for such are considered for Malta. However, there is a growing volume of more anecdotal evidence, particularly emerging from the local media sources, that more than suggest emerging climate change and the resulting impacts on the current Maltese tourist economy. For example, Mercieca (2012a) has indicated that Malta is among the ten poorest countries globally in terms of water resources per inhabitant (172 out of 180) stating that nowhere else in Europe is water more scarce. He suggests that there are only some 60 m^3 of potable water per inhabitant and that nowhere in Europe is water such an acutely scarce resource where the European minimum norm is 1000 m^3. Osbourne (2014) also confirms such a concern, suggesting that water shortages are one of the most significant dangers to the economic well-being of the Mediterranean region. Clearly this also has profound resource and environmental implications for continued and sustainable tourism growth in Malta.

As early as 2010 Deidun expressed concerns regarding changing ecology with a report headlining *'Tis the season to be jelly*, which reported increased incidents of jellyfish blooms. Tremlett (2013) has also tracked the changing ecology of the islands, which he suggests is now significantly affecting tourism. He suggests that the record surge in jellyfish blooms in 2013 was not only transforming local Maltese ecosystems but also now threatening the health of tens of thousands of tourists. Similar assessments by Piraino *et al.* (2014) and summarized by Tremlett (2013) show that the island of Lampedusa (some 160 km from Malta), which receives some 300,000 visitors per year, has only one swimming week a year free from jellyfish and that the socio-economic impact on tourism will result in the loss of millions of Euros per annum. He goes on to estimate that over 150,000 tourists are now treated annually for severe jellyfish stings. In the same year the global news media organization CNN reported through its travel blog on the dramatic increase in jellyfish populations in the Mediterranean and Black Sea and warned of the potential impacts to tourists (CNN, 2013; Fig. 11.1). Other hazards were reported by Mercieca (2012b) and Chetcuti (2012), who highlight the abnormal rise in numbers of victims bitten by the 'Asian Tiger Mosquito' and the rise in incidence of severe attacks during recent summer seasons. This in turn was reported on social media and online travel sites such as Trip Advisor, which raised warnings and concerns relating to a 'mosquito problem' across the islands. In this respect the power of media and its impact to quickly highlight and distribute such information globally and the impact this may have on local tourist economies should not be under-estimated.

Added to this, there has also been a growing discourse on the increased frequency of severe weather events and severe storm damage. In this context recent reports by, for example, Fritz (2014), *The Times of Malta* (2014) and *Malta Today* (2014) highlighted the severity of recent autumnal storms which were likened to tropical hurricanes with winds in excess of 95 mph. These storms alone resulted in over 250 storm damage reports through local flooding and damaged infrastructure and resulted in the suspension of flights in and out of Malta International Airport, suspension of all ferry services to and from the islands and inter-island links, general power outages within tourism resorts and damage to the electricity grid. These incidents of course, in turn, created major disruption and difficulties for tourism operations and in turn the tourism economy across the islands. Again such incidents were reported by online travel sites such as *The Times of Malta* (2014, 2016), thus emphasizing the potency of online media in distributing global information of this nature.

The recognition of heat stress and prolonged durations of high temperatures has also been another phenomenon recently reported as unusual climatic events. Such phenomena reported, by for example Sansone (2012) and Muscat (2014a), have not only increased the growth of heat-related illness but also had consequences for public utilities and the disruption to water and power supplies across the islands.

Less tangible but still perceived as a potential real-time threat is also the growing

Fig. 11.1. Jellyfish warnings, Balluta Bay, Malta.

awareness of sea level rise and the potential flood risk to existing Maltese tourism resorts and beaches. These potentially include, for example, the beaches and resorts of Mellieha Bay, Ramallah and Golden Bay, Sliema (ferries), St Julians and Spinola Bay, Marsaxlokk and Birzebbugia and the beaches and bays of Xendi and Ramla on Gozo (Micallef, 2012; Sansone, 2012; Muscat, 2014a, b). The loss of already limited beaches and associated tourism facilities would indeed be a dramatic impact for the islands. In turn such occurrences will potentially have significant implications for local infrastructure and the local tourist economy. Again, Muscat's (2014b) illustration and visual projection of how Manoel Island and The Strand, Sliema, would look with a 0.5 m sea level rise is a dramatic case in point where substantial tourism infrastructure and tourism operations could be severely compromised (Fig. 11.2).

These issues present some real evidence of the current situation of where strains are emerging. Given such evidence, current predictions present some stark warnings for tourism in Malta and the associated tourism economy. Evidence would suggest that the summer season will become less sustainable due, particularly, to heat and periods of prolonged high temperatures, water shortages, high solar exposure and increased risk from ecological hazards. The additional economic costs and security questions for the continuity in supply for public utilities such as electricity and water are also a growing concern. Such issues would suggest that seasonality and shift in tourism numbers to shoulder- and off-season tourism periods might be a valid actuality in the medium to long term. Whilst predictions of this nature are not an exact science this will in turn provide challenges to the existing tourist economy and environment and the way in which the social and economic well-being of the islands can be sustained. A move to more specialized niche markets away from traditional sun and sea markets will inevitably be a major challenge. In turn, the need to adapt the market and economy to less numbers of tourists during the summer and the need to promote better quality, higher spending tourists that do not wholly rely on the sun and sea factors during the spring, autumn and winter should perhaps be a starting point in this respect. The availability of summer seasonal workforce (particularly the availability of casual labour) and additional more highly trained personnel presents a quandary and may also lead to some sizeable difficulties and shifts for human resource management. The need to increasingly move toward new opportunities for sustainable tourism niche market development will in turn present new challenges for the Maltese tourism industry in this context.

Future Resilience: Malta at a Crossroads

Clearly evidence from the literature, thus far, shows that threats derived from climate change will ultimately impact upon the long-term future viability of tourism environments and, of course, ultimately their continued survival. From the evidence presented thus far, the future for tourism in Malta appears to be at a crossroads.

As such, over the last decade two factors have clearly emerged. One suggests that tourism is having a major environmental impact on many established tourism destinations in Malta and the second suggests that potential threats from environmental change are likely to create considerable adverse impacts to tourist economies of the islands unless managed effectively. Thus we can begin to infer that a clear juxtaposition and paradox has emerged between, on the one hand, tourism, itself, creating many undesirable impacts at tourism destinations and on the other, environmental and by association climate change threatening to adversely impact on tourism infrastructure, ultimately threatening the very nature, character and socio-economic well-being of many tourist destinations. This duel disadvantage and the resulting implied threats to the Maltese tourist economy are again very real in the context.

From the existing literature it looks as if several complex dynamics, tensions and pressures now present some real tangible

Fig. 11.2. Coastal vulnerabilityat St Julians and Sliema, Malta: (a) Balluta Bay, St Julians; (b) storm damage at Balluta Bay, St Julians; and (c) storm surges at the promenade, Sliema.

challenges to the tourism industry. These are largely concerned with the synergies (or lack of) between the often disparate stand-points and interests that attempt to recognize predictions for climate change, meet economic and social needs from tourism growth demands and meet the ever-growing aspirations of the diverse tourism stakeholders. It is probably still too early to speculate on definitive outcomes, but the interaction and relationship between these issues will, as time will tell, ultimately determine the future sustainability and viability for Maltese tourism and the future prospects for tourism growth. Despite some government efforts to fully recognize climate change impact, a better coordinated and strategic approach to problem recognition and management solutions is now certainly overdue and one that presents a real challenge for the continuation of tourism growth prospects across the Maltese islands.

From the literature reviewed, such complex relationships raise several questions on the continued need for tourism destinations, including those in Malta, to address key management issues. These, in summary, would appear to include a collection of what would appear to often be quite disparate interests that require a more strategic and coordinated approach in order to address both the impacts of climate change and the need to sustain tourism economic well-being.

The synergies between these factors would appear to be critical if Malta is to sustain its tourist economy. However, the implications from such varied discourses suggest that it has become increasingly important to nurture and develop destination management principles and practices that in essence reflect better strategic policy approaches. These, as already stated, are often complex and vary according to localities and destination characteristics. A one-size-fits-all solution to setting strategic objectives is not perhaps the final answer. A flexible solution that can adapt to the recognition of problems, that can be adaptable to finding and implementing solutions and in turn meet stakeholder expectations (both nationally and within specific tourism localities) would seem to be fundamental. The key challenge, in this context, is one of 'balance' and as such is primarily one which requires better synergy between all the interested parties within tourist economic and social structures. This is especially pertinent for each Maltese tourist locality, resort or destination.

Conclusion: Risk Hazard and Resilience – Getting the Balance Right

From the issues drawn from the literature potential impacts of predicted climate change particularly for the Maltese islands will undoubtedly pose a significant threat to natural coastal environments, tourism infrastructures and the tourist 'communities' at specific localities in Malta. Contemporary evidence from the literature illustrates that perceived and actual threats are, indeed, real although accurate predictions and current assessments still remain at best ambiguous and at worst suffer from vagueness, apathy and to some extent 'media' hyperbole. Again, social media can act as a powerful deterrent in the promotion or demotion of a tourism destination where issues, risks and hazards associated with climate change can be increasingly played out on a 'virtual' global platform. In conclusion, this review presents a broad assessment of key contemporary threats, key challenges and possible resolutions to current threats to Maltese tourism from perceived climate change predictions.

In summary, it seems fair to state that there still remains continuing uncertainty regarding the science of climate change and the validity of current predictions. However, as already stated, the very recent findings and conclusions from the IPCC report (2014) would confirm worst assertions. In this regard, general perceptions from the literature demonstrate that there is recognition of adverse climatic events such as ecological change, resource depletion, heat stress, increased incidence of storm surges and a general rise in sea level. Perceptions also recognize predicted erosion and structural damage that will result from such incidents. However, there are mixed responses when resolutions, responses and actions are considered. Here, knowledge gaps and disagreement can frustrate options to take ameliorative action. In essence there remains a complex relationship of interrelated processes that combine at differing levels and stages of the tourism destination management hierarchy. This dynamic involves processes associated with several parameters including tourism destination management, processes of problem recognition, balancing strategic policy decisions, meeting stakeholder expectations and providing solutions to the key challenges, problems and threats currently at hand. At many Maltese tourism localities, these are ongoing processes, the dynamics of which are often complex, consisting of multi-layers of diverse stakeholder interests. Synergies between these diverse groups would still appear to remain in their infancy with little strategic direction being taken. More disappointing perhaps is that these dynamic processes, particularly the predictions for adverse climate change vis-à-vis ever ongoing strides to promote further tourism growth, are increasingly and potentially in conflict or at odds with one another. The current growth strategy from the private sector to boost tourism visitors to over 2 million per annum is a case in point. One could argue that this does not bode well for the future prospects for Maltese tourism in the short term.

Despite remaining uncertainties, there is evidence to suggest that there is an unequivocal necessity to maintain strategic momentum for all tourism stakeholders, and this includes both public, private, business, user and community representatives to engage and integrate more fully with decision making and policy processes. Pertinent to this is engagement with appropriate long-term policy implementation measures which connect more closely with existing environmental, governmental, legal, financial and technological frameworks. These are themes now commonly explored in the literature, for example by UNEP/OECD (2010), Jones and Phillips (2011), Hall *et al.* (2012) and Sing (2012). In essence such discourse means creating or ensuring structures for management and decision making that are fit for purpose. At present evidence suggests that such structures and organizational frameworks remain lacking. The future success in managing tourism, and this is particularly pertinent to the Maltese islands, is a need to promote effective measures that support a strategic cyclic process of problem recognition, addressing challenges and implementing effective resolutions at destinations.

In this respect there are the beginnings of change in Malta too. Evidence from projects such as the environmental management initiatives implemented at the Radisson Blu Hotel complex and resort at Golden Bay and the Hilton hotel at St Julians demonstrate that environmental management systems that conserve resources and reduce impacts (particularly in carbon) can be economically sustainable. The more comprehensive 'ECO GHAWDEX' project on Gozo, which promotes broader sustainable tourism practices, also provides a good starting platform. In Malta and elsewhere, however, these tend to remain largely a loose uncoordinated patchwork of schemes and unfortunately, because of this, their wider influence and impact remains somewhat marginal. This would still seem to be the case across the wider tourism sectors in Malta.

The evidence is now clear. Maltese tourism will inevitably have to adapt to changing patterns of tourism growth with shorter summer seasons and longer spring, autumn and winter seasons. The impact on existing enviro-socio-economic patterns for tourism will also inevitably need to adapt and adjust. The challenge also remains one which requires action in the form of better-coordinated policy initiatives. Without such measures the risks from climate change and the predicted negative impacts this will have on the Maltese tourism economy will inevitably escalate, threatening the very nature, economic well-being and future sustainability of the industry. None the less, these are threats and challenges that all the tourism stakeholders in Malta still need to fully recognize.

References

Anon (2014) MHRA suggests setting maximum carrying capacity for Malta. *Times of Malta* 17 June. Available at: www.timesofmalta.com/articles/view/20140617/local/mhra-suggests-setting-maximum-carrying-capacity-for-malta-as-it-calls-for-new-tourism-strategic-plan.523836 (accessed 10 October 2014).

Austin, R. (2012) The real Malta: a rubbish tip. *The Times of Malta* 28 November.

CCCA (2010) *National Climate Change Adaptation Strategy – Consultation Report*. CCCA, Malta, November.

Chetcuti, K. (2012) Malta under attack from mosquitos. *The Times of Malta* 3 October. Available at: www.timesofmalta.com/articles/view/20121003/local/Malta-under-attack-from-mosquitoes.439394 (accessed 30 May 2016).

CNN (2013) Coming to Europe This Summer – Jellyfish. CNN Travel. Available at: http://travel.cnn.com/coming-europe-summer-jellyfish-600901 (accessed 30 May 2016).

Deidun, A. (2010) 'Tis the season to be jelly. *The Times of Malta* 30 May, Valletta.

Dodds, R. (2007) Malta's tourism policy, standing still or advancing to sustainable tourism? *Island Studies Journal* 2(1), 47–66.

European Environmental Agency (EEA) (2012) *Climate Change, Impacts and Vulnerability in Europe 2012*. EEA 12/2012, Copenhagen.

Fritz, A. (2014) Wicked Mediterranean storm whips up 95 mph wind gusts in Malta. *Washington Post* 7 November, Washington, DC.

Hall, M., Gossling, S. and Scott, D. (2012) *Tourism and Climate Change: Impacts, Adaptation and Mitigation*. Routledge, Abingdon, UK.

Intergovernmental Panel on Climate Change (IPCC) (2007) *Climate Change 2007, the Fourth IPCC Assessment Report*. UNEP, Cambridge University Press, Cambridge.

Intergovernmental Panel on Climate Change (IPCC) (2014) *Fifth Assessment Report on Climate Change 2014*. UNEP, Cambridge University Press, Cambridge.

IUCN, MedPlan and WWF (2012) *A Changing Mediterranean Marine Environment, under Predicted Climate Change Scenarios*. IUCN, Gland, Switzerland.

Jones, A.L. and Phillips, M. (eds) (2011) *Disappearing Destinations: Climate Change and the Future Challenges for Coastal Tourism*. CAB International, Wallingford, UK.

Malta Independent (2009) Climate change makes for a precarious future for Malta. *Malta Independent* 16 August. Available at: www.jc.um.edu.mt/__data/assets/pdf_file/0003/79095/Climate_change_makes_for_a_precarious_future_for_Malta.pdf (accessed 3 November 2014).

Malta Today (2014) Over 250 reports of storm damage lodged with Enemalta. *Malta Today* 8 November, Valletta.

Malta Tourism Authority (MTA) (2012) *Tourism Policy for the Maltese Islands 2012–2016*. MTA, Valletta.

Malta Tourism Authority (MTA) (2015) *Tourism in Malta Edition 2015*. MTA, Valletta.

Mercieca, F. (2012a) Averting the imminent water crisis. *Sunday Times of Malta* 1 April.

Mercieca, F. (2012b) More fall victim to Asian Mosquito. *The Times of Malta* 12 October.

Micallef, M. (2012) Wrecked by lashing winds. *The Times of Malta* 11 March.

Ministry for Resources and the Environment (2004) *First Communication of Malta to the UNFCCC*. Ministry for Resources and the Environment, Valletta.

Ministry for Tourism (2015) *Draft National Tourism Policy 2015–2020 – Ensuring Sustainable Growth*. Ministry for Tourism, Valletta.

Muscat, C. (2014a) We will all be touched by climate change. *The Times of Malta* 7 April.

Muscat, C. (2014b) Predicted sea level rise 'will hit whole of Malta'. *The Times of Malta* 8 June. Available at: www.timesofmalta.com/articles/view/20140608/local/predicted-sea-level-rise-will-hit-whole-of-malta.522437 (accessed 30 May 2016).

NASA (2013) GISS Surface Temperature Analysis: Global Maps. Available at: http://data.giss.nasa.gov/gistemp/maps (accessed 30 May 2016).

Osbourne, H. (2014) Climate change water shortage threatens US and Mediterranean. *International Business Times* 2 January.

Piraino, S. *et al.* (2014) *Pelagio benovici* sp.nov. (Cnidaria Scyphozoa), a new jelly fish in the Mediterranean sea. *Zootaxa* 3974(3), 455–468.

Sansone, K. (2012) It's officially a heatwave. *The Times of Malta* 13 July.

Sing, T.V. (2012) *Critical Debates in Tourism*. CH13, Channel View Publications, Clevedon, UK.

The Times of Malta (2014) Flights diverted as sudden storm lashes Malta, some damage reported. *The Times of Malta* 5 October, Valletta. Available at: www.timesofmalta.com/articles/view/20141005/local/update-5-flights-diverted-as-sudden-storm-lashes-malta-some-damage.538552 (accessed 4 June 2017).

The Times of Malta (2016) Storm over Malta diverts flights, disrupts feasts, suspends fair. *The Times of Malta* 24 June, Valletta. Available at: www.timesofmalta.com/articles/view/20160624/local/storm-over-malta-diverts-flights-disrupts-feasts.616624 (accessed 4 June 2017).

Tremlett, G. (2013) Jellyfish surge in the Mediterranean threats environment and tourists *The Guardian* 3 June.

UNEP and OECD (2010) *Climate Change and Tourism Policy in OECD Countries*. UNEP, OECD.

United Nations World Tourism Organization (UNWTO) (2014) *Tourism Highlights, 2014 Edition*. UNWTO, Madrid.

12 Case Study Iceland: Climate Change and Tourism Sustainability and its Effects on Icelandic Coastal Destinations

I. Jenkins*
University of Iceland/Háskóli Íslands, Iceland

Introduction

This chapter focuses on Iceland, aiming to encapsulate the climate change debate within the context of coastal tourism destinations. It will look at evaluating adaptation and responses to climate change and also compensations provided by climate change. Iceland is relatively 'new' to tourism even though it has had a moderately small tourist market for some time that, prior to 2008, had remained at a stable level for many years. It has always been on the tourist horizon but was regarded as a somewhat unusual destination and not one found on the main tourism mind-maps. That said, since the crash of the Icelandic banks in 2008 and the volcanic eruption of Eyjafjallajökull in 2010, there has been an exponential growth in tourism in Iceland; somewhat of an irony given the crisis that 2008 caused in terms of a general decline in global tourism receipts and trips. Furthermore, it is something of a conundrum to ascertain exactly why tourism in Iceland has become so popular and why it is now one of the island's major exports, nearly matching that of the fishing industry for export earnings (ITB, 2014a).

Eyjafjallajökull and the Icelandic Banking Crash: Success out of Natural and Economic Disasters?

Although climate change is now considered inevitable, the actual rate of change is a critical factor relating to rising sea levels and extreme weather conditions (IPCC, 2007). However, it is not just sea levels that tourism destinations need to be concerned about, but also changes to weather patterns such as the increased risk of flooding and more violent storms and heat/cold intensities. These changes will inevitably disrupt tourism flows and Iceland has been a good example of global tourism disruption through natural phenomenon. For example, the 2010 eruption of Eyjafjallajökull in Iceland caused serious disruption to air traffic in the northern hemisphere. Whilst not part of climate change 'per se' it illustrated the force and effect of nature and the impacts it can have on global transport networks, which are key features of any tourism product. This eruption captured the world's attention and imagination and it is asserted that it also greatly affected tourism in Iceland, in a positive sense. Iceland now has upward of 1.2 million tourists

*E-mail: jenkins@hi.is or dr.isjenkins@gmail.com

per year (2015) and this has grown exponentially from 360,392 in 2004 (ITB, 2014a) resulting in greatly increased earnings from tourism (Figs 12.1, 12.2). Eyjafjallajökull is just one of 30-plus active volcanoes in Iceland and illustrates the power of nature to affect human transport and destinations. The volcanic eruption of 2010 cost airlines at least £130 million per day (Wearden, 2010) and caused much disruption and cancellation to travel, which can be seen as a negative effect on world tourism. However, closer examination of the effects show that perhaps this natural occurrence helped to actually increase tourism in Iceland. Perhaps this could be an allegory for climate change and its ability to influence global tourism.

Conceivably, the lesson or maxim here is that change is inevitable and that disaster can, in the future, create higher tourist flows to certain destinations. Iceland has been said to be beyond the 'edge of the world' (Clements, 2015, p. 137) and in terms of tourist development has, as mentioned in the introduction, had a nominal tourist industry for some time. Until the eruption and the crash of the banks, Iceland had small volumes of tourists and it is suggested that what has

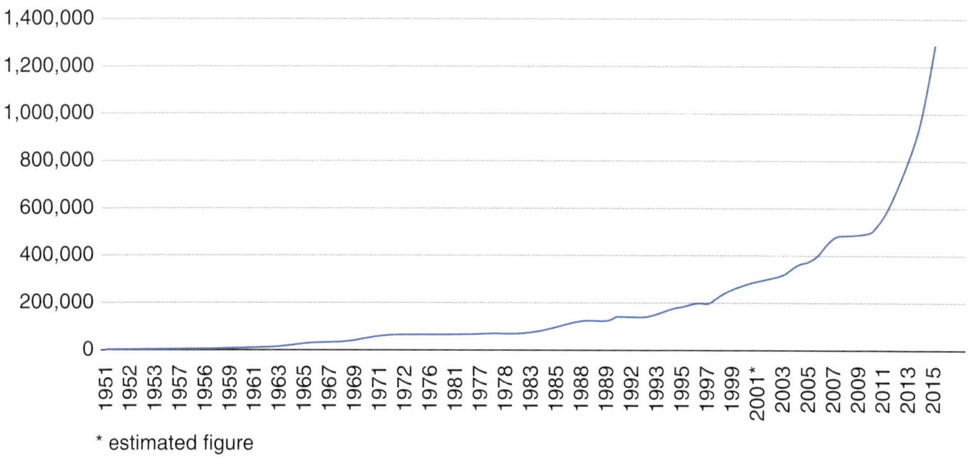

Fig. 12.1. Foreign visitor arrivals by air and sea to Iceland 1949–2015. Number of visitors (ITB, 2014b).

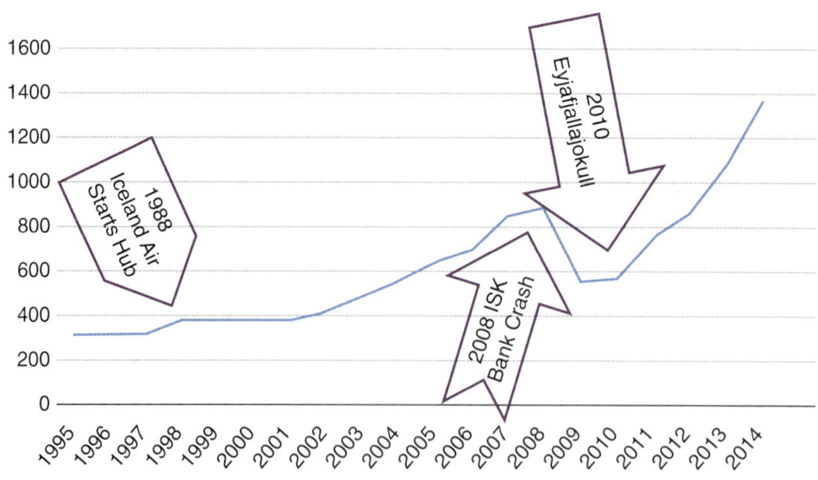

Fig. 12.2. International tourism receipts (current US$) Iceland (World Bank Data, 2014).

happened in Iceland is a redirecting of tourists from other destinations towards this new tourist milieu.

Icelandic tourism is based primarily upon nature (ITB, 2014b) and therefore the topic of climate change is extremely important to its tourism product; it is therefore an interesting case for examining the consequences of climate change and tourism development. As already stated, its tourist market has exponentially grown over the last 7 years and Iceland seems to have greatly benefited from both the economic and natural shocks of 2008 and 2010.

Other Tourism Factors

Although it can been asserted that Eyjafjallajökull, together with the collapse of the Icelandic banking system, was an excellent marketing tool for Iceland, other factors have also collaborated to give Iceland this exponential growth. Although the collapse of the Icelandic banks focused world attention on Iceland, from a utilitarian perspective it also meant a devaluation of the Icelandic Krona, reducing prices by a half. This had serious effects domestically, but helped to make Iceland less expensive. In addition, the Icelandic Tourist Board implemented a sustained marketing campaign which they believe has been effective in enticing tourists to Iceland. Other factors have also helped, such as the excellent air links with Iceland; Iceland was one of the first countries to offer air transport as LCC (Low Cost Carriers) and the Island has also developed as an excellent stop-over destination for international flights. Keflavik has become an international hub for many carriers and it has capitalized on this stop-over market, together with an increase in legacy airlines and new LCCs. In 2015 the airport had 4.86 million passengers (KIA, 2016b), a 25% increase from 2014, and in 2014 had a total of 22 carriers operating throughout the year (KIA, 2016a) (although the number of carriers differed during different seasons). In response to this tourism demand, Iceland has increased the number of LCCs landing at Keflavik, which now includes its own LCC airline WOW. These LCCs run flights during the peak season on an extensive basis and some run throughout the year, such as EasyJet. Given the size of the Icelandic population (320,000), this is a considerable success.

Furthermore, there is a perception of Iceland being an extremely safe destination, with low crime rates and little risk of terrorism. In addition, it remains a fairly unexplored destination for tourists and has developed quite specialized products for exploring nature, capitalizing upon the short-break market, especially for long weekends.

Tourism generates ISK302,667bn (ITB, 2015) and is now almost on par with the fishing industry, which is Iceland's mainstay with regard to export earnings. That said, the exponential growth of tourism has had negative effects and this can be seen in terms of infrastructure developments and also aspects of waste disposal and recycling. Sustainable energy, at a low carbon cost, is not really problematic, as most of Iceland's energy comes from either geothermic or hydroelectric power generation.

However, there are major challenges ahead for Iceland: it is difficult and expensive to maintain the road network with such a small population and such a large island, coupled with the increasing number of tourists. As with any accelerated increase in tourism, certain locations are becoming extremely stressed and have begun to exhibit signs of degradation in terms of the environment and society. Certain routes can now be counted as mass tourism circuits, especially destinations on the Golden Circle, which is probably the top tourist circuit in Iceland. It has been difficult for tourism planners to keep up with this sudden growth and currently they seem to be reacting rather than being proactive towards tourism growth.

Geography of Iceland

Iceland is situated almost mid-way between Europe and America, located just within the Arctic Circle and its northernmost parts are on latitude 66°. 'The population of Iceland is just over 300,000, with almost two-thirds of the population living in the capital, Reykjavík, and surrounding areas. Iceland has an area of 103,000 km, and is the second largest island in Europe' (MfEI, 2010, p. 7). In fact both European and American continental plates meet in the middle of Iceland and the division can be seen at Þingvellir National Park, one of UNESCO's designated world heritage sites and a major Icelandic tourist attraction, just some 40 km north-east of Reykjavík. Iceland is more or less entirely volcanic and was affected by the last glaciation period, which has left many landscape phenomena similar to other locations, such as Scotland. This has created a strange and unique topography which tricks the eye into thinking that it has had a multitude of different rock formations during its geological history, but in fact it is a fairly new landscape of some 15 million years (Grimsson, 2007). The variety of landscapes is extraordinary and this is one of the reasons why tourists are attracted to Iceland. On many of the visitor surveys, nature is given as a key reason for travelling to this location (ITB, 2014b).

In terms of Iceland's geographical location relative to the centres of urban populations, it has been described as 'on the edge of the world', which is a fairly accurate description of its isolated location in terms of transport routes and its position in the middle of the Atlantic. The interior (highlands) of the island is uninhabited and the spatial pattern of settlement can be seen as distributed around the coastal regions (Fig. 12.3), similar to Australia, but in this case, Iceland is surrounded by glaciers not sand! This again is reflective of geomorphology. The highlands contain the major glaciers and landscapes that are in many cases alien to the tourist lens. This is perhaps one of the reasons why many films and television programmes have been made in Iceland including James Bond, *Die Another Day* (Film in Iceland, 2016), *Fortitude* and *Game of Thrones*.

There are three National Parks in Iceland and one is UNESCO designated, Þingvellir National Park, situated to the northwest of

Fig. 12.3. Basic geography of Iceland (Landmælingar Íslands, 2016).

Reykjavík. Two of these are destinations with glaciers, the other is a lowland park that has both physical and cultural elements important to its designation.

Prior to road transport, transport by sea was the main access to areas of population. However, it is now possible to circumnavigate the island by road but this has only been feasible since 1974 when the road link, 'Route 1, ... 1,332 kilometres (828 miles)' (Iceland Magazine, 2014), was completed in the southern part of the island, linking at Skeiðarársandur, which joined the road to Reykjavík and the other major centres of population (Iceland Magazine, 2014). Reykjavík contains almost two-thirds of Icelanders, with Akureyri, located in the north of Iceland (population 20,000), being the second largest city. The most efficient transport system between population centres is air transport and there are a number of airports all of which can take large jet aircraft. However, two airports are the main hubs for transport in Iceland: Keflavik, which serves Iceland's international requirements, and Reykjavík Domestic Airport (in Reykjavík) connecting domestic flights within the island and some flights to the Faroe Islands and Greenland.

Icelandic Sustainability and Global Sustainable Tourism Council

Sustainability is part of the Icelandic Strategic Plan for Tourism and a strategic report was published in 2013 relating to tourism developments in the country. Sustainability is also linked to climate change in the Environmental Plan of 2006 (MfEI, 2002, 2006) where the focus is primarily based on reducing emissions, but with no mention of the contribution of air transport, which is the primary transport carrier for tourists. A number of reports emanating from the Ministry of the Environment and Industry mention sustainable developments as very important and focus on aspects of sustainability development (MfEI, 2006). It is also recommended that Iceland should consider the Global Sustainable Tourism Council (GSTC) criteria for destinations which would confirm that its sustainability is measured on a global stage. Since its introduction (GSTC, 2015), GSTC has made an attempt at standardizing the aspects of tourism sustainability at destinations and companies.

Clearly, the exponential growth of tourism in Iceland has instigated much debate on how tourism can be managed and maintained relative to sustainability. There is also evidence that some key elements of sustainability are beginning to be enacted in Iceland's capital, Reykjavík. But Iceland still has a long way to go; one study ranked Reykjavík 14th out of 16 cities on the Nordic Sustainability Index (PKF, 2013). However, it does have the potential to become more sustainable in terms of environmental impacts with the access of geothermal energy and the progress towards refuse management and recycling. On comparing Iceland with top locations such as Switzerland, there are many aspects of sustainability that Iceland needs to develop and emulate, in order to improve its sustainability scores and help in mitigating global warming. As mentioned earlier, Iceland is relatively new to the tourist business and it is suggested that the exponential growth in tourism has meant a new focus on strategic policies regarding sustainability and climate change.

Iceland does have its own quality grading system called Vakinn, which uses a metric to award the quality grading of hotels and other establishments within the tourism industry. That said, a closer evaluation of the criteria sees little evidence of sustainable measures except perhaps in terms of 'cultural knowledge of the area' and 'towels being washed or continually used' (Vakinn, 2010). There is a separate measure (award) for sustainability but one would have expected that quality would be meshed with sustainability. There appears to be little if any evaluation of a hotel's direct environmental impact, which needs to be evaluated in the grading system; the system requires changes to incorporate a sustainability section ensuring that hotels adhere rigorously to environmental principles.

An assessment of the recycling and waste policies gives an insight into Iceland's strategy for improving environmental policies and those connected with sustainability. There are a number of documents available, but many seem dated and need revision as the strategy available seems passé and not linked to the climate change debate (MfEI, 2002). There are municipal recycling facilities and visible recycling bins throughout the centre of Reykjavík, but in terms of 'composting' there seems to be little effect, with most organic waste placed into general waste handling, which ends up in landfill sites (Fischer, 2013). This is clearly an area that needs to be improved and enhanced given the increase in tourist numbers that is predicted for the future: 1.75million by 2030 (PKF, 2013). The Icelandic landfill does produce climate change gases such as methane, but some Icelandic companies utilize these landfill sites and collect the gases (methane) for local bus services and other transport vehicles (Mannvit, 2017).

Perhaps the principal question for Iceland (reflective of other similar destinations) is: how to manage the carbon emissions of aircraft that carry tourist visitors to Iceland? The Icelandic Government has developed strategies for sequestration and mitigation of their CO_2 emissions but it is not evident what they will do with this 'by-product' given the exponential growth in tourism. As is evidenced by research, air transport accounts for 2–7% of global emissions of CO_2 (depending upon which source is used) and is clearly an important element in climate change (Belobaba *et al.*, 2016). It is also interesting to note that the Paris COP21 agreement excluded tackling air industry emissions.

For Iceland this is a conundrum. Tourism is a valuable export earner but not without environmental impacts and not solely limited to Iceland. There need to be some polices for addressing aircraft emissions and how to offset the carbon. Although Iceland benefits from tourists, tourists also benefit from Iceland. There needs to be an equitable solution to who is responsible for the emissions and it can be argued that both Iceland and tourists are jointly responsible for these. Therefore, it necessarily follows that the offset and sequestration should be made by both actors. This issue never appears to be fully dealt with by tourist organizations, tourist board policies or, indeed, by the World Climate Talks. A further argument could also be used in terms of airlines; it is suggested that it may well be that airlines should be made responsible for ensuring offsets just like other industries. ICAO will be meeting in 2016 to discuss this very issue and it will be interesting to see what the likely reduction of carbon emissions will be (ICAO, 2016). Whatever the arguments for and against ultimate responsibility, there needs to be planned action to reduce emissions or at least to offset them. It could also be argued that because Iceland's main sources of energy are thermal and hydroelectric, this could offset any air transport emissions so why does Iceland need to address this issue, as it already has low carbon energy levels compared to other countries?

Climate Change Effects on Iceland

The Icelandic Government has a number of policy documents and instruments relating to climate change and how Iceland intends to tackle this. The documents are to be found with the Ministry for the Environment and Natural Resources. A strategic plan was produced in 2007 (MfEI, 2007) which primarily dealt with the idea of a reduction of CO_2 emissions and how the Icelandic Government will act upon the Kyoto Agreement. The current plan is dated 2007 and the Government is intending to update this following the Paris Climate deal in 2015 (Iceland Monitor, 2015). However, currently a number of factors have been identified which are likely to affect Iceland. These are as follows.

Rising sea levels

If temperatures continue to rise then coastal areas of Iceland will be exposed to increased risk of flooding caused by sea level rises. Most of the population areas of Iceland are close to the coastline although there are combinations of levels where population areas are to be found. Certainly, aspects of flooding can be envisaged as increasing tourist numbers will inevitably impact upon sewers and water runoff. It is suggested that there needs to be some significant planning in terms of new builds, especially the development of tourism facilities such as tourist accommodation. With a predicted increase of tourist numbers to 1.75 million by 2030 (PKF, 2013) there will have to be a significant development in hotel builds and other tourist infrastructure, which will put pressure on the current water system and runoff. There is also a need to consider storm proofing hotels from flood damage perhaps by building flood restriction/protection structures into hotels and tourist accommodation, especially on lower ground levels.

Increase in precipitation

Planning also needs to be made for the likelihood of increased precipitation. Although not necessarily an immediate concern, there is a need to plan for this and construct extra capacity into storm and drain systems, especially with increased water usage and drainage of tourist developments.

Increased temperature changes

The predictions are for increased temperature, and localized temperature figures confirm this. This will particularly affect glacial melts and as Iceland has some of the largest glaciers in Europe this will cause reduction in glacier size and increased risk of flooding in certain areas. In fact the prediction is that all the glaciers will disappear within the next 150–200 years (IfSS, 2012). That said, Iceland has had to regularly cope with this type of hazard as many glaciers are situated on volcanoes that erupt sub-glacially, causing large flows of glacial water, known as jökulhlaups, into the drainage systems (Björnsson, 2003). Currently Icelandic glaciers are retreating and this will be exponential if global temperatures continue to rise.

Another problem may well be changes in the Gulf Stream and its effects on Iceland. Iceland is relatively warm given its high latitude and if the Gulf Stream stops or is diverted, this could mean that Iceland will be plunged into Arctic temperatures which will be disruptive and incur extra infrastructural costs, resulting in the need to change current designs to meet extreme changes in temperature. It may also affect the seasonal flows of tourists.

Extreme weather

An increase in extreme weather could also affect Iceland. Iceland regularly has extreme weather warnings which may increase in frequency and velocity. In particular, 'wind speed', which can cause extremely serious damage and appears to be one of the main reasons for emergency service callouts and rescues, is a particular problem in Iceland, which already experiences storms with incredibly fierce winds. Tourists frequently get caught in extreme weather and have to be rescued by the emergency services. Again, there needs to be consideration on how adaptation can be met by the introduction of policies and plans. Structural design changes will certainly be required for the construction of new buildings in order to meet the increasing force of winds and storms.

Resilience

This is perhaps one of the key questions related to climate change (Adger *et al.*, 2011). How will a community's tourist economic system and nature cope with changes to the climate? It is evident that this is not a new question as human existence came about through adaptation to climate change. It is asserted that effective measures of resilience reflect spatial scale and ultimately it is the locale and community that will have to absorb and adapt to changes to ensure that economic and social systems function effectively.

The definition of resilience is 'the ability of the system to withstand a major disruption within acceptable degradation parameters and to recover within an acceptable time and composite costs and risks' (Haimes, 2009, p. 498).

When tourism is considered, it is possible to see resilience not just in terms of a loss of tourists to a destination. An analogy might be made with Iceland where resilience can be seen with regard to its current tourist flows and the future conditions where it is predicted that Iceland will, by 2030, receive 1.75 million tourists (PKF, 2013) (this could be seen as a major disruption for a population of only approximately 320,000). If the effects of climate change are added on, there is a potential recipe for extreme systems stress in Iceland. Already signs of stress are appearing as tourist growth has been extremely fast and exponential (20% per year).

Iceland is not without experience in resilience; in fact one could use the analogy of Norse and Viking voyaging as a form of resilience. It is suggested that one reason for the Vikings settling in Iceland was because of past social or climate change which meant that the Vikings migrated (perhaps an original climate refugee!). Iceland certainly does not have the most hospitable of climates and there was a definite need for resilience if settlers were to survive. It seems that Icelanders are now facing different kinds of pressure and change due to this new growth in tourism but will need to remain as resilient as they were in the past; Iceland needs to plan for these shocks and incorporate policies and detailed plans to deal with them.

Throughout the centuries, Iceland has had to cope with many major disasters, especially on an island where there are more than 30 active volcanoes. It has certainly learned to plan for disasters relating to natural phenomenon and has a specialized disaster team for dealing with volcanic eruptions. It would therefore not seem beyond the realms of possibility to transfer this expertise on disaster planning to climate change. As already stated, Iceland already copes with extreme weather, in particular winds, which are perhaps the most damaging factor to Icelandic infrastructure next to glacial floods, which occur after a volcanic eruption. It can be suggested that perhaps the Icelandic mentality is one of resilience and that adapting to climate change is simply another factor in the survival of Iceland. However, there will need to be careful planning and clear strategies for adaptation, which is already beginning to happen at governmental level.

What need to be put in place are actual targets related to climate change adaptation rather than simply focusing on mitigation. Mapping areas and identifying high-risk zones is a necessary aspect of Icelandic resilience planning.

Adaptation

The overall inference relating to Icelandic adaptation is that of complacency and that, at least in the short term, 'things will work out'. Hence the level of planning and policy instruments is extremely low. Iceland has considered aspects of climate change and how to deal with adaptation and in 2015 announced a 16 Point Plan to deal with Climate Change (MBL.IS, 2015). But it seems to lack any urgency, which perhaps is because currently, tourism is booming. Yet it is the very destinations where tourism is most successful that will most probably be affected by climate change. These are the population centres located on the coast, which will be most prone to sea-level rises and extremes of weather. Up until 2015 there has been very little impetus to deal with adaptation to climate change in terms of physical action to structures and land use, furthermore a significant majority of the Icelandic public do not feel immediately threatened by climate change (IfSS, 2012). Added to that, the general process of destinations applying adaptation to climate change shows much diversity in approach (Füssel, 2007).

However, what is not clear is how adaptation will be enacted. With accelerated tourism growth there is a need to incorporate these adaptation changes (land use, infrastructure and building design) into a number of destinations of the tourism market.

Examples of adaptation

There are few, if any, clear examples of adaptation to climate change in Iceland. The subject has been discussed but it is not necessarily seen as urgent and, given the increase in tourism, many believe there is in fact an element for positivity as climate change may well benefit Iceland, so why adapt? (IfSS, 2012). This is somewhat short-sighted as the population centre of Iceland is focused on the south-west where most tourism is located and it is here that the likely predicted effects of climate change will start to take effect.

As mentioned earlier, the main hazard is likely to be flooding, especially in low-lying areas around Reykjavík. As ISS states: 'Relevant information and tools, such as maps and databases, are needed to develop strategies and action plans. Lacking maps of flooding areas makes it for instance difficult to enforce building regulations and plan how cultural heritage may be protected' (IfSS, 2012). Currently, one can guess where flooding may occur and strengthening harbour defences may help, but planning needs to involve a comprehensive survey of topography and runoff, together with a strengthening of coastal defences.

There is evidence of some additional adaptation planning for climate change in other areas of Iceland such as the 'Coast Adapt Programme', which ran for 3 years between 2009 and 2012 and has detailed the likely flood risks for at least two communities. Two areas of Iceland, Árborg and Mýrdalshreppur, which are in southern Iceland, have been identified as risk areas for potential flooding due to a number of factors including climate change. These two areas contain the urban areas of Selfoss and Vik, which are also considered tourist locations, especially Vik, which 'rests on a narrow rim of flatland between the sea shore and the mountains. The coastal area built with volcanic sediments from the Katla eruption in 1918 has since 1971 eroded at a rate of 8–10 m per year' (CoastAdapt, 2012). There has been some climate change adaptation with the building of dykes, management of beach erosion and the provision of organizations to deal with civil defence relating to extreme weather conditions.

Market Changes and Effects on Iceland

The climate will change (Scott *et al.*, 2012) and so too will destinations, everything is 'flux' but the key is the speed. There will clearly be climate winners and losers and there appears to be a change towards destinations with new and more specialized tourist products. Iceland seems to fit this change as a postmodern tourist destination; its exponential growth since 2008 has been phenomenal. It seems to reflect changing aspects of the tourism market especially given its location, but whether this is the beginning of 'climate refugee tourists' is a moot point (see Chapter 2). Iceland's geographical location is clearly at the 'edge of the world', which has meant for a millennium, isolation from the mainstream changes of global economic frames and systems. However, the exponential growth in air travel has clearly assisted in relocating Iceland and perhaps explains some of the current growth of tourism in Iceland.

There is clearly a structural change in tourist markets, reflecting the post-modernist aspect of fracturing and perhaps hyper-reality (Urry, 2002, p. 149). Even though currently post-modernism seems somewhat of a pejorative term in certain tourism frames, it has agency in terms of explaining market changes and forces. For example, the market has clearly fractured and splintered, which has changed the mass tourism market into smaller and smaller segments. The tourist is also far more footloose and able to change destinations at the earliest signs of trouble or simply on a whim. In addition, Iceland seems to offer aspects of hyper-reality, seen within the context of film tourism and nature tourism such as whale watching, geyser gazing, blue pooling and 4×4ing.

Furthermore, the risk and destructive natural forces of active volcanoes, that can release such immense natural energy, seem to be an attractive tourist product. Where else can you walk on volcanoes buried beneath layers of ice, and within easy reach of

600 million plus populations (Western Europe and America)?

Mowforth and Munt (2009) clearly address this possible change and give a good treatise on the new forms of tourist that are emerging with an expanding middle class, that is beginning to leave the 'Costas' behind and search for newer forms of tourism to enjoy. This seems to support the proposition that they may be venturing to Iceland.

Conclusion

Iceland has long had to endure the vagaries of climate change and weather. This 1000-year-old Norse settlement has developed a certain resilience to living in what is quite an inhospitable climate. Hence, Iceland's response to climate change and adaptation, although seemingly somewhat ponderous, given its past, may be understandable. Consideration has been given to climate change but this is mainly within the context of emissions mitigation rather than actual physical adaptation to climate change. Tourism has now become the new export earning currency for Iceland and it is difficult to attribute this to definitive factors. However, there is clearly a 'sea change' in the tourist market and Iceland seems to be benefiting from perhaps changes in climate. None the less, many of Iceland's tourism developments are close to the coast or related to climate changes (glaciers and wildlife) and are therefore likely to be affected by predicted changes in weather caused by climate change. There has been some consideration of climate change and its effects on Iceland but little related to adaptation. Adaptation will be needed if Iceland is to continue to benefit from tourist developments. The predicted weather changes for Iceland could be seen simply as an exaggerated profile of current weather conditions, which are considered to be even now 'extreme at times', therefore adaptation to cope with increases in frequency and velocity of extreme weather would not necessarily be too difficult to adopt.

In addition, there is clearly still much to be done in terms of waste management and sustainability policies for hotels and other tourist facilities. An incorporation of an environmental quality section for grading hotels (Vakinn) would seem an essential step in trying to mitigate climate change. This might also enhance changing perceptions towards climate adaptations for hotels. There is also the issue of air transport and who should be responsible for the emissions of tourism. In view of the issue being avoided by the Paris Talks COP21 in 2015, there seems little impetus for Iceland to deal with this and it will be interesting to see the outcome of the ICAO talks in 2016.

Iceland is benefiting from a growth in tourism and perhaps climate change may well continue to enhance the number of tourist visits to Iceland, as they change from other destinations. However, in order to maintain the tourist monies it will also be necessary for Iceland to consider climate change adaptation policies and more sustainable policies for a pristine destination 'beyond the edge of the world'.

References

Adger, W.N., Brown, K., Nelson, D.R., Berkes, F., Eakin, H. *et al.* (2011) Resilience implications of policy responses to climate change. *WIREs Climate Change* 2, 757–766.

Belobaba, P., Odoni, A. and Barnhart, C. (2016) *The Global Airline Industry*, 2nd edn. Wiley, Chichester, UK.

Björnsson, H. (2003) Subglacial lakes and jökulhlaups in Iceland global and planetary change. *A Planetary Perspective* 35(3–4), 255–271.

Clements, J. (2015) *A Brief History of the Vikings. The Last Pagans or the First Modern Europeans?* Robinson, London.

CoastAdapt (2012) *Adapting to Climate Change in Iceland.* Institute for Sustainability Studies, University of Iceland. Available at: http://ssf.hi.is/sites/ssf.hi.is/files/Skyrslur/adapting_to_climate_change_in_iceland.pdf (accessed 15 June 2017).

Film in Iceland (2016) Made in Iceland. Available at: www.filminiceland.com/made-in-iceland (accessed 15 June 2017).

Fischer, C. (2013) Municipal Waste Management in Iceland ETC/SCP February. Available at: www.eea.europa.eu/publications/managing-municipal-solid-waste/iceland-municipal-waste-management (accessed 17 December 2015).

Füssel, H.-M. (2007) Adaptation planning for climate change: concepts, assessment approaches, and key lessons review article. *Sustain Science* 2, 265–275.

Grimsson, F. (2007) The Miocene floras of Iceland origins and evolution of fossils floras from North-West and Western Iceland 15 to 6 Ma. PhD Thesis, Institute of Earth Sciences Sigillum Universitatis Islandiae.

GSTC (2015) Global Sustainable Tourism Council Criteria. Available at: www.gstcouncil.org/en/gstc-criteria/sustainable-tourism-gstc-criteria.html (accessed 17 December 2015).

Haimes, Y.Y. (2009) Perspective on the definition of resilience in Systems. *Risk Analysis* 29(4), 498–501.

ICAO (2016) ICAO's Global Aviation Dialogues (GLADs) on Market-Based Measures to Address Climate Change 2016. Available at: www.icao.int/Meetings/GLADs-2016/Pages/default.aspx (accessed 15 February 2016).

Iceland Magazine (2014) The Ring Road around Iceland Completed Forty Years Ago, Travel. Available at: http://icelandmag.visir.is/article/ring-road-around-iceland-completed-forty-years-ago (accessed 11 February 2016).

Iceland Monitor (2015) Iceland Announces Climate Change Plan. Politics and Society. Available at: http://icelandmonitor.mbl.is/news/politics_and_society/2015/11/25/iceland_announces_climate_change_plan (accessed 11 August 2017).

IfSS (Institute for Sustainability Studies) (2012) *Adapting to Climate Change in Iceland CoastAdapt Report*. University of Iceland, Reykjavik.

IPCC (2007) Projections of Future Changes in Climate. Available at: www.ipcc.ch/publications_and_data/ar4/wg1/en/spmsspm-projections-of.html (accessed 17 December 2015).

ITB (Icelandic Tourist Board) (2014a) Tourism in Iceland in Figures – April 2014. Available at: www.ferdamalastofa.is/static/files/ferdamalastofa/Frettamyndir/2014/mai/toursim_in_icland_infigf2014.pdf (accessed 11 February 2016).

ITB (2014b) Foreign Visitors to Iceland 1949–2014 (Excel-document). Available at: www.ferdamalastofa.is/en/recearch-and-statistics/numbers-of-foreign-visitors (accessed 15 February 2016).

ITB (2015) Tourism in Iceland in Figures April 2015. Available at: www.ferdamalastofa.is/static/files/ferdamalastofa/Frettamyndir/2015/mai/tourism-in-iceland-in-figures_15.pdf (accessed 15 February 2016).

KIA (Kelflavik International Airport) (2016a) Facts and Figures 2014. Available at: www.kefairport.is/resources/Files/Facts_and_figures_2014_2.pdf (accessed 11 February 2014).

KIA (Kelflavik International Airport) (2016b) Passenger Statistics. Available at: www.kefairport.is/English/About-us/Facts-and-figures (accessed 11 February 2016).

Landmælingar Íslands (2016) Overview map 1:2 000 000. Available at: www.lmi.is/en/wp-content/uploads/2011/09/is2000.pdf (accessed 1 December 2016).

Mannvit (2017) Methane Gas Processing in Akureyri. Available at: www.mannvit.com/projects/methane-gas-processing-in-akureyri (accessed 15 June 2017).

MBL.IS (Iceland Monitor) (2015) Iceland announces climate change plan. *Iceland Monitor*, Reykjavik. Available at: http://icelandmonitor.mbl.is/news/politics_and_society/2015/11/25/iceland_announces_climate_change_plan (accessed 15 June 2017).

MfEI (Ministry for the Environment) (2002) *Welfare for the Future Iceland's National Strategy for Sustainable Development 2002–2020*. MfEI, Reykjavik.

MfEI (The Ministry For The Environment in Iceland) (2006) *Welfare for the Future Iceland's National Strategy for Sustainable Development*. MfEI, Reykjavík.

MfEI (Ministry for the Environment) (2007) *Iceland's Climate Change Strategy*. MfEI, Reykjavík.

MfEI (Ministry for the Environment) (2010) *Iceland's Fifth National Communication on Climate Change Under the United Nations Framework Convention on Climate Change*. MfEI, Reykjavík. Available at: http://unfccc.int/resource/docs/natc/isl_nc5_resubmit.pdf (accessed 17 December 2015).

Mowforth, M. and Munt, I. (2009) *Tourism and Sustainability: Development, globalisation and new tourism in the Third World*. Routledge, London.

PKF (2013) *Promote Iceland: Long-term Strategy for the Icelandic Tourism Industry*. PKF, London.

Scott, D., Hall, C.M. and Gössling, S. (2012) *Tourism and Climate Change: Impacts, Adaptation and Mitigation*. Routledge, London.

Urry, J. (2002) *The Tourist Gaze*, 2nd edn. Sage, London.
Vakinn (2010) Hotels, Evaluation Criteria. Available at: www.vakinn.is/static/files/Enska/Accommodation-criteria/hotel-critetia-1.7.2014.pdf (accessed 5 October 2015).
Wearden, G. (2010) Ash cloud costing airlines £130m a day. *Guardian* 16 April. Available at: www.theguardian.com/business/2010/apr/16/iceland-volcano-airline-industry-iata (accessed 17 December 2015).
World Bank Data (2014) International Tourism, Receipts (Current US$). Available at: http://data.worldbank.org/indicator/ST.INT.RCPT.CD (accessed 15 June 2017).

13 Case Study Barbados: Policy, Practice and Science: Perspectives on Climate Change and Tourism in Barbados – Conflict or Congruence?

J. Cumberbatch,* L. Nurse and K. Francis
University of the West Indies, Barbados

Introduction

Barbados is the most eastern of the Caribbean islands with a population of 277,821 in 2010 on a land area of 430 km^2 (Barbados Statistical Service, 2013; Fig. 13.1).

Approximately 66% of the population, as well as the majority of the tourism infrastructure, are located in the coastal areas in the west, south and south-east. The tourism on the west coast has long been described as the luxury product and the majority of the hotels, restaurants, villas and condominiums lie directly on the shoreline. These structures are at risk from the anticipated effects of global climate change, i.e. sea level rise, beach erosion, saltwater intrusion, increased flooding and storm surges. This is of grave concern since tourism has been the leading engine of economic growth in Barbados over the past five decades (Worrell *et al.*, 2011). Figure 13.2 shows sustained growth from 1956 onward with interruptions caused by three international economic recessions: the 1981–82 Gulf War; in the aftermath of the 11 September 2001 attack on the World Trade Center in New York; and since the 2008 world recession.

Currently, the sector continues to grow, although it is sluggish. According to the World Travel and Tourism Council (WTTC), the direct contribution of Travel and Tourism to GDP was US$505.4 million (10.8% of total GDP) in 2014. However, prospects are positive and tourism's contribution to GDP is forecast to rise by 1.8% in 2015, and to rise by 3.3% per annum, from 2015 to 2025, to US$710 million (12.3% of total GDP) in 2025. As expected, the industry contributes considerably to employment; in 2014, the total contribution of travel and tourism to employment, including jobs indirectly supported by the industry, was 35.7% of total employment (45,000 jobs) (WTTC, 2015).

As the country slowly emerges from the 2008 recession the projected effects of global climate change conjure worrying prospects. Not only is global climate change expected to have significant negative impacts, it will require considerable investment to build resilience to withstand the effects. This chapter presents the most recent climate model projections for Barbados, focusing on the implications for tourism planning. The knowledge, attitudes and practices of tourism stakeholders on the west coast are analysed

*Corresponding author, e-mail: janice.cumberbatch@cavehill.uwi.edu

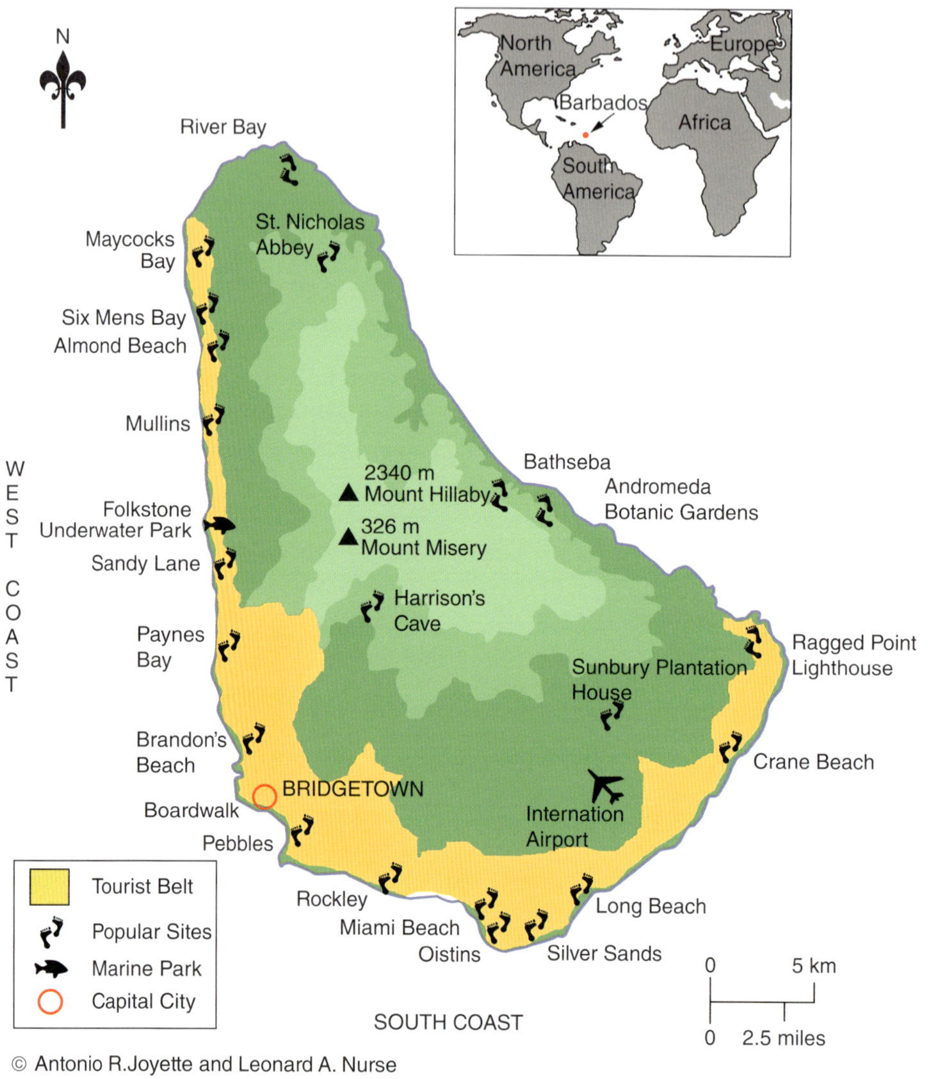

Fig. 13.1. Map of Barbados.

to determine the extent to which they are ready, willing and able to adapt to climate change. Current tourism policy is examined to ascertain whether it is congruent with the climate projections and sufficient to support the needs of the industry.

Climate Change: Assessing Problems and Challenges

Given the volume of greenhouse gases already emitted into the atmosphere and the associated inertia in the climate system, it is certain that global temperatures will continue to rise for at least another century (IPCC, 2007, 2013). Under RCP[1] 4.5 W/m^2 (intermediate low scenario), temperatures in the Caribbean are projected to rise by between 1.2°C and 1.5°C above the 1986–2005 mean (IPCC, 2013). Regional surface air temperatures would exceed these values under the higher (but plausible) RCP scenario of 8.5 W/m^2. Analysis of the observed data also shows that while the percentage of warm days (i.e. ±30°C) has increased, the

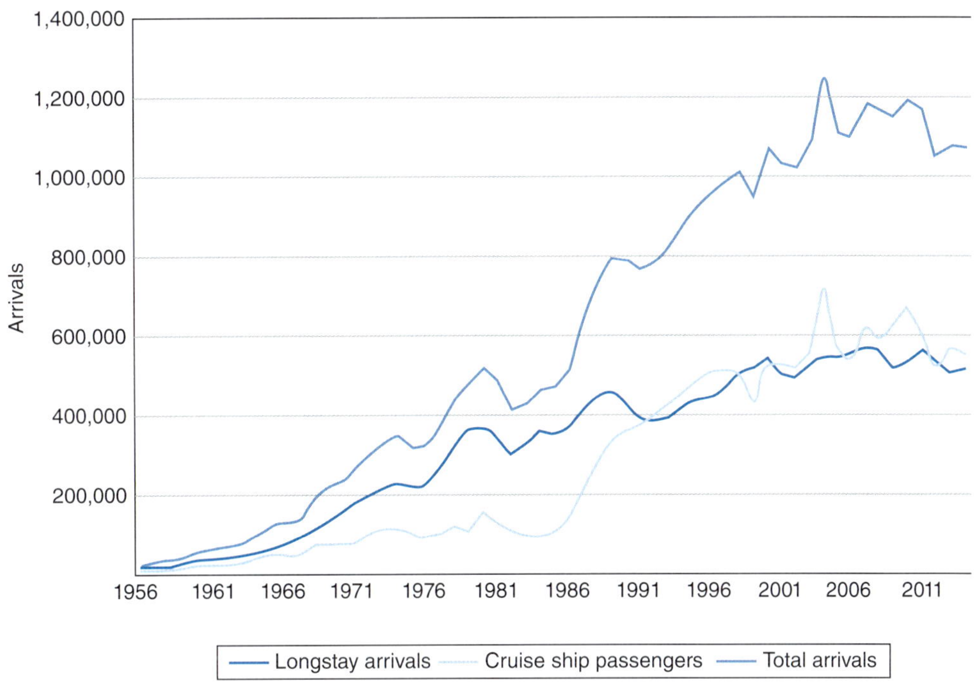

Fig. 13.2. Visitor arrivals to Barbados 1956–2014.

number of 'cool' days has decreased (Peterson *et al.*, 2002), evidently indicative of the hemispheric warming that has been occurring ever since the 1900s. This trend is projected to continue until at least the end of the present century.

Downscaled HadCM3[2] climate model projections for Barbados under SRES A2 and B2 scenarios,[3] suggest that in coming decades the island is likely to experience reduced mean daily, monthly and annual rainfall amounts relative to the 1960–1999 baseline, in both the dry and wet seasons (Charlery and Nurse, 2010). While the decline is projected to be lower in the dry season (21% and 12% for the A2 and B2 scenarios, respectively) than in the wet season (61% and 31% for the A2 and B2 scenarios, respectively), it represents a significant threat to many economic sectors, including tourism (Charlery and Nurse, 2010). These findings are consistent with projections for the Caribbean region as a whole, where on average the mean annual rainfall reduction has been approximately 0.18 mm/year since 1900 (IPCC, 2013). However, while the rainy season is projected to contract, the number of *heavy rainfall events* is expected to increase, with potential adverse consequences such as flooding and heavy sedimentation via runoff.

Sea level rise is another key climate change indicator that is of considerable interest and concern to small, low-lying coastal states like Barbados. Over the 20th century, the observed rate of sea level rise in the Caribbean has been approximately 1.8 mm/year, slightly higher than the global average (IPCC, 2013). Although no long-term sea-level data exist for Barbados, the geologic record suggests that a rate of change of this magnitude would be reasonably representative for the island. The most recent long-term estimates for the Caribbean suggest that by the last two decades of the present century, sea level would have risen by 0.5–0.6 m relative to the 1986–2005 mean (IPCC, 2013). Recent sea level projections for Barbados under RCP scenarios 2.6 W/m^2, 4.5 W/m^2 and 8.5 Wm^2 show that water levels around the island will continue to rise at an accelerating rate until the end of the present century (Fig. 13.3).

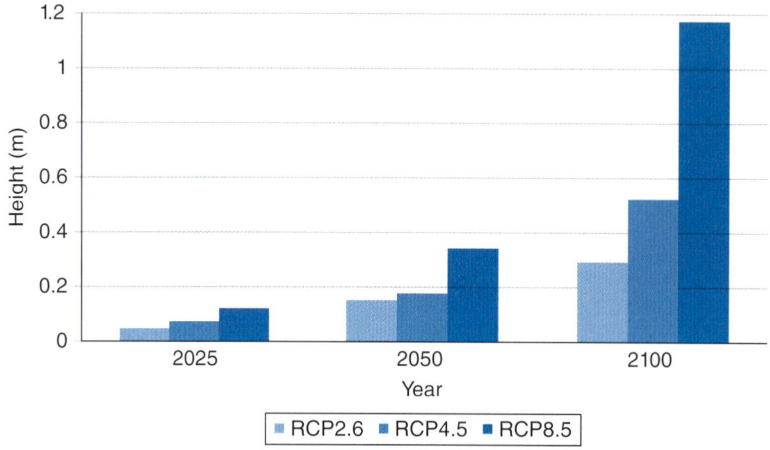

Fig. 13.3. Recent sea level projections for Barbados, using RCP 2.6, 4.5 and 8.5 W/m^2 (GFDL-CM3).

In addition, while there is no clear evidence that the total number of Atlantic hurricanes will increase as a result of atmospheric and oceanic warming, the *intensity* (and possibly frequency) of systems especially at the higher end of the Saffir-Simpson scale (categories 3–5) is expected to increase by 2–11% by 2100 (Knutson *et al.*, 2010; Emanuel, 2013). Hurricanes generate storm surges, as the rapidly falling pressure and strong onshore winds associated with their passage produce unusually high water levels at the coast. Storm surge is a major threat to coastal areas and is most often responsible for the extensive beach and infrastructure damage caused by hurricanes. The recent work of Knutson *et al.* (2010), Lin *et al.* (2012) and Emanuel (2013) whose findings show a direct causal relationship between atmospheric warming and hurricane intensity, must therefore be of enormous concern to low-lying nations like Barbados, whose GDP is highly dependent on the country's coastal assets.

Climate change implications for the Barbados tourism sector

The tourism sector has already begun to face and will continue to be confronted with many direct and indirect challenges as a consequence of global climate change. In economic terms, Moore (2010) points out that both the 'supply' side (physical plant and attractions) and 'demand' side (shift in visitor behaviour in response to change) of the product will be impacted. For most of the Caribbean, in particular countries such as Barbados where >90% of the tourism accommodation and related physical infrastructure is located at or within 2.0–3.0 km of the shoreline, sea level rise is a direct and existential threat (Pulwarty *et al.*, 2010). Equally at risk from elevated sea level are other facilities including the island's only airport, its deep-water port and marinas, major highways linking the tourism corridor on the west and south coasts, and the main electrical power generation facility. In addition, the combination of projected higher temperatures (increased evaporation), reduced rainfall (lower infiltration and aquifer recharge) and sea level rise (salinity intrusion) will have consequential adverse effects on the groundwater resources of Barbados, a country already classified as water scarce. Like all other sectors, tourism cannot be sustainable without adequate freshwater availability, thus appropriate adaptation and related management responses need to be implemented as a matter of urgency. The commissioning of a reverse osmosis facility in the year 2000[4] to augment the existing freshwater supply and a project to replace leaking mains[5] are two welcome interventions.

It is now well established that elevated sea surface temperatures pose one of the most serious threats to the region's coral reefs (Donner *et al.*, 2005; Manzello *et al.*, 2007; Oxenford *et al.*, 2008). In Barbados, >65% (Hoetjes *et al.*, 2002) and 70% (Oxenford *et al.*, 2008) of the island's corals were bleached during the intense El Niño events of 1997–98 and 2005–2006, respectively, when sea surface temperatures were at least 1°C above the seasonal mean. It must therefore be of some concern that recent climate model outputs confirm that warming of the Caribbean Sea is consistent with global and hemispheric trends (IPCC, 2014) and the rate is expected to accelerate in coming decades (Nurse and Charlery, 2014). The experiments suggest that under the IPCC SRES A2 and B2 scenarios, Caribbean Sea surface temperatures (SSTs) will experience a mean increase in warming of approximately 0.13°C in the period 2000–2029, to around 0.31°C in the decades 2030–2059 and exceed a mean of 0.41°C during the years 2070–2099 (Nurse and Charlery, 2014). Moreover, the research also indicated that over the same period the *range* in SST between the warmest and coolest months would contract, so that the region's corals would be exposed to consistently high water temperatures throughout the year. In such circumstances, corals would be at considerably greater risk of bleaching and associated morbidity and mortality, as there is contraction of the traditional cool season that is favourable to recovery (Nurse and Charlery, 2014). The coral reef degradation that is expected to accompany increasingly more frequent and longer bleaching episodes would have negative indirect impacts on the Barbados tourism industry.

There is also great concern that as the world's oceans become increasingly acidic as a consequence of absorption of higher levels of CO_2, ocean acidification will heighten the threat to corals and other calcifying organisms such as shellfish. Observational evidence confirms that since the start of the Industrial Revolution (circa 1760) ocean acidity has increased by approximately 30%, consistent with the observed acceleration of greenhouse gas emissions since that time. Consequently, a 14–30% decline in calcification is projected by 2050 (Fabricius *et al.*, 2011), so that erosion rates are expected to exceed the rate of reef accretion over the next several decades (Hoegh-Guldberg *et al.*, 2007; Pandolfi *et al.*, 2011; IPCC, 2013). In the case of the Caribbean Sea, data from NOAA Ocean Acidification Product Suite show that aragonite saturation state (Ωarg), a reliable proxy for net calcification rate, declined from approximately 4.15 in 1988 to about 3.9 in 2012.[6] This trajectory is expected to be maintained and most likely accelerate, given present and projected rates of global greenhouse gas emissions. Thus, the climate change-related risks of bleaching and acidification combined with other anthropogenic stressors would place Barbados' coral reefs at considerably great risk with consequential negative impacts on the quality and sustainability of the tourism product.

Coral reefs also provide the resource base for the important scuba diving industry, a key element of the tourism product offered across the CARICOM region (Table 13.1). The industry generates significant revenue for the region, amounting to at least US$1.2 billion annually (UNEP, 2005). Scuba diving is highly sensitive to environmental changes, owing to its dependence on healthy coral reefs which constitute the industry's primary resource base. While corals continue to face threats from various anthropogenic and natural sources (e.g. land-based contaminants, anchor damage, mining, storms and swell waves), it is now widely acknowledged that climate change poses one of the most significant risks to their sustainability and will amplify existing threats (Donner *et al.*, 2005; Hoegh-Guldberg *et al.*, 2007; IPCC, 2007, 2014; Carpenter *et al.*, 2008).

While it is unclear at this stage that climate change has had a permanent negative effect on visitor preferences, it has been clearly demonstrated that the manner in which the market perceives the attractiveness of a destination, including cost of accommodation, may be shaped by climate-related resource degradation (Buzinde *et al.*, 2010; Moore, 2010; IPCC, 2014). In the Caribbean region, Uyarra *et al.* (2005), for instance, have drawn attention to the negative effect that beach

Table 13.1. Revenue from scuba diving for selected Caribbean countries.

Country	Estimated revenue (million US$)	Year	Source
Antigua & Barbuda	0.6–1.0	2002	Ivor Jackson and Associates, 2002
Barbados	0.9–1.1	2013	Peterson *et al.*, 2014
Bonaire	50.0	2012	University of Amsterdam, 2013
Belize	30.0–37.0[a]	2007	Cooper *et al.*, 2008
Grenada	3.1		IABIN-OAS, 2009
St Lucia	7.3	2000	Barker and Roberts, 2002
St Maarten	9.6	2010	Bervoets, 2010

[a]Includes sport fishing revenues.

erosion and coral bleaching would have on visitors' willingness to return to Barbados and Bonaire, respectively. In the case of Barbados, the study suggested that tourism revenues could decline by as much as 46% with a significant reduction in beach area (Uyarra *et al.*, 2005). Similarly, in a study of coastal squeeze on the island of Martinique, Schleupner (2008) concludes that accelerated sea level rise would adversely impact the quality of the coastal amenity as an important tourism attraction. It was further noted by Parsons and Thur (2008) that changes in the 'quality' of scuba diving sites in Bonaire, defined by the variables *visibility*, *species diversity* and *percentage coral cover*, significantly affect diver preference. Diver surveys showed that a decline in the quality of the reefs at the time of the survey to good, medium and poor, represented a mean per person revenue loss of approximately US$45, US$142 and US$192, respectively. The study also suggested that if reef quality were to deteriorate from medium to poor, divers would be inclined to select a different dive destination altogether (Parsons and Thur, 2008). While Bonaire was the focus of this investigation, authorities in Barbados may wish to analyse the implications of these findings for the local dive industry, since both countries draw heavily from largely similar markets.

Research Context

The science is clear: global temperatures will continue to rise for at least another century (IPCC, 2007, 2013). As a consequence, the tourism sector will continue to be confronted with many direct and indirect challenges. How therefore is the industry responding? In 2012, a knowledge, attitudes and perceptions (KAP) survey was conducted with tourism stakeholders along the western coastline of Barbados to understand better their perceptions of climate change, its associated risks and its potential impact on tourism. The targeted businesses were guest houses, beach apartments, hotels, villas, condominiums, restaurants and bars. A total of 158 establishments were identified in the study zone and 82 were surveyed, representing approximately 50% of the population.

Results confirmed that all of the respondents knew the term 'global climate change' and they were aware of its potential impacts on the tourism industry. Despite the fact that 56% of the sample said that they did not know enough or hardly anything about the topic, 78% managed to provide an acceptable definition or explanation of what global climate change is about, either: providing responses corresponding to the IPCC definition of climate change; making mention of some of the acknowledged environmental impacts of climate change such as melting ice caps, sea level rise, beach erosion and floods and droughts; or by highlighting some of the key contributors to climate change, e.g. emissions from industries and factories, the burning of fossil fuels and deforestation.

While 82% of the respondents believed that climate change is a very important factor in the tourism industry in Barbados, 76% stated that the industry was not prepared to handle any extreme climate event and a further 12% were uncertain. This situation is exacerbated by the fact that when asked to indicate measures that had been put in place by the government to deal with a list of known effects of climate change, between 38% and 66% were consistently unable to identify any measures. The only initiative that was known by 62% of the respondents was the Holetown Waterfront Improvement Project, which addresses beach erosion and property protection along the west coast. They had a greater level of awareness of the support offered by the Barbados Hotel and Tourism Association (BHTA) in the form of public awareness programmes. In their opinion, more public awareness and education was required about the potential threats of climate change to the industry, the island and the region. They also felt that greater environmental planning and management, upgraded infrastructure, and disaster planning and management are needed to enhance preparation for the effects of climate change.

The west coast of Barbados is prone to beach erosion and coastal flooding and the respondents confirmed that they had experienced these events repeatedly over the years. Therefore, it was not surprising that 90% of the participants saw beach erosion as a substantial threat to their business and 83% had the same level of concern about coastal flooding. Similar levels of concern were expressed over sea level rise (84.1%), hurricanes (82.9%) and flooding (82.9%). Overall, therefore, more than 80% of the respondents perceive the effects of climate change impacts to be high-level threats to their businesses. Despite this, 72% said that they were only moderately concerned about the threat of global climate change, 24% were very concerned and 2% each were unconcerned or uncertain. The majority, 72% of respondents, also believed that the government was primarily responsible for addressing climate change. However, 62% agreed that the responsibility was also that of businesses and industry and 58% said all citizens should be involved in addressing the threat of climate change.

More than half of the respondents (63%) have begun to insert measures into the planning and management of their establishments to deal with the effects of climate change. Most of the emphasis is being placed on addressing flooding, hurricanes, high winds and water shortages rather than on beach erosion, storm surges and sea level rise. Measures include insurance, purchasing pumps, clearing drains, storing sand bags, planting trees, maintaining hurricane supplies, installing shutters, installing water tanks, structural reinforcement and developing a hurricane plan. Those that have not put any measures in place say that they are hindered by a lack of expertise, financial difficulties, the unavailability of assistance or the lack of technology.

In an attempt to determine how serious respondents were about protecting their businesses, they were asked what percentage of their profit they would be willing to spend to protect their business from the threat of global climate change. More than half (60%) of the respondents failed to provide an exact percentage of the profit they were willing to allocate. Of the 40% who provided an actual percentage, 14% indicated that they would spend less than 5% of their profit, while 4% were willing to spend between 26% and 36% of their profit.

Based on these results it was determined that while the respondents were knowledgeable about global climate change there are gaps in their knowledge. They also have an expectation that the government will take the lead in putting measures in place to deal with the effects of global climate change, even though they are unaware of many of the programmes already in place. In addition, some respondents have already begun to put measures in place to protect their properties but they require technical and financial support.

Climate Change – Managing Solutions

In light of the fact that the tourism industry is relying on the government for support in the drive towards climate change adaptation the study examined the current policies and programmes that have been implemented in the public sector. The White Paper on Tourism Development in Barbados was finalized and approved by Cabinet in 2012. The document acknowledges the importance of sandy beaches, reefs and coastal ecosystems to the success of the industry, as well as the fact that the coastline is an extremely critical asset of the island's economy. As a consequence, the government has pledged to preserve and manage these precious resources from the damage caused by natural disasters and the risks associated with climate change (Strategic Solutions Inc, 2012). To this end there is a commitment to initiating a multi-sectoral response to mitigate the negative impacts of global warming and climate change by:

- Requiring the utilization of sustainable green building materials and other conservation measures for all new developments or renovation projects in the design, construction, operation and maintenance of existing spaces;
- Promoting the use of green technology;
- Enforcement of environmental legislation, mandating environmental impact assessments; and
- Supporting national environmental awareness programmes targeting schools, businesses and the general public to promote sustainable practices and eliminate littering and other undesirable behaviours; and implementing recommendations from the Caribbean Community Climate Change Centre (CCCCC) pilot study on the effect of climate change on tourism (Strategic Solutions Inc, 2012, p. 102).

Further evidence that the government is committed to safeguarding the tourism industry from the threat of climate change is found in the projects and programmes that have already been implemented. For example, the government has borrowed US$30 million from the Inter-American Development Bank (IADB) and is contributing a further US$12 million for a Coastal Risk Assessment and Management Program, which is intended to focus on coastal risk assessment and monitoring; construction and pre-investment studies for beach stabilization works; and institutional strengthening for coastal risk management and climate change adaptation. This is the most recent in a series of projects undertaken with such funding to protect the coastline of the country. Previous projects included a Coastal Infrastructure Programme (2002–2009) – US$24.2 million which focused on shoreline stabilization, beach enhancement and improved public access in three areas of significance to the Barbados tourism industry.

Significant advances have also been made in the energy sector. The Government of Barbados has established an Energy Smart Fund to provide financial and technical support to renewable energy efficiency projects in Barbados. To capitalize the fund, the government obtained a loan of US$10 million from the IADB under an agreement for the Sustainable Energy Investment Programme known as the 'Energy Smart Fund'. The fund consists of a package of financial instruments and technical assistance to address the main market failures that prevent the country from adopting renewable energy and energy-efficient technologies on a large scale. The fund not only provides subsidized loans for the implementation of viable renewable energy and energy-efficient projects, it also provides grants to businesses for funding pre-investment studies and assessing the technical and financial viability of these projects (www.energy.gov.bb/web/energy-smart-fund).

Efforts have also been underway to make improvements in the water management sector. As indicated above, salinity intrusion from sea-level rise will have adverse effects on the groundwater which affects both local residents as well as the tourism sector. To this end the Government of Barbados obtained a US$50 million loan from the IADB to implement a water and sanitation systems upgrade which is intended to improve water resources management in Barbados and sustainable water and wastewater service provision by the Barbados Water Authority (BWA). Specifically, the project will support the Government of Barbados' efforts to modernize the institutional setting of the water and sanitation sector; improve water production infrastructure and the efficiency of the operations of the BWA by reducing unaccounted-for-water and implementing adequate cost-recovery mechanisms (www.iadb.org).

In addition to the White Paper, the Government of Barbados also approved a Tourism Masterplan 2014–2023 in January 2015 (Government of Barbados, 2015). One of the many strategies in the Masterplan requires that resources be invested in the implementation of a climate change awareness programme. In offering a rationale for promoting this strategy, the Masterplan supports the findings of the 2012 study by stating that the adaptive capacity within the tourism sector in Barbados is low and that properties and enterprises are not sufficiently engaged in the implementation of appropriate measures to prepare themselves for the increasing adverse impacts of climate change. Further, it is suggested that while there is some indication that financial constraints play a part in this lack of engagement, the major reason is the lack of technical ability to design and undertake the necessary adaptation initiatives (Government of Barbados, 2015, p. 132).

In this regard it is important to note that there are on-going initiatives to address the technical challenges amongst the practitioners in the industry. One such project is the Caribbean Hotel Energy Efficiency Action (CHENACT) project, which seeks to implement an energy-efficiency model and a clean energy policy for the hotel sector and improve the competitiveness of small and medium-sized hotels under 400 rooms, through the improved use of energy, with the emphasis on renewable energy and micro-generation. During the pilot phase detailed energy audits were conducted in 32 hotels in Barbados and walk-through assessments were completed in 30 hotels. Ozone Depleting Substance audits were also undertaken in Barbados (Loreto Duffy-Mayers, CHENACT Barbados, 2015, personal communication).

It is clear that the Government of Barbados has developed a strong policy and institutional framework to protect the integrity of ecosystems and build resilience. Moreover, millions of dollars in loans have been invested in a diversity of coastal, water resource management and renewable energy programmes that can help the tourism sector adapt to the effects of global climate change. With so much already invested in all of the logical areas – public awareness and education, environmental planning and management, upgraded infrastructure, and disaster planning and management – the question is what can be done differently to better engage the practitioners in the process of climate change adaptation.

Conclusions: Future Expectations

The logical recommendations for increasing the participation of tourism stakeholders in climate change adaptation are typically create and enforce legislation, invest in infrastructure and implement public awareness programmes. All of this is being done in Barbados, so what can be added to the mix? Perhaps more attention is needed to understanding the process of change, i.e. how to change the mind-sets of the individuals and to successfully nurture organizational change in support of climate change adaptation. It may well be time to bring change specialists and marketers into the climate change arena. Even as governments and international agencies invest millions in awareness and education programmes and in climate change adaptation and mitigation measures, the practitioners in the tourism industry still claim to not fully understand climate change, its effects and what is required of them. Those that do comprehend seem to find the financial investment beyond their capacity, even with Smart Energy Funds and other such initiatives in place. Therefore, it is time for change specialists to analyse the situation and design processes geared at moving individuals and organizations to a new state of being. For example, one of the leading theories on change recommends five essential steps:

1. Awareness of the need for change.
2. Desire to support and participate in the change.
3. Knowledge on how to change.
4. Ability to implement the required skills and behaviours.
5. Reinforcement to sustain the change (www.prosci.com).

The evidence presented in the Barbados case supports the fact that time and money has been invested in steps 1, 3 and 4. However, less time has been invested in understanding how to kindle the desire for the change and also in developing programmes to support the change once it has occurred. This is the domain of change specialists. Moreover, there is a need for marketers who can design the appropriate messages to reach the various target audiences. Without this type of input the Government of Barbados will continue to incur millions of dollars of debt in support of climate change adaptation without a corresponding response from the industry.

Notes

[1] Representative Concentration Pathways Scenarios (RCPs).

[2] HadCM3 – Hadley Centre Coupled Model version 3, is a coupled climate model that has been used extensively for climate prediction, detection and attribution, and other climate sensitivity studies.

[3] IPCC Special Report on Emissions Scenarios are used to make projections of possible future climate change.

[4] The 30,000 m^3/day facility was built under a 'build-own-operate' contract arrangement between the Government of Barbados and the private sector.

[5] This is a component of a US$35.6 million 'Water and Sanitation Systems Upgrade Project', jointly funded by the Government of Barbados and the IDB.

[6] See: http://coralreefwatch.noaa.gov/satellite/oa/saturationState_GCR.php (accessed 10 November 2015).

References

Barbados Statistical Service (2013) *2010 Population and Housing Census: Vol. 1*. Barbados Statistical Service, Bridgetown, Barbados.

Barker, N. and Roberts, C. (2002) *Preliminary results from reef valuation study, Saint Lucia, West Indies 2000-2001*. Environment Department, University of York, York, UK, 5 pp.

Bervoets, T. (2010) *Working Paper on the Economic Valuation of Country St. Maarten's Coral Reef Resources*. Nature Foundation, Philipsburg, St Maarten, 25 pp.

Buzinde, C.N.D., Manuel-Navarrete, E.E.Y. and Morais, D. (2010) Tourists' perception in a climate of change: eroding destinations. *Annals of Tourism Research* 37(2), 333–354.

Carpenter, K.E.M., Abrar, G., Aeby, R.B., Aronson, S., Banks, A. *et al.* (2008) One-third of reef-building corals face elevated extinction risk from climate change and local impacts. *Science* 321(5888), 560–563.

Charlery, J. and Nurse, L. (2010) Areal downscaling of global climate model outputs: an approach that avoids data remodeling. *Climate Research* 43, 241–249.

Cooper, E., Burke, L. and Bood, N. (2008) *Coastal Capital: economic contribution of coral reefs and mangroves to Belize*. World Resources Institute, Washington, DC, 53 pp.

Donner, S.D., Knutson, T.R. and Oppenheimer, M. (2005) Model-based assessment of the role of human-induced climate change in the 2005 Caribbean coral bleaching event. *Proceedings of the National Academy of Sciences USA* 104(13), 5483–5488.

Emanuel, K.A. (2013) Downscaling CMIP5 climate models shows increased tropical cyclone activity over the 21st century. *Proceedings of the National Academy of Sciences USA* 110(30), 12219–12224.

Fabricius, K.E., Langdon, C., Uthicke, S., Humphrey, C., Noonan, S. *et al.* (2011) Losers and winners in coral reefs acclimatized to elevated carbon dioxide concentrations. *Nature Climate Change* 1(3), 165–169.

Government of Barbados (2015) *Final Report Barbados Tourism Master Plan 2014–2023 Report II: The Environment* 145.

Hoegh-Guldberg, O., Mumby, P.J., Hooten, A.J., Steneck, R.S., Greenfield, P. *et al.* (2007) Coral reefs under rapid climate change and ocean acidification. *Science* 318(5857), 1737–1742.

Hoetjes, P., Kong, A.L., Juman, R., Miller, A., Miller, M. *et al.* (2002) Status of coral reefs in the eastern Caribbean: the OECS, Trinidad and Tobago, Barbados and the Netherlands Antilles. In: Wilkinson, C. (ed.) *Status of Coral Reefs of the World: 2002*. Global Coral Reef Monitoring Network, Australian Institute of Marine Science, Queensland, pp. 325–342.

IABIN-OAS (Inter-American Biodiversity Information Network) (2009) Economic Valuation of Goods and Services Derived from Coral Reefs: Results from the South Coast of Grenada. PowerPoint presentation made to ReefFix Exercise Workshop, Grenada, 2009.

Intergovernmental Panel on Climate Change (2007) *Climate Change 200: Impacts, Adaptation and Vulnerability: Contribution of Working Group II to the Fourth Assessment Report of the IPCC*. Cambridge University Press, Cambridge, p. 976.

Intergovernmental Panel on Climate Change (2013) *The Physical Science Basis. Contribution of Working Group I to the Fifth Assessment Report of the Intergovernmental Panel on Climate Change*. Cambridge University Press, Cambridge, p. 29.

Intergovernmental Panel on Climate Change (2014) *Impacts, Adaptation, and Vulnerability. Contribution of Working Group II to the Fifth Assessment Report of the Intergovernmental Panel on Climate Change* (Barros, V.R., Field, C.B., Dokken, D.J., Mastrandrea, M.D., Mach, K.J. *et al.* (eds)). Cambridge University Press, Cambridge, pp. 1613–1654.

Ivor Jackson and Associates (2002) *Yachting Study Antgua/Barbuda*. Prepared for the Economic Commission for Latin America and the Caribbean (ECLAC) Sub-Regional Office, Trinidad and Tobago, 99 pp.

Knutson, T.R., McBride, J.L., Chan, J., Emanuel, K., Holland, G. *et al.* (2010) Tropical cyclones and climate change. *Nature Geoscience* 3(3), 157–163.

Lin, N., Emanuel, K., Oppenheimer, M. and Vanmarcke, E. (2012) Physically based assessment of hurricane surge threat under climate change. *Nature Climate Change* 2(6), 462–467.

Manzello, D.P., Brandt, M., Smith, T.B., Lirman, D., Hendee, J.C. and Nemeth, R.S. (2007) Hurricanes benefit bleached corals. *Proceedings of the National Academy of Sciences USA* 104(29), 12035–12039.

Moore, W.R. (2010) The impact of climate change on Caribbean tourism demand. *Current Issues in Tourism* 13(5), 495–505.

Nurse, L.A. and Charlery, J.L. (2014) Projected SST trends across the Caribbean sea based on PRECIS downscaling of ECHAM4, under the SRES A2 and B2 scenarios. *Theoretical and Applied Climatology*, DOI: 10.1007/s00704-014-1346-1.

Oxenford, H., Roach, R., Brathwaite, A., Nurse, L., Goodridge, R. *et al.* (2008) Quantitative observations of a major coral bleaching event in Barbados, Southeastern Caribbean. *Climatic Change* 87, 435–449.

Pandolfi, J.M., Connolly, S.R., Marshall, D.J. and Cohen, A.L. (2011) Projecting coral reef futures under global warming and ocean acidification. *Science* 333(6041), 418–422.

Parsons, G.R. and Thur, S.M. (2008) Valuing changes in the quality of coral reef ecosystems: a stated preference study of SCUBA diving in the Bonaire national marine park. *Environmental and Resource Economics* 40(4), 593–608.

Peterson, A., Turner, R., Gill, D., Forster, J., Fitzsimmons, C. *et al.* (2014) *Future of Reefs in a Changing Environment: An Ecosystem Approach to Managing Caribbean Coral Reefs in the Face of Climate Change*. CERMES Technical Report No. 69, University of the West Indies, Barbados, 44 pp.

Peterson, T.C., Taylor, M.A., Demeritte, R., Duncombe, D.L., Burton, S. *et al.* (2002) Recent changes in climate extremes in the Caribbean region. *Journal of Geophysical Research: Atmospheres (1984–2012)* 107(D21), ACL-16.

Pulwarty, R., Nurse, L. and Trotz, U. (2010) Caribbean islands in a changing climate. *Environment: Science and Policy for Sustainable Development* 52(6), 15–27.

Schleupner, C. (2008) Evaluation of coastal squeeze and its consequences for the Caribbean Island Martinique. *Ocean & Coastal Management* 51(5), 383–390.

Strategic Solutions Inc (2012) *White Paper on the Development of Tourism in Barbados*. Ministry of Tourism, Barbados, 250 pp.

UNEP (2005) *Caribbean Environment Outlook*. United Nations, New York, p. 75.

University of Amsterdam (2013) *The Total Economic Value of Nature on Bonaire*. IVM Institute for Environmental Studies, Amsterdam, 69 pp.

Uyarra, M.C., Cote, I.M., Gill, J.A., Tinch, R.R.T., Viner, D. and Watkinson, A.R. (2005) Island-specific preferences of tourists for environmental features: implications of climate change for tourism-dependent states. *Environmental Conservation* 32(01), 11–19.

World Travel and Tourism Council (2015) *Travel and Tourism Economic Impact 2015 Barbados*. WTTC, London, p. 24.

Worrell, D., Belgrave, A., Grosvenor, T. and Lescotta, A. (2011) An analysis of the tourism sector in Barbados. *Economic Review* XXXVII(1), 49–75.

14 Case Study Mexico: Riviera Maya – How is the Riviera Maya Tourism Industry Dealing with Climate Change? An Overview of Non-climatic Stressors that Determine the Destination's Vulnerability to Climate Change

R. Santos-Lacueva,* S. Anton Clavé and Ò. Saladié
Department of Geography, Rovira i Virgili University, Spain.

Introduction: The Riviera Maya

The Riviera Maya is a well-known Caribbean coastal destination in the state of Quintana Roo (Mexico), as shown in the location map (Fig. 14.1). It has a wealth of natural and cultural resources. The Mesoamerican Barrier Reef (the second largest in the word), the Sian Ka'an Biosphere Reserve, the Tulum National Park and, in general, the white-sand beaches, underground rivers, mangroves and rainforests all form a rich collection of biodiversity. Its Mayan heritage, which can be observed at various archaeological sites (Fig. 14.2) and experienced as part of the living culture, also makes the Riviera Maya especially attractive.

Quintana Roo was one of the most underdeveloped and marginalized territories in Mexico until the late 1960s (Torres and Momsen, 2005) when Cancun was selected to be the site of the first planned tourism development in Mexico (Ambrosie, 2015). Riviera Maya, previously known as the Cancun-Tulum corridor, emerged in the middle of the 1980s as a response to the massive development of Cancun. It was conceived to provide nature and exclusivity, and benefit from the pulling power of Cancun, which by then had become well established. It started out as an integrated destination in 1997, with the creation of the Fideicomiso de Promoción Turística de la Riviera Maya (Riviera Maya Destination Marketing Office). Within a few years it became one of the most famous coastal destinations and a major source of income for the state. In 2014 it was host to more than 4 million tourists (43.4% of the total for Quintana Roo); it had 394 hotels (42.3% of Quintana Roo) and a total of 42,011 rooms (47.6% of Quintana Roo). According to the report *Barómetro turístico de la Riviera Maya* (unpublished) by the Riviera Maya Destination Marketing Office in 2015, the average rate of occupancy was 81.6%.

Riviera Maya consists of two municipalities – Solidaridad and Tulum – both of which are made up of several localities. Playa del Carmen is the capital city of Solidaridad, and Tulum village is the capital of Tulum. A total of 65.7% of all the hotels and 84.9% of all the rooms in Riviera Maya are in Solidaridad. Most of the hotels are in

*Corresponding author, e-mail: raquel.santos@urv.cat

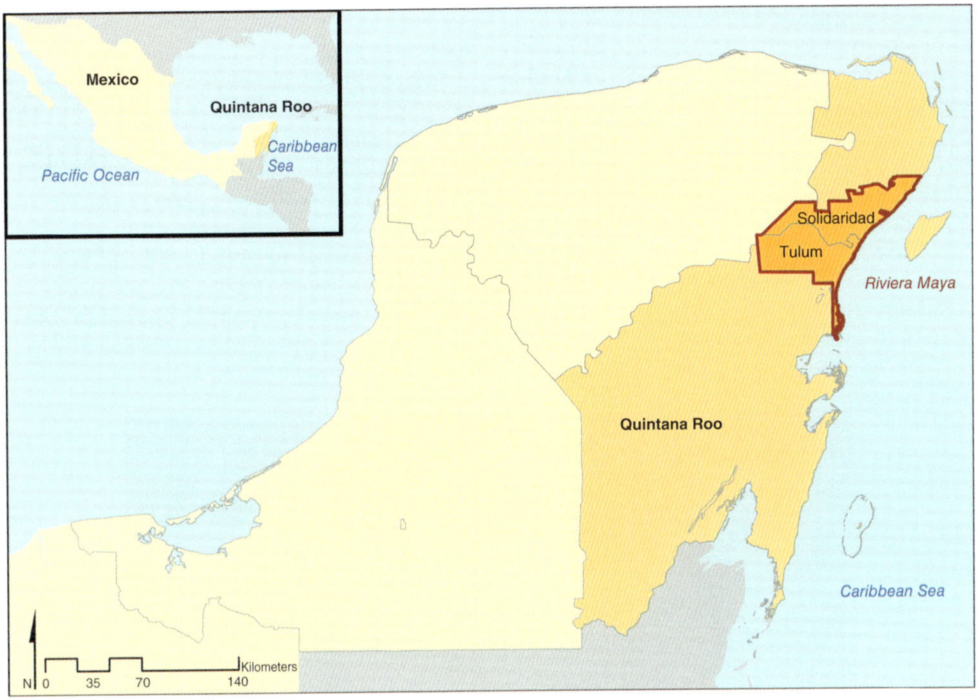

Fig. 14.1. Location map of Riviera Maya.

Fig. 14.2. Tulum archaeological site. Photo by Raquel Santos-Lacueva.

Playa del Carmen (43.1%) even though they only represent 18.2% of the rooms. In Riviera Maya, 80% of the rooms (24% of the hotels) are in the five-star or special category. A total of 73 hotels (75.3% of the rooms) offer all-inclusive plans. Most rooms (83%) are in the larger hotels (more than 100 rooms) (*Barómetro turístico de la Riviera Maya*).

The valuable natural resources of Riviera Maya are highly vulnerable to climate change. Although Riviera Maya has a tropical climate with nearly 250 days a year of sun, its geographical position exposes it to extreme meteorological events during the rainy season. Additionally, its competitiveness also depends on the quality of its natural resources – the coral reef, the underwater biodiversity, the pristine white beaches, the jungle and the mangroves – all of which are potentially affected by global warming. Quintana Roo also has a special hydrological system made up of underground rivers and cenotes (Fig. 14.3), which, being an important tourist attraction, complicates water and waste management.

Fig. 14.3. Tourist swimming in cenote. Photo by Raquel Santos-Lacueva.

Otherwise, one of the most distinctive features of Quintana Roo's society is its duality. In fact, rapid economic and urban growth have reinforced inequality (Torres and Momsen, 2005). Throughout the state the rural settlements of the native Mayan inhabitants predominate and contrast with the few urban tourist poles that have developed along the coast. The former are marginalized and lack basic services and infrastructures, while the latter concentrate foreign and national immigrants, investment and modern infrastructures and services. Duality is also a feature in cities, where the tourist areas are clearly differentiated from the rest of the urban areas. These no-tourist areas lack even the most basic of services and are undergoing rapid growth.

Despite these situations, Riviera Maya has been recognized as a committed destination that promotes sustainability in tourism planning and it has been recognized by the Global Sustainable Tourism Council in 2014 as one of the fourteen Early Adopter destinations in the world.

Research Context

Research on vulnerability to climate change has evolved from a focus on the biophysical impact of mitigation policies and hazards (Thywissen, 2006; Füssel, 2010; Hinkel, 2011) to an analysis of adaptation strategies from economic policies and resilience scope (Eakin and Luers, 2006; Füssel and Klein, 2006). In order to be evaluated, there have been increases in the number of local studies, more analyses of the implication of stakeholders and a greater focus from the social sciences on the significance of non-climatic factors (Füssel, 2007; Moreno and Becken, 2009; O'Brien and Wolf, 2010; Huebner, 2012).

The present research has reviewed policy reports and undertaken 16 interviews with stakeholders to study the extent to which stakeholders perceive policies and strategies on climate change to be factors that determine the ability of a destination to adapt to and mitigate it from a contextual perspective (Füssel, 2010; O'Brien and Wolf, 2010). The case of Riviera Maya is relevant because perceptions, decisions and actions are heavily influenced by socio-economic and political factors, and because of the physical characteristics that increase the destination's exposure to natural disasters and climate change-related hazards.

The interviews were held in person between November 2014 and January 2015. Participants were selected for their relevance, due to their role at the destination, their position and affiliation, and also their policy area, administrative level, professional profile and gender (Table 14.1). The interviews combined open and closed questions in four blocks of information: climate change and tourism; obstacles to successful adaptation and mitigation; role of stakeholders in adaptation and mitigation; and coordination of decision making between the tourism and environment authorities.

Climate Change: Assessing Problems and Challenges: From Scientific Reports to Stakeholders' Perceptions

Climate change in Riviera Maya: scientific reports

The Government of Mexico is aware that climate change has had an impact on coastal destinations (SEMARNAT and PNUMA, 2006; INE, 2008). SEMARNAT *et al.* (2011) and SEMARNAT (2012) point out that Mexico is particularly threatened by extreme meteorological events such as hurricanes, cyclones (IMTA and SEMARNAT, 2009), floods and heatwaves. Natural disasters caused 154 deaths per year on average between 1999 and 2011 and this century the cost of damage is expected to increase from 3.2% to 6% of GDP. This is because extreme events are more intensive and more frequent and because human settlements and economic activities are more exposed (SEMARNAT, 2012). Between 1970 and 2009 hurricanes increased in both frequency and intensity, especially in the Gulf of Mexico and on the Caribbean coast (SEMARNAT, 2012).

Table 14.1. Relevance of participants.

Position	Affiliation
Director of Tourism Planning and Development	Secretariat of Tourism, Quintana Roo State
Director of Climate Change and Environmental Management	Secretariat of Ecology and Environment, Quintana Roo State
Head of Environmental Risk Department	Impact and Environmental Risk Institute, Quintana Roo State
General Director of Tourism	Municipality of Solidaridad
General Director of Environmental and Urban Planning	Municipality of Solidaridad
General Director of Economic Development and Attracting Investments	Municipality of Solidaridad
Director of ZOFEMAT (Federal Maritime Land Area)	Municipality of Solidaridad
Director of Tourism	Municipality of Tulum
General Director of Urban Development and Ecology	Municipality of Tulum
General Director	Riviera Maya Destination Marketing Office
Executive Director	Riviera Maya Hotels Partnership
General Manager of Sustainable Hospitality Initiative of MARTI (Meso American Reef Tourism Initiative)	Riviera Maya Hotels Partnership
Director	Amigos de Sian Ka'an
Deputy director of the Climate Change Programme	Amigos de Sian Ka'an
Founder and Director	Sustainable Riviera Maya
Professor of Sustainable Tourism	Caribe University

Riviera Maya is located in the area with the greatest risk of hurricanes and storm surges (Fig. 14.4) (Government of Quintana Roo, 2013). Since tourism became a flourishing industry there, four hurricanes have impacted: Gilbert (1988), Roxanne (1995), Emily (2005) and Wilma (2005). The last one was the most severe and caused damage valued at more than 18 million pesos, mostly due to losses in tourism (IMTA and SEMARNAT, 2009). Additionally, in Riviera Maya a temperature increase of 1–1.1°C is expected by 2020, 1.9–2°C by 2050 and 2.9–3.3°C by 2080 (SECTUR, 2013). SECTUR (2013) also argues that precipitation in Riviera Maya will decrease and envisage that sea level will rise by 3.1 cm/year and states that coastline erosion between 2004 and 2012 was 1.22m/year. This report qualifies physical vulnerability as being very high.

The potential effects of these changes are (SECTUR, 2013): floods; beach erosion; damage to urban areas and populations, infrastructure (housing, communications, and energy and water supplies) and the environment; economic losses; fires; health problems (increased heart rate and an increase in dengue, cholera and paludism); a scarcity of potable water and waste accumulation due to extreme weather; deterioration of cenotes and proliferation of invasive species, such as Sargasso algae (Fig. 14.5).

Climate change in Riviera Maya: stakeholders' perceptions

Stakeholders influence decision and non-decision making and determine policies, strategies and plans to deal with climate change. So, what they think about this issue is important because their perceptions of risk might not be the same as the perceptions of scientists. The stakeholders interviewed consider that climate change influences tourism in Riviera Maya 4.13 out of 5. The worst effects they identify are with the rise in sea level, the extreme meteorological events and coral bleaching (Table 14.2). Interestingly, most of them think that climate change events are already having a negative effect on tourism in Riviera Maya (Table 14.3).

As can be seen in Table 14.4, stakeholders point out that the main difficulty of responding to the problems identified is the scarcity of economic resources, the lack of awareness, and the poor coordination between different public authorities, and

Fig. 14.4. Tourism infrastructure affected by storm surges. Photo by Raquel Santos-Lacueva.

Fig. 14.5. Beach affected by Sargasso alga. Photo by Raquel Santos-Lacueva.

between the authorities and the private sector. There are also significant differences in answers according to the profile – lobbyist or politician – and the policy area – environment or tourism – of the respondents. Non-governmental participants are more critical of the scarcity of economic resources than politicians and the difference in their opinions about limitations of human resources is even greater. The opinions of lobbyists and politicians about awareness are also quite opposite. The problems of poor coordination between institutions and organizations are also regarded as greater by those who are not politicians. As far as the policy area is concerned, those participants from the environmental sector are most negative about economic resources, private sector support, public sector capacity and coordination between the public authorities and the private sector. Finally, the tourism sector considers that the greatest problems are those of awareness.

Table 14.2. Answers to the question 'What are the worst effects of climate change in Riviera Maya? (Select three)'

Effect of climate change	Mentions (%)
Sea level rise	29.2
Extreme meteorological events (heatwaves, storms, hurricanes, etc.)	27.1
Coral bleaching	20.8
Rise in temperature	12.5
Migration of species	6.2
Acidification of the ocean	4.2
Increase in organisms, illness, insects, etc.	0.0
Drought	0.0
Others	0.0

Climate Change: Managing Solutions: Policy Context, Key Actors, Roles and Coordination

Policy context

The first steps to establishing a climate-change policy took place in Mexico with the signature of the United Nations Convention in 1992. In 2005 the federal government established the Inter-Secretarial Commission for Climate Change, the aim of which was to construct the policy framework for adaptation. This institution developed the National Strategy for Climate Change (ISCCC, 2007) and the Special Programme on Climate Change for the period 2009–2012 and 2014–2018 (ISCCC, 2009, 2014). The National Development Plan 2007–2012 (Government of Mexico, 2007) included climate change for

Table 14.3. Perception of the risk of climate change impacts in Riviera Maya.

	Mentions (%)		
Effect of climate change	It does not affect tourism and it will not affect it	It is already affecting tourism	It does not affect tourism yet but it will affect it in (years according to average answers)
Sea level rise	6.2	87.6	6.2 (15 years)
Extreme meteorological events (heatwaves, storms, hurricanes, etc.)	6.4	81.2	12.4 (7 years)
Coral bleaching	0.0	62.5	37.5 (10 years)
Increase in temperature	18.7	62.6	18.7 (15 years)
Migration of species	18.7	56.3	25.0 (13 years)
Acidification of the ocean	18.7	50.0	31.3 (14.5 years)
Increase in organisms, illness, insects, etc.	6.2	50.0	43.8 (9 years)
Drought	68.7	18.7	12.6 (6 years)
Others	62.5	31.3	6.2 (10 years)

Table 14.4. Affirmative answers about problems according the participant's profile and policy area.

Problem	Affirmative answer (%)				
	Profile		Policy area		
	Lobby	Politician	Tourism	Environment	Total
Economic resources are limited	85.7	66.7	66.7	85.7	75
There is not enough awareness	100.0	44.4	77.8	57.1	68.7
There is a lack of coordination between decision makers and the private sector	85.7	44.4	55.5	71.4	62.5
Techniques and knowledge about the issue are limited	57.1	55.6	55.6	57.1	56.2
Human resources are limited	85.7	22.2	55.5	42.8	50.0
Data and information are insufficient	85.7	22.2	55.5	42.8	50.0
Future implications are uncertain	42.9	44.4	42.9	44.4	43.7
There is a lack of coordination between public authorities	85.7	11.1	42.8	44.4	43.7
Public sector capacity is limited	57.1	33.3	22.2	71.4	43.7
Private sector support is not enough	42.8	11.1	11.1	42.8	25.0

the first time, and it is still a feature of the current National Development Plan 2013–2018 (Government of Mexico, 2013). The Climate Change Advisory Council developed the Policy Framework for Medium Term Adaptation (Climate Change Advisory Council, 2010) with a time-horizon up to 2030. Moreover, other federal programmes include issues related to climate change (e.g. the Sectorial Environment and Natural Resources Program 2007–2012, 2013–2018) (SEMARNAT, 2007, 2013).

In 2012 Mexico approved the General Law of Climate Change (2012), which grants states and municipalities the competency to take adaptation and mitigation actions. In 2010 Quintana Roo created the State Commission on Climate Change and drew up the State Program of Action on Climate Change (2013), under the State Development Plan 2011–2016 (Government of Quintana Roo, 2011), which makes climate change one of the priority projects in the Quintana Roo's Axis Green. The state also drafted instruments such as the REDD+ (Reduction of Emissions of Deforestation and Degradation) Technical Advisory Council of Quintana Roo (2012), the Law of Climate Change Action (2012), the Law of Ecological Equilibrium and Environmental Protection (2001), the Quintana Roo Inventory of Greenhouse Gas Emissions (2013) and the Quintana Roo Centre for Climate Change Education (2011). Other state programmes reinforce this policy: for example, the State Urban Development Program, the State Program for Territorial Management, the State Territorial Zoning Program and the Environmental Preservation and Natural Resources Sector Program. Moreover, Quintana Roo has entered into innovative regional agreements for the country, such as the Declaration of Campeche, by which Yucatan and Quintana Roo agree to combat climate change together (2011).

The General Law of Climate Change expects municipalities to draw up a Municipal Action Plan for Climate Change. Although Solidaridad and Tulum have not drawn up their plans yet, they have other instruments that help to cope with the effects of climate change even though they were not designed with this purpose (for example, municipal regulations and plans about civil defence, urbanism and construction).

Key actors and roles

The policies implemented by the administration aside, the stakeholders interviewed for this research consider that the most important organizations fighting against climate change in terms of adaptation and mitigation are: Amigos de Sian Ka'an, an NGO focused on environment protection throughout Quintana Roo; the Riviera Maya Hotels Partnership (AHRM), a non-governmental organization (NGO) that represents the hotel sector in Riviera Maya; and the Secretariat of Ecology and Environment (SEMA), a Quintana Roo governmental organization. This gives some idea of the relative importance of NGOs, the state administration and environmental policy for the stakeholders involved.

Therefore, in Riviera Maya NGOs play an important role in promoting significant strategies, which then are sometimes taken up by the public authorities. One example is the contribution of Amigos de Sian Ka'an to the State Action Plan for Climate Change (SAPCC) in Quintana Roo (2013), coordinated by the SEMA. Quintana Roo University was also involved. Amigos de Sian Ka'an developed the strategies for adaptation and mitigation concerning the tourism sector. In fact Quintana Roo was the first Mexican state to incorporate a section specifically on tourism. Amigos de Sian Ka'an and MARTI, in alliance with the Secretariat of Tourism of Quintana Roo, also launched the Guide for Planning, Design and Sustainable Construction in the Mexican Caribbean (Amigos de Sian Ka'an *et al.*, 2012). This led to the Mexican law that established that new tourism urbanizations had to be sustainable and respect requirements such as design, building, management and cessation of activity.

The AHRM also plays a key role. It is in charge of the Sustainable Management Programme for Accommodation of MARTI initiative. The programme started in 2007 and aims to give advice on and encourage good environmental practices in hotels, such as saving energy or improving waste management. AHRM is one of the organizations with the most experience in the destination. Ever since the beginning of Riviera Maya as a tourism project promoted by the Government of Quintana Roo, it has proved the strength of the private sector as a lobby.

A final example is Riviera Maya Sostenible, an NGO that seeks to get companies, government, society, other NGOs and academia to work together. It is closely connected with the Global Sustainable Tourism Council. In fact, in December 2014 both organizations arranged a workshop on the global criteria of sustainable tourism which resulted in the creation of a council to promote the sustainability of Riviera Maya. This council has become a strong lobby that represents more than 50 entities and they meet once a month.

As well as SEMA, which coordinated the SAPCC and gathers the Direction of Climate Change of the state, there are other important government actors. Among them, the Riviera Maya Destination Marketing Office plays an important role in fostering the image of Riviera Maya as a destination committed to sustainability. Two federal institutions – the Secretariat of the Environment and Natural Resources and the Secretariat of Tourism – are also involved. The latter has carried out recent studies on the vulnerability to climate change of ten strategic destinations in Mexico including Riviera Maya (SECTUR, 2013). Nevertheless, stakeholders interviewed do not recognize it as a key actor in dealing with climate change.

Coordination

The stakeholders regard the coordination between organizations as the main barrier to dealing effectively with climate change in Riviera Maya. In fact, more than 60% state that the coordination between tourism and environmental areas does not work properly. We also asked them to say to what extent stakeholders from the environmental area take into account tourism issues, and vice versa. Even though differences are slight, environmental authorities take more notice of tourism (3.4 out of 5) than tourism authorities of the environment (2.7 out of 5). On the other hand, most tourism stakeholders think that they take the environment into account (3.1 out of 5) and most environmental stakeholders think that they take tourism into account (2.9 out of 5).

Conclusions: Future Expectations

Mexico is highly exposed to natural disasters, which are expected to become more intense and frequent due to climate change.

This has led to climate change being added to the political agenda and a federal normative and policy framework being developed. This includes the General Law of Climate Change, which is encouraging the involvement of states and municipalities. Nowadays, the Quintana Roo state has already assumed its role but local limitations prevent the Municipal Action Plans from being put into practice. In Riviera Maya, the main obstacles are socio-economic and political (for example, the lack of awareness, the extreme economic dependence on tourism, the scarcity of economic resources and the weakness of the local public administration).

Our research reveals some key points that policy makers need to consider if they are to reinforce adaptation and mitigation strategies in Riviera Maya.

1. Some NGOs are sufficiently important to lead actions not only because of the role they play at the destination, but also because of their long expertise and knowledge in the field.
2. The lack of awareness in the tourism and local political spheres must be corrected if coordination is to be improved between the tourism and environmental sectors and the local institutions are to get more involved.
3. Stakeholders consider that the most urgent impacts that need to be dealt with are the rise in sea level rise, extreme meteorological events and reef bleaching.

Finally, this research reveals that it is of the utmost importance to take heed not only of scientific reports on physical features and climate projections but also stakeholders' perceptions if the power dynamics, priorities and difficulties destinations have in coping with climate change are to be fully understood.

References

Ambrosie, L.M. (2015) Myths of tourism institutionalization and Cancún. *Annals of Tourism Research* 54, 65–83.

Amigos de Sian Ka'an, MARTI and Secretariat of Tourism of Quintana Roo (2012) Guide for Planning, Design and Sustainable Construction in the Mexican Caribbean[in Spanish]. Available at: http://sedetur.qroo.gob.mx/marti/index.php (accessed June 2017).

Climate Change Advisory Council (2010) *Policy Framework for Medium Term Adaptation*. Climate Change Advisory Council and Secretariat of Environment and Natural Resources, Mexico, 55 pp [in Spanish]. Available at: http://gia.imta.mx/geoportal/docs/Marco_Politicas_Adaptacion_Mediano_Plazo.pdf (accessed June 2017).

Eakin, H. and Luers, A.L. (2006) Assessing the vulnerability of social-environmental systems. *Annual Review of Environment and Resources* 31(1), 365–394.

Füssel, H.-M. (2007) Vulnerability: a generally applicable conceptual framework for climate change research. *Global Environmental Change* 17(2), 155–167.

Füssel, H.-M. (2010) Review and quantitative analysis of indices of climate change exposure, adaptive capacity, sensitivity, and impacts. In: *Word Development Report 2010*. World Bank, Washington.

Füssel, H.-M. and Klein, R.J.T. (2006) Climate change vulnerability assessments: an evolution of conceptual thinking. *Climatic Change* 75(3), 301–329.

Government of Mexico (2007) *National Development Plan 2007–2012*. Government of Mexico, Mexico DF, 323 pp [in Spanish].

Government of Mexico (2013) *National Development Plan 2013–2018*. Government of Mexico, Mexico DF, 184 pp (In Spanish). Available at: www.itesa.edu.mx/pdf/PND.pdf (accessed June 2017).

Government of Quintana Roo (2011) *State Development Plan 2011–2016*. Government of Quintana Roo, 104 pp [In Spanish]. Available at: www.cidge.gob.mx/wp-content/uploads/2013/05/QUINTANA-ROO2.pdf (accessed June 2017).

Government of Quintana Roo (2013) *State Program of Action on Climate Change*. Government of Quintana Roo, INECC-AECID and University of Quintana Roo, Chetumal, 116 pp [in Spanish]. Available at: www.ccpy.gob.mx/pdf/agenda-qroo/PEACC_2013_ESPAnOL.pdf (accessed June 2017).

Hinkel, J. (2011) Indicators of vulnerability and adaptive capacity: towards a clarification of the science–policy interface. *Global Environmental Change* 21(1), 198–208.

Huebner, A. (2012) Public perceptions of destination vulnerability to climate change and implications for long-haul travel decisions to small island states. *Journal of Sustainable Tourism* 20(7), 939–951.

IMTA and SEMARNAT (2009) *Análisis de la frecuencia e intensidad de los ciclones tropicales para prevenir los efectos, presentes y futuros, debidos a la variabilidad y al cambio climático en México*. Mexican Institute of Water Technology (IMTA) and Secretariat of Environment and Natural Resources (SEMARNAT), Mexico City, 96 pp. Available at: www.inecc.gob.mx/descargas/cclimatico/2009_estudio_cc_va4 (accessed June 2017).

INE (2008) *Evaluación de la afectación de la calidad del agua en cuerpos de agua superficiales y subterráneos por efecto de la variabilidad y el cambio climático y su impacto en la biodiversidad, agricultura, salud, turismo e industria*. National Institute of Ecology (INE), Mexican Institute of Water Technology (IMTA) and Secretariat of Environment and Natural Resources (SEMARNAT), Mexico DF, 108 pp. Available at: www.inecc.gob.mx/descargas/cclimatico/ev_calidad_agua_cc.pdf (accessed June 2017).

ISCCC (2007) *National Strategy for Climate Change*. Government of Mexico and Secretariat of Environment and Natural Resources, Mexico DF, 160 pp [in Spanish]. Available at: www.inecc.gob.mx/descargas/cclimatico/2012_est_nal_cc.pdf (accessed June 2017).

ISCCC (2009) *Special Program on Climate Change 2009–2012*. Government of Mexico and Inter-Secretarial Commission for Climate Change, Mexico DF, 118 pp [in Spanish]. Available at: www.semarnat.gob.mx/archivosanteriores/programas/Documents/PECC_DOF.pdf (accessed June 2017).

ISCCC (2014) *Special Program on Climate Change 2014–2018*. Government of Mexico, Mexico DF, 151 pp [in Spanish]. Available at: www.semarnat.gob.mx/sites/default/files/documentos/transparencia/programa_especial_de_cambio_climatico_2014-2018.pdf (accessed June 2014).

Moreno, A. and Becken, S. (2009) A climate change vulnerability assessment methodology for coastal tourism. *Journal of Sustainable Tourism* 17(4), 473–488.

O'Brien, K.L. and Wolf, J. (2010) A values-based approach to vulnerability and adaptation to climate change. *Wiley Interdisciplinary Reviews: Climate Change* 1(April), 232–242.

SECTUR (2013) *Estudio de la vulnerabilidad y programa de adaptación ante la variabilidad climática y el cambio climático en diez destinos turísticos estratégicos, así como propuesta de un sistema de alerta temprana a eventos hidrometeorológicos extremos*. Secretariat of Tourism, Mexico DF. Available at: www.sectur.gob.mx/programas/planeacion-y-politica-turistica/ordenamiento-turistico-sustentable/cambio-climatico/estudio-de-vulnerabilidad-al-cambio-climatico-en-el-sector-turistico (accessed June 2017).

SEMARNAT (2007) *Programa Sectorial de Medioambiente y Recursos Naturales 2007–2012*. Government of Mexico, Mexico DF, 138 pp. Available at: www.semarnat.gob.mx/archivosanteriores/Documents/PROMARNAT%202013-2018.pdf (accessed June 2017).

SEMARNAT (2012) *Adaptación al cambio climático en México: visión, elementos y criterios para la toma de decisiones*. Secretariat of Environment and Natural Resources (SEMARNAT), Mexico DF, 186 pp. Available at: http://biblioteca.semarnat.gob.mx/janium/Documentos/Ciga/libros2009/CD001364.pdf (accessed June 2017).

SEMARNAT (2013) *Programa Sectorial de Medioambiente y Recursos Naturales 2013–2018*. Government of Mexico, Mexico DF, 138 pp. Available at: www.semarnat.gob.mx/archivosanteriores/Documents/PROMARNAT%202013-2018.pdf (accessed June 2017).

SEMARNAT and PNUMA (2006) *El cambio climático en América Latina y el Caribe*. Secretariat of Environment and Natural Resources (SEMARNAT) and United Nations Environmental Program (UNEP), 140 pp.

SEMARNAT, INE and CCTA (2011) *Diagnóstico de las Tendencias actuales de fenómenos meteorológicos extremos y proyección de su actividad al clima futuro cercano 2030 y clima futuro lejano 2080. Considerando los efectos del cambio climático global y otros efectos locales*. Secretariat of Environment and Natural Resources (SEMARNAT) and National Institute of Ecology (INE), 85 pp. Available at: www.inecc.gob.mx/descargas/cclimatico/2011_creaas_inf_final.pdf (accessed June 2017).

Thywissen, K. (2006) *Components of Risk. A Comparative Glossary*. United Nations University Institute for Environment and Human Security, Bonn.

Torres, R. and Momsen, J. (2005) Planned tourism development in Quintana Roo, Mexico: engine for regional development or prescription for inequitable growth? *Current Issues in Tourism* 8(4), 259–285.

15 Case Study Dubai: A Theme Park Approach to Climate Change

Angela Anthonisz[1]* and Tim Heap[2]
[1]*University of Northampton, UK;* [2]*The University of Derby, Buxton, UK*

Introduction

Despite its semi-desert climate, Dubai is one of the fastest-growing tourist destinations in the world, with visitor numbers increasing from 374,000 in 1982 (Henderson, 2006) to 14.9 million in 2015 (Gulf News, 2017) and projections to reach 20 million plus by 2020 (DTCM, 2017). With an economy previously reliant on its oil resources the obvious limitations in terms of a range of attractions, lack of infrastructure and the natural resources associated with creating visitor demand appear to have been overcome via extensive inward investment designed to propel tourism forward as a pillar of the economy (Henderson, 2006). The government's obvious commitment to tourism growth is evidenced by the rapidly increasing infrastructure, the growth in the hosting of major sports events such as the Dubai Classic and the Dubai World Cup and as hosts of the EXPO2020. While tourism is used by governments to boost the economy, improve image and regenerate coastal cities it would seem unusual for a destination such as Dubai to adopt a strategy such as this given the existing problems at the destination in terms of the deteriorating environment in the region (Madichie, 2013), the concerns raised by the local population on tourism's negative impacts on the local culture (Odeh, 2010) and the real threats associated with climate change and rises in sea levels. At the micro-level Dubai's long-term sustainability can be called into question when it comes to its accelerated growth in infrastructure which appears to contradict the sustainable agenda of the EXPO and to a certain extent implies a lack of sustainable thinking. At the macro-level Dubai, as a coastal city that sits only 16 m above sea level, also needs to concern itself with the environmental challenges associated with global warming and rising sea levels. The UAE has nearly 1300 km of coastline. Approximately 85% of the population and over 90% of the infrastructure of the UAE is located within several metres of sea level in low-lying coastal areas (ERAS, 2005).

This chapter considers the nature of the accelerated development of Dubai and discusses the possible approaches being considered as Dubai changes the shape of its coastline in a bid to accommodate new attractions to maintain its position as an international tourist destination. Rather than seeing the future as a series of threats Dubai is embracing the necessary changes

*Corresponding author, e-mail: angela.anthonisz@northampton.ac.uk

and incorporating them in future plans for the 'Theme Park Destination' that it has become. The introduction of the term 'Theme Park Tourism Destination' is one that commits the destination to a brand strategy of continuous change in response to the market as the major proviso for that change, but also a destination that anticipates, and in Dubai's case influences or leads, those market changes. From the desert snow palace of 'Frozen' to the largest swimming pool in the world via diverting the sea into the desert, this hyper-real world of Dubai is the biggest theme park in the world.

Dubai's Cultural Environmental Agenda and the Myth of Sustainability

Dubai aims to increase visitor numbers to 20 million by 2020 and the sub-theme of the EXPO was meant to drive forward a number of initiatives associated with environmental management. While recent awareness of environmental issues has escalated since 2013, the initial push for sustainable practices was launched by Dubai's Department of Tourism and Commerce Marketing (DTCM) in 2009 when they introduced the Dubai Green Tourism Award. In 2015 the Dubai Supreme Council for Energy announced its Demand Side Management strategy that is designed to support green implementation methods in building regulations, reuse of water and use of outdoor light (Navdar, 2015). The strategy has set ambitious goals for Dubai with aims to reduce water consumption by 15%, energy usage by 20% and CO_2 emissions by 20% by 2030. It is difficult to understand this in the context of the doubling of both the number of bed spaces in Dubai by 2020 and size of the airport to accommodate the 20 million visitors (90% of whom will arrive by air). This is a direct challenge to Dearden and Manopawitr (2011), who suggested that 'In the future it is likely that air travel will rise considerably in cost as governments seek ways to reduce carbon footprints. Taxes on air fares will probably increase and increased scarcity of fossil fuels will lead to further increases. The overall result is likely to be less international travel, especially long haul international travel' (p. 151).

According to Ruschmann (2001), environmental sustainability must include a planning process that aims to preserve and allow for rational use of the natural resources as they relate to tourism activities at the destination. Compare this with the deterioration of the environment in the region which is a major challenge in terms of sustainability with one of the millennium development goals (MDG) for the region focused on how this may be achieved (Madichie, 2011). While some cities such as Abu Dhabi have made some headway in this respect with the creation of Masdar ecological city and the first 'positive energy building', the city of Dubai with its focus on leisure tourism appears to have concentrated efforts on constructing attractions such as Ski Dubai, an indoor sports centre where the indoor air temperature has to be maintained at 1°C and 30–40 tonnes of fresh snow is produced each night, and new hotels and infrastructure designed to cater to the anticipated 20 million visitors in the year of the EXPO. As the number-one family visitor attraction in the city, Ski Dubai currently maintains its appeal by theming the attraction around one of the latest Disney offerings, the movie *Frozen*.

The conquering of the desert has become a reflection of the modern in that new techniques and management can roll back the sands (Ryan and Stewart, n.d.). The Islamic culture in this context is important as it conceives of Paradise as a green garden with running waters and fountains. Thus the environmental sustainability merges with the cultural imperatives and is linked through the visitor or tourist. Green is also the colour traditionally associated with the prophet. Just as, within a Zayedist ideology, the presence of oil is perceived as a gift from God, so too the ability to transform the desert into a green pasture is an expression of God's creative will (Ouis, 2002). Ouis also identified the extensive greening of the city as consisting of planting 20 million date palms and bringing 200,000 ha of land into cultivation, a pattern which had appeared to be continuing given World Economic Forum figures on forest cover change (WEF, 2013/14). The greening of the desert, while artificially constructed, could be considered as a means of showing progress in the same way that creating the world's busiest airport, designing the tallest building in the world (Burj Khalifa) and winning the bid to host EXPO2020 were sending global messages in terms of Dubai being a modern, thriving destination. The key issue with regard to these developments should perhaps be whether the impact on the environment is sustainable. As Ouis (2011) points out, 'how is it possible that the mega-projects in the UAE are presented as environmentally friendly in the public discourse?' (p. 1409). The answer to this may lie within the observation made by Gaffney (2013) in that legacy is often a reflection of the host's understanding of sustainability as a concept.

It is very difficult to gain data on the environmental impacts in Dubai especially when those impacts need to be jointly quantified, such as from transport to air pollution. The link is proven and the increase in traffic in Dubai is visible, but the official figures prove elusive. This is a critical issue considering that tourism policies, promotion and their inclusion in the elaboration of development plans, are often supported with very little information about the nature of the benefits and costs deriving from such policies (Jones and Munday, 2004). One simple example of environmental degradation according to the UNWTO (2010) is the link between air pollution and tourism that is exampled as being especially relevant in regions that have witnessed exponential growth over a short time period and are predicted to have continued growth. In Dubai's case this is compounded by a climate that produces high year-round temperatures and is geographically located within one of the most powerful high-pressure elements of any Hadley cells in the world; the likelihood is that global warming will reinforce the longevity of the cell rather than disperse it. This is then further exacerbated by the lack of wind to help disperse the pollutants. The latest figures for air quality in Dubai are for 2009 and rank the city as 77th in the world for CO_2 intensity and 104th in the world for the concentration of particulate matter (WEF, 2013/14). The rate of increase in key pollutant indicators was modelled by Saenz de Miera and Rosselló (2014) in Mallorca, where the results showed that a 1% increase in tourist numbers can be related to up to a 0.45% increase in PM10 levels. This does not appear to have been factored into any strategic plans for Dubai in their quest for a 100% increase in tourist numbers on existing levels by 2020. Added to this is the consequent increase in pollutants related to the building of the extended tourist provision.

It could be assumed that key inhibitors for growth may therefore come to the fore in the coming years, closely linked to the issues of sustainability where water depletion, sea level rises and air quality are significant causes for concern. Tourism is often pointed to as a strategic sector in economic and social development, but the premise of increased tourist numbers is rarely linked to the generation of factors such as the increases in the proportion of air pollution related to environmental degradation and consequent negative effects on the economic balance sheet. In 2008 the Abu Dhabi Environment Agency commissioned the Stockholm Environment Institute to conduct a preliminary assessment of the impact of climate change on the UAE. Focusing on the coastal cities of the Emirates, the report identified several scenarios linked to varying levels of sea rise from 1 m up to 3 m by 2100. In the worst case scenario of a 3 m rise by 2100, Dubai's elaborate infrastructure would probably be the shortest lived in human history with almost the entire city submerged (Table 15.1). To ignore these scientific projections is illogical, but the race for growth by 2020 means that Dubai has no choice but to ignore them, or adapt to them, which is possible within the short life cycles of the 'theme park rides'. They can easily be replaced as the history of development in Dubai has shown over the past 30 years.

This is probably an extension of Gaffney's (2013) 'host understanding of sustainability' to 'host choosing not to understand', leading to an 'illusion of sustainability' as Dubai in itself is illusory in its constructed, commodified, hyper-real 'fantasy world' of

Table 15.1. Dubai – Zones of inundation in line with sea level rise.

Land use	1 m rise	3 m rise	9 m rise
Barren lands	2	3	4
Built-up urban	7	133	200
Others (education, hospital, cemetery, etc.)	1	3	5
Transportation (roads, port, airport)	1	3	4
Grass	0.4	1	1
Plantation/orchard	1	1	2
Woody vegetation (*Prosopis*)	0.3	1	2
Shrub (*Leptadenia*)	0.3	1	1
Coastal/saline	0.04	0.1	0.2
Mangrove	0.02	0.03	0.03
Water bodies	0.2	0.3	0.3
Wetlands (tidal flat, mudflat, creek)	1	1	1
Total	14	147	221

snow in the desert, and it is this that sells and sustains the destination. The visitor is happy to walk into this illusory world of the constructed environments and accept it as being green and environmentally sustainable as they will view nothing that suggests otherwise. This is Mandelartz's (2012) 'Tourist Bubble', which he explains is not solely a physical place but is extended into the social and cultural environment of each individual. He further explains that visitors are captured in a bubble of their social and cultural norms, which in this case is the hyper-reality of snow in the desert. Perhaps this then leads to a 'Host Bubble' for the Emirati, which, as outlined by Giddens (1990), provides 'ontological security' as the confidence or trust we have in the surroundings in terms of the physical environment and the social environment consisting of subjects and people that share our lives, provide stability and a continuity to our identity.

In this case the built environment, the product that is Dubai, changes the society, but the illusion is a 'continuity of identity' and the denial of new 'social and cultural environments'. Interviews with Emiratis suggest they bemoan the changes to the environment and culture, but they then support the acceleration of that change. It is the fact that the host bubble and the tourist bubble overlap and that as the tourism product is successful that gives justification to both parties. However, in Fay's (1996) cultural bubble there is an implication that risk has to be included in this exposure to change. Membership of groups and identification whether ethical, racial, gender or religious often have only survived by their cultural and social individuality and they want to remain distinct as a way of resisting the domination of more powerful groups. These subcultures do not want to be acculturated into the changed world, so the Emirati's denial of change on the altar of economic success may be being absorbed into that acculturated world.

The problem is when this illusion meets reality in the form of environmental disaster as the Gulf becomes super-saturated with the untreated sewage of 20 million visitors (and 4 million residents) as there is not enough water to treat that sewage! Dubai is a Middle East South Sea Bubble business built upon illusion, linked to sustaining investor confidence and subject to a suggested environmental Armageddon, and this does not suggest economic sustainability. The indication is that rather than follow the logic of these arguments we suggest that Dubai has to ignore them (economically and environmentally) as it builds a future destination that accepts sea level rises as a way of increasing capacity and re-enforces the theme park concept of a world of illusions. This confirms Jones (2011), who comments that 'whereas there is growing awareness about current climate change threats and the need to act, there is a continued strong drive to sustain and develop tourism infrastructure and facilities'. The financial importance of tourism is paramount in an economy that, alongside everything else, plans to build a new education hub and wellness and health tourism city to ensure tourist capacity continues to build sustainability for the hotel sector in terms of bed occupancy levels required for the return on investment by the international hospitality industry. Something that is entirely at odds with Jenkins (2011, p. 69), who says that 'disaster threats may lead to the emergence of new tourism products and a more responsible and "greener" industry'.

Dubai's Economic Agenda

In contrast to what appears to be a strategy of redefining or ignoring the environmental agenda there is, in contrast, an imperative to follow the economic agenda associated with developing and sustaining Dubai as a premier tourist destination, a strategy that involves extending the city via land reclamation outwards from the coast and expansion inland to make way for new theme parks and larger attractions such as the new snow dome which will dwarf the existing Ski Dubai facility, which is currently Dubai's number one visitor attraction. This is classical seaside resort zoning, but the new build cannot be segregated in these terms as the coastal areas are considered the domain of the 5- to 7-star provision and initially limited geographically in terms of development, but Dubai's answer is to bring the sea into the desert as opposed to keeping it out and to build in the capacity to absorb 2–3 m sea level rises within the development. The answer is simply to rebuild the existing coastline in the guise of promenades and replenish the beaches whilst you are extending the concept inland (Figs 15.1, 15.2).

Jenkins (2011) comments that 'undoubtedly economics are at the heart of the climate

Fig. 15.1. Beach nourishment project on Dubai promenade.

Fig. 15.2. Dubai creek extension (www.skyscrapercity.com/showthread.php?t=822856&page=11 (accessed 1 March 2016)).

change and tourism debate', but this was used within the context of environmental sustainability. In Dubai's case it is the major driver and indicator of success and the other pillars of sustainability are left to adapt. The reasoning behind this imperative does appear to follow Jenkin's 'amelioration of 'existing political risk models'. In Dubai's case the politics are entirely linked to the development strategy and model of economic sustainability as it is the Emirati that entirely control and drive the real-estate development necessary for growth. This political expediency therefore ignores Jenkins' view that 'climate-tourism models for destinations' are necessary for continued tourism investment (p. 69).

As a tourist destination Dubai's economy has benefited from a number of 'core resources and attractors' (Brent-Ritchie and Crouch, 2003) that have made a significant contribution to the growth of the city. For Dubai, as a city in a desert landscape, the coastline and winter sun created an initial draw for sun, sea and sand with much of the development extending along the coastline and around the Creek area in the 1980s and 1990s. The success was sustained by the brand identity of a 5-star resort combined with 5-star attractions and 5-star prices.

This, over time, led to a more complex set of products that today is 'brand Dubai or theme park Dubai' and it is this that fuels the continual demand for more, bigger, better and newer products within the brand that seem to appear organically alongside the existing 5-star portfolio. This demand-led brand then feeds or demands growth whether environmentally, socially or economically sustainable or not and perhaps investors realize that the long-term returns are dependent upon that growth. There can be no steady state within Dubai.

The success of the tourism industry in Dubai was, therefore, facilitated by strategic location, a reputation for safety and extensive marketing campaigns which boosted the image of Dubai in terms of consumerism and luxury. The current and projected growth will increase hotel provision, and create a more balanced product portfolio, but will attract new markets in the more accessible budget category and displace the existing luxury demographic that make up the majority of tourists at the destination. The new build 4-star may exceed some of the existing 5-star provision in terms of product/price market, which in turn would require refurbishment of the older build with inherent costs. The economic arguments are complex, but the 'theme park tourist' requires continuous change and this has been noted by the hoteliers in Dubai. The industry has identified that the tourist demographic (they term it 'generation') changes every 3 years (a long way from generations x, y and z), meaning that refurbishment, continuous renewal and product development is essential in recognizing those changes. After 10 years a hotel may lose out on two changes of consumer preferences; the response by the resort is to pre-empt this by offering the changes on a continuous basis. For Dubai, landmark hotels such as the Atlantis and the Burj al Arab have traditionally met with consumer demand, but as the destination develops these older products now compete with a number of international and locally grown lifestyle hotel offerings that have been added into the hotel portfolio, including Versace, Vida, W and Manzil. The roller coaster not only stops to allow people to get off and new ones to get on, but it also must find time to change the nature of the ride at the same time.

These prosperity and lifestyle choices in developed countries have historically contributed to global warming in terms of greenhouse gas emissions (World Bank, 2010). The urbanization of Dubai, driven by demand for economic growth and from the consumer, means that emissions will also increase dramatically, but there appears to be no alternative possible for the destination as it is locked in to this growth through investment portfolios promising exponential returns and a global market demand for such destinations.

The key to overall sustainability in the literature often appears linked to the management of the destination in terms of the quality of life for the local population. The fact that the Emirati form only 10–15% of the overall population of Dubai, and that they are protected against those negative factors, may play a big part in the lack of environmental management of the destination. The counter to the environmental problems in Dubai's terms are that the environmental quality which assumes negative impact on the attractiveness of a destination for the tourist, but that the continual rejuvenation of the destination hides the negative factors outlined by much of the literature. It is the 'hall of smoke and mirrors' that suggests alternatives for every shape and size of individual. The role of the environment as the main factor of competitiveness has already been called into question by Porter (2000), who noted that it is very difficult to separate the effects of tourism from other processes in the area, and by Lim and Cooper (2009), who criticize the concept of sustainable tourism being almost exclusively linked to the environment.

Ruschmann (2001) emphasizes that sustainability is a more complex issue and needs to be thought of along four dimensions, including economic, environmental, socio-cultural and political-institutional when planning for sustainability. But Dubai seems more in line with Aronsson (2000), who argues that the concept of sustainability is nothing more than an illusion and that the best-case scenario is only to make tourism more sustainable than it was before, or in Dubai's case an illusion of more sustainability. This situation could be confirmed by studies such as Cater (1995) and UNEP (2011) that show that advances in sustainability are becoming a drain on a destination's core economic progress, making it difficult, if not impossible, to implement measures that will achieve greater levels of sustainability. Ultimately, destination success in terms of sustainability must be linked to competitiveness as, identified by Brent-Ritchie and Crouch (2010), otherwise the destination becomes less attractive, tourist numbers decline and economic viability is questioned. Brent-Ritchie and Crouch also include the proviso that government strategies implemented to ensure the quality of the environment needs to meet with the tourists' expectations and improve the quality of life for its residents. In Dubai's case the illusions are complete for both parties even though the theme park may be built on sand.

The UAE government's participation in international treaties and conferences such as UN Climate Change conferences, and the initiation of energy forums such as Dubai Global Energy Forum where the climate change subject was introduced, combined with the intention to have a diversification of energy sources 'oil, solar and nuclear' are positive signs of the good intention towards a sustainable future (Al Rustamani, 2014).

Conclusion

As the site for the EXPO2020, Dubai is investing heavily in an illusory world that contains a future tourism infrastructure in an already overdeveloped city to ensure that visitor spend increasingly acts as a key pillar of economic growth. This vision with a suggestion of economic sustainability with positive effects on employment, levels of disposable income and quality of life, seems far removed from what are the suggested outcomes if we consider the science of global warming and its potential impact upon the destination. Dubai is being built upon a fragile environmental base; a society that is constructed within a 'wolkenkuckucksheim' and an economy that presently provides US$27 billion per annum income, but that requires continuous growth and investment streams to maintain and then double that visitor income. Only by a growth strategy can the returns on investment be satisfied. Dubai historically has borrowed its way out of debt by increasing its debt/income ratio through cultural and family ties, but is this really sustainable in the post 'Arab spring' world and the reaction to the 'Arab spring' including the rise of a re-invented extreme Islamism? This is further questioned in terms of the economic debate surrounding the fluctuations in the oil price, which though forecast to rise in the longer term, is producing uncertainty in the investment markets. In simple terms this predominantly economic view of sustainability does not acknowledge the complexities of what makes the destination attractive, and ultimately competitive in the long term as the future seems oriented around creating a giant theme park rather than 'greening' the desert. Dubai, for example, has the financial ability to develop wholesale measures that could focus on the use of solar power, on creating reduced carbon properties such as the EXPO site, but it does not appear to be concentrating on these. The cost of keeping clean the photoelectric cells in a destination with high pollution levels and windborne sand from the desert perhaps nullifies the potential of this provision, so the lack of progress is probably expedient and not deliberate.

On the other hand, in this chapter we introduce a Dubai that seems to be reflecting another aspect of the climate change debate than the one prevalent in current academic debate, by embracing change in terms of adapting physically to the potential implications of that change. It appears confident in its ability to effect continuous change into the future simply because it has a legacy and experience of accelerated change and nothing so far has suggested it needs to question that confidence. To build a new canal bringing the sea into Dubai within an 18-month timeframe again confirms its ability to respond quickly. It hides the illusion of 'business as usual' within the 'theme park' of the destination offer and the tourist continues to buy into the brand as it satisfies the perceptions from within the safety of their tourism bubble and the developers from within the safety of their 'destination bubble'. The sustainable illusion of Dubai is, therefore, an oxymoron as in its unsustainability it becomes sustainable. It uses change as the meter for sustainability, which may be possible as long as it can provide the return on investment.

References

Al Rustamani, Z.A. (2014) Impacts of climate change on urban development in the UAE: The case of Dubai. Thesis, Paper 29. Available at: http://scholarworks.uaeu.ac.ae/cgi/viewcontent.cgi?article=1029&context=all_theses (accessed 4 June 2017).

Aronsson, L. (2000) *The Development of Sustainable Tourism*. Continuum, London.

Brent-Ritchie, J.R. and Crouch, G.I. (2003) *The Competitive Destination: A Sustainable Tourism Perspective*. CAB International, Wallingford, UK.

Brent-Ritchie, J.R. and Crouch, G.I. (2010) A model of destination competitiveness/sustainability: Brazilian perspectives. *Brazilian Public Administration Review (RAP) Special Issue Rio De Janeiro* 44(5) 1049–1066. Available at: www.scielo.br/pdf/rap/v44n5/v44n5a03.pdf (accessed June 2017).

Cater, E. (1995) Environmental contradictions in sustainable tourism. *The Geographical Journal* 161(1), 21–28. Available at: ftp://ftp.puce.edu.ec/Facultades/CienciasHumanas/Ecoturismo/ArticulosTurismo/Art%C3%ADculos%20cient%C3%ADficos/Turismo%20sostenible/contradicciones_ambientales_turismo_sostenible.pdf (accessed 1 March 2016).

Dearden, P. and Manopawitr, P. (2011) Climate change – coral reefs and dive tourism in South West Asia. In: Jones, A. and Phillips, M. (eds) *Disappearing Destinations: Climate Change and Future Challenges for Coastal Tourism*. CAB International, Wallingford, UK.

Department of Tourism and Commerce Marketing (DTCM) (2017) Tourism Vision 2020. Available at www.visitdubai.com/en/department-of-tourism/about-dtcm/tourism-vision-2020 (accessed 4 June 2017).

ERAS (2005) cited in *Climate Change. Impacts, vulnerability and adaptation*. Environmental Agency, Abu Dhabi (2009). Available at www.ead.ae/Documents/RESEARCHERS/Climate%20change%20impacts%20-%20Eng.pdf (accessed June 2017).

Fay, B. (1996) Contemporary philosophy of social science: a multicultural approach. In: Mandelartz, P. (ed.) Tourist Behavior during and after Times of Crisis. Insights into Attitudes, Perceptions and Response Mechanisms. PhD Thesis, University of Derby, UK.

Gaffney, C. (2013) Between discourse and reality: the un-sustainability of mega-event planning. *Sustainability* 5, 3926–3940.

Giddens, A. (1990) The consequences of modernity. In: Novelli, M. (ed.) *Niche Tourism, Contemporary Issues, Trends and Cases*. Butterworth Heinemann, Wallingford, UK.

Gulf News (2017) 14.9 million overnight visitors for Dubai in 2016. Department of Tourism statement says emirate is 'firmly on course' to hit 20 million visitors by 2020. *Gulf News*, 6 June 2017. Available at: http://gulfnews.com/business/sectors/tourism/14-9-million-overnight-visitors-for-dubai-in-2016-1.1974824 (accessed 4 June 2017).

Henderson, J.C. (2006) Tourism in Dubai: overcoming barriers to destination development. *International Journal of Tourism* Research 8, 87–99.

Jenkins, I. (2011) Climate change – risk management issues and challenges. In: Jones, A. and Phillips, M. (eds) *Disappearing Destinations: Climate Change and Future Challenges for Coastal Tourism*. CAB International, Wallingford, UK.

Jones, A. (2011) UK coastal tourism destinations – assessment of perceived climate impacts: issues for destination management, local governance and public policy making. In: Jones, A. and Phillips, M. (eds) *Disappearing Destinations: Climate Change and Future Challenges for Coastal Tourism*. CAB International, Wallingford, UK.

Jones, C. and Munday, M. (2004) Evaluating the economic benefits from tourism spending through input-output frameworks: issues and cases. *Local Economy* 19(2), 117–133.

Lim, C. and Cooper, C. (2009) Beyond sustainability: optimising island tourism development. *International Journal of Tourism Research* 11(1), 89–103.

Madichie, N.O. (2011) IRENA – Masdar City (UAE) – exemplars of innovation into emerging markets. *Foresight: The Journal of Future Studies, Strategic Thinking and Policy* 15(4), 278–293.

Madichie, N.O. (2013) Is the Middle East the land of the future? It is not a given! *Foresight* 15(4), 321–333.

Mandelartz, P. (2012) Tourist behavior during and after times of crisis. insights into attitudes, perceptions and response mechanisms. PhD Thesis, University of Derby, UK.

Navdar, P. (2015) Dubai Municipality's Plan for a Sustainable City. Available at: www.hoteliermiddleeast.com/24041-dubai-municipalitys-plans-for-a-sustainable-city/ (accessed 1 March 2016).

Odeh, M. (2010) The wicked problem of tourism: economic benefits versus socio-cultural impacts. *WIT Transactions on Ecology and the Local Environment* 139, 345–357. Available at: http://library.witpress.com/pages/PaperInfo.asp?PaperID=21234 (accessed 14 July 2014).

Ouis, P. (2002) Greening the Emirates: the modern construction of nature in the United Arab Emirates. Sage Publications. Available at: http://journals.sagepub.com/doi/abs/10.1191/1474474002eu252oa (accessed 14 July 2014).

Ouis, P. (2011) Engineering the Emirates: The Evolution of a New Environment. Available at: http://dspace.mah.se/dspace/bitstream/handle/2043/13973/EngEmOuis.pdf?sequence=2 (accessed 14 July 2014).

Porter, M. (2000) Location, competition and economic development: local clusters in a global economy. *Economic Development Quarterly* 14, 15–34.

Ruschmann, D.V.M. (2001) *Turismo e planejamento sustentável: a proteção do meio ambiente* In: Da Cunha, S.K. and da Cunha, J.C. (eds). (2005) Tourism cluster competitiveness and sustainability: proposal for a systemic model to measure the impact of tourism on a local development. *Brazilian Administration Review* 2(2), 47–62.

Ryan, C. and Stewart, M. (n.d.) The Paradoxical Social Construction of Eco-Tourism – The Case of the Luxury Desert Lodge at Al Maha, Dubai. Available at: http://pc.parnu.ee/~htooman/EuroChrie/Welcome%20to%20EuroCHRIE%20Dubai%202008/papers/The%20Paradoxical%20Social%20Construction%20of%20Ecotourism.pdf (accessed 14 July 2014).

Saenz de Miera, O. and Rosselló, J. (2014) Modelling Tourism impacts on air pollution: The case study of PM10 in Mallorca. *Tourism Management* 40, 273–281.

The National (2015) Dubai targets 9% annual growth in tourist numbers as calculation method altered. *The National*, 4 May 2015. Available at: www.thenational.ae/business/travel-tourism/dubai-targets-9-annual-growth-in-tourist-numbers-as-calculation-method-altered (accessed 1 March 2016).

UNEP Yearbook (2011) Emerging Issues in Our Global Environment. Available at: http://wedocs.unep.org/handle/20.500.11822/8276 (accessed 4 June 2017).

UNTWO Annual Report (2010) A Year of Recovery. Available at: http://media.unwto.org/sites/all/files/pdf/finalannualreportpdf.pdf (accessed 20 July 2014).

World Bank (2010) World Development Report 2010: Development and Climate Change. Available at: https://openknowledge.worldbank.org/handle/10986/4387 (accessed 4 June 2017).

World Economic Forum (WEF) (2013/14) The Global Competitiveness Report: Full Data Edition. Available at: www3.weforum.org/docs/WEF_GlobalCompetitivenessReport_2013-14.pdf (accessed 4 June 2017).

16 Case Study Vietnam: Climate Change Impacts on UNESCO World Heritage – The Case of Hoi An Ancient Town

Huong T. Bui[1]* and Tuan-Anh Le[2]
[1]*Ritsumeikan Asia Pacific University (APU), Japan;* [2]*Southern Cross University, Australia*

Introduction

With a coastline of over 3000 km, Vietnam is likely to be one of the countries most affected by climate change. Many tourist destinations, cities and urban centres in Vietnam are located along the coastal zones and rivers. Owing to increases in temperature, sea levels are projected to rise from 57 to 73 mm by 2100 and without appropriate adaptation actions, this increase could result in the inundation of 39% of the coastal land area in the Mekong River Delta in the south, 10% of the Red River Delta area in the north and 2.5% of the area in the Central Coastal Region, causing severe impacts on infrastructure, the population and economic activities including tourism (UN Habitat, 2014).

Hoi An Ancient Town is located in Vietnam's central coastal region of Quang Nam Province, approximately 30 km from Da Nang City and 812 km from Hanoi to the north and 845 km from Ho Chi Minh City to the south. As a UNESCO World Heritage Site (WHS), the ancient city of Hoi An is the most important tourist destination of Quang Nam Province. This WHS is not only an attractive blend of cultures but is also a favourable seaside resort destination in the Cua Dai and An Bang area (Fig. 16.1). In the 5 years between 2007 and 2011, revenue from tourism increased 149.53% from US$39,695,776 to US$59,356,769 in 2011. Between 2004 and 2011, the economic growth rate was 11.5% per year (UN Habitat, 2014). A tourism-based economy has created opportunities to develop new livelihoods and has created jobs and income for 5000 direct labourers and 12,500 indirect labourers. Altogether, income from tourism and related trade and services contributes 64% to the GDP of the city (Hoi An Center for Monuments Management and Conservation, 2008).

The city is located between 0 and 15 m above sea level, on the north bank near the mouth of the Thu Bon River flowing to the sea at Cua Dai, about 7 km from the city centre. Both protected and buffer zones are characterized by low-lying elevations, with unstable topographic foundations that are prone to erosion. Hoi An consequently has a large and growing tourist infrastructure that urgently needs support from improved urban environmental management and protection from the impacts of erosion, flooding and salinity intrusion, which are likely to be exacerbated by climate change (Asian Development Bank, 2014).

*Corresponding author, e-mail: huongbui@apu.ac.jp

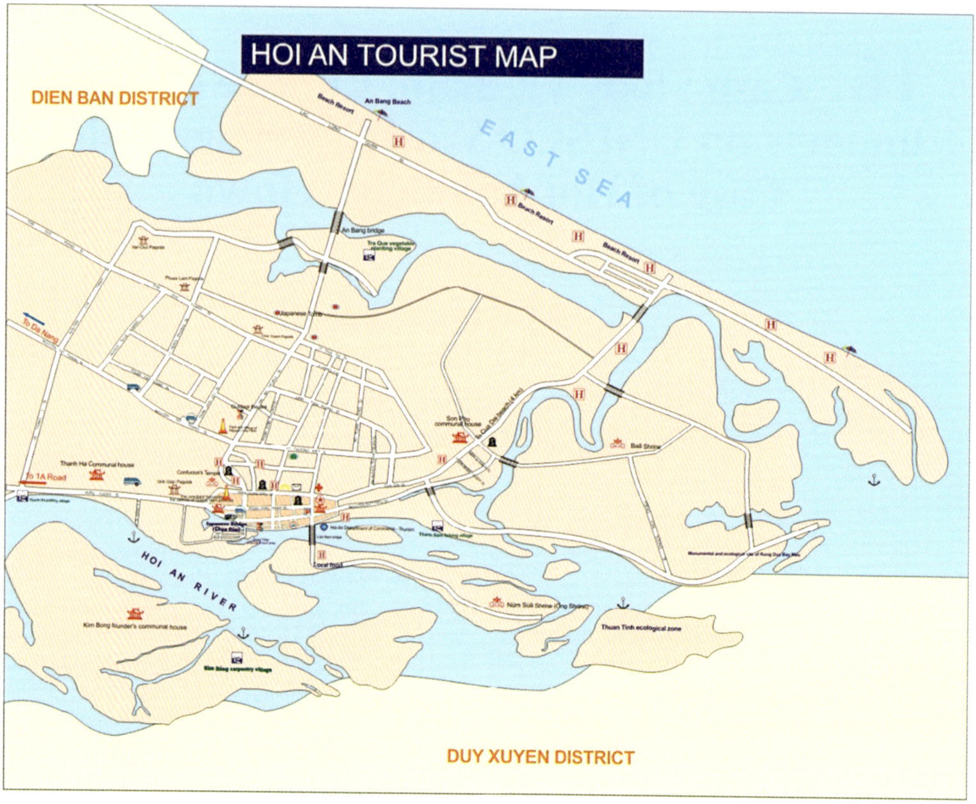

Fig. 16.1. Hoi An tourist map (courtesy of Hoi An Department of Trade and Tourism).

The establishment of the National Target Program to Respond to Climate Change in December 2008 responded to the call for an action plan at national level. The programme has determined the consequences of climate change and has established national priorities including coordination between national, provincial and community-level authorities for effectively combating climate change. According to the programme, the People's Committees of provinces, cities and districts are responsible for preparing and implementing climate change response action plans. However, the extent to which the action plans are effective in combating climate change remains in question. The current case study therefore analyses impacts of climate change on the tourism industry in Hoi An before discussing the effectiveness in response to the change taken by various sectors.

Research Context

Climate change affects tourist destinations in four ways: direct impacts from the changing climate systems, indirect impacts on cultural heritage and the environment, indirect impacts associated with societal change, and impacts induced by climate change mitigation and adaptation in other sectors (Scott *et al.*, 2012). This case study identifies and examines three types of impacts: (i) direct impacts of climate change on the coastal region of Cua Dai; (ii) indirect impacts on the cultural heritage site of Hoi An Ancient City; and (iii) induced impacts on the tourism industry resulting from climate change mitigation. Research data were derived from various reports of national and international organizations assessing climate change vulnerability, the tourism industry, and newspaper articles reporting current aspects of the city under threat by climate change. In addition to drawing on secondary data sources, the authors interviewed visitors, government officers and business owners in the summer of 2015. Interview data are used to illustrate the impacts of climate change on Hoi An. The authors highlight direct impacts of the rising sea level, which causes coastal erosion, indirect impacts from increased severity of floods and typhoons affecting the heritage assets of the ancient city, and induced impacts of action mitigating the effects of climate change on the tourism-based economy of Hoi An.

Climate Change: Assessing Problems and Challenges

Sea-level rise is one of the most salient impacts of climate change. The rising sea has caused erosion and inundation, has impeded drainage and has increased the risk of river flooding and coastal habitat loss, as well as resulting in higher water tables, which can adversely affect the stability of foundations of infrastructure (Scott *et al.*, 2012). During the last 50 years, Vietnam's sea level along the coastline has risen by approximately 20 cm (Asian Development Bank, 2014). Along the coastline of the Cua Dai area, between 10 and 20 m of land on the shoreline are eroded every year, resulting in coastal erosion of 150 m in 8 years. Several major causes of erosion have been identified: (i) loss of sand owing to sand mining on the Vu Gia and Thu Bon Rivers for civil construction; (ii) loss of sand owing to construction of a series of reservoirs; (iii) construction of resorts and protection structures on the beach; (iv) the natural evolution of the river mouth and its flood-tide delta and ebb-tide delta; and (v) dredging of a navigation channel on the estuary area of the Vu Gia–Thu Bon river system (Asian Development Bank, 2014).

Coastal erosion of the Cua Dai area is the top concern of the public and private sectors as well as of tourists. The public sector is concerned about severity of the erosion at a pace that has never been witnessed before, as noted by the head of the city's natural resources and environment office:

> It is the worst situation of its kind that I have ever seen. The beach is dramatically disappearing. It stretched out 200 m just 10 years ago, but now it has narrowed to 40 m. The rest is underwater now.... The erosion has even approached a main section of the 1.5-km road connecting the beach with Hoi An. We can often see waves crashing heavily against it (interview, *Vietnam News*, 2014).

The private sector has invested heavily in beach resort development in the area, and is expressing concern regarding the proportion of resort-front beaches that would be lost to inundation and accelerated erosion, as beaches would simply disappear. A staff member at a beach resort who has witnessed the coastal squeeze over the years commented:

> The resort was launched in 2000 with a beach that stretched out for 40 m, but it is much smaller now, no wider than 10 m during the summer. During summer, the view is mostly just waves crashing against the resort's dike (interview, *Vietnam News*, 2014).

International tourists to the region have recognized the impacts of climate change on the coastal line of the city. Tourists perceived beach-erosion control measures as aesthetically unpleasant, but seemed to understand the measures were necessary to protect the beach and were part of on-going work to restore it through beach nourishment:

> I think the erosion was caused by climate change. I heard about tropical storms that dragged sands away from the shore. Personally, I think the scenery remains beautiful even though the long sandbag and cement dike near the pier visually polluted the natural landscape. My friends recommend An Bang beach. But it is crowded there, not as quite as this place (Cua Dai) (interview conducted by authors, June 2015).

Erosion has already damaged housing in the coastal area. As a notable example, the entire village of Phuoc Trach was relocated because of erosion affecting nearly 200 households. If no action is taken, the expectation is that within the next 3 to 5 years around 55 houses in the area will be seriously affected by erosion (UN Habitat, 2014).

In addition, the rising sea level has intensified the severity of floods, particularly as large areas of the city around the Thu Bon and De Vong Rivers are 2 m below sea level and therefore especially sensitive to flooding effects. According to the Hoi An People's Committee, flooding in 2007 caused damage to roads estimated at VND2842 billion (approximately US$136 million). Built heritage may be damaged as a result of climate change-related flooding, sea-level rise and associated coastal flooding and erosion, along with increased exposure to extreme events. These conditions can be particular threatening to heritage structures such as buildings, which are already in a fragile state (McIntyre-Tamwoy, 2008). The ancient buildings were designed for a specific climate, and are less isolated from the ground than newer buildings that utilize geomembranes and other technologies to limit exposure to water and soil processes. Changes in water levels or salinity may accelerate deterioration of foundations of ancient structures. In addition, changes in wet and dry cycles can induce crystallization and dissolution of salts and affect other heritage assets (UNESCO, 2007).

The World Heritage site of Hoi An comprises many ancient wooden houses, which are subject to regular floods that have been worsened by increasing storm surges in the region. Hoi An has two distinct seasons: a dry season, which runs from January to July, and a rainy season that extends from August until December. On average, between three and five major floods occur every year, with an average inundation level of 3 m. The three most severe floods occurred in 1964, when the water level reached 3.4 m above sea level, in 1998 (2.99 m) and in 1999 (3.21 m) (as marked in Fig. 16.2). More than 100 flood events were recorded in Quang Nam from 1979 to 2010 and the frequency of these events is increasing: 1980–1989 recorded 21 typhoons and 22 floods, 1990–1999 recorded 17 typhoons and 26 floods, and 2000–2009 recorded 26 typhoons and 49 floods along with a significant increase in natural disasters (Asian Development Bank, 2014).

Storms in Hoi An usually occur during the rainy season, between September and December. They normally consist of strong winds and heavy rains and can cause storm surges. Effects of the storms include damage from high winds, local landslides, and flash and coastal flooding. Approximately 25% of all storms that reach Vietnam in the rainy season affect Hoi An. The most severe storms usually occur in years that have abnormally high rainfall, in excess of 2400 mm/year. During storms, wind speeds can reach 110 km/h with unpredictable changes in direction, causing damage to buildings, including the culturally and economically important historic buildings found throughout the city (UN Habitat, 2014).

Coastal erosion, increased flooding and strong storms are transforming coastal tourism in the region, with implications for property values, insurance costs, destination competitiveness and marketing. These effects pose a significant challenge to the economy of the region, where tourism represents a mainstay of the economy (Scott *et al.*, 2012). Although statistics of visitor arrivals to Hoi An in the last 5 years indicate a continuous trend of visitation to the WHS of Hoi An, after the beach erosion in 2014 the number of visitors to Cua Dai declined by approximately 8% (Table 16.1). Informal discussions with an officer of the Hoi An Department of Trade and Tourism revealed a further decline in visitation to this coastal resort area in the first half of 2015. Tourists are leaving Cua Dai in preference for the An Bang beach about 4 km from the city centre.

A survey conducted by the European Union-funded Environmentally and Socially Responsible Tourism Capacity Development Programme (ESRT, 2014) found that among major tourist destinations in Vietnam, Hoi An has the longest stay for international tourists with an average of 4.2 nights. Consequently, these tourists' whole-trip expenses were the highest, with US$395.8 in Hoi An. As Table 16.1 shows, the number of international tourists exceeds the number of domestic tourists, particularly in the coastal

Fig. 16.2. Water levels in Tan Ky house (the boy is 1.3 m in height).

Table 16.1. Tourist arrivals to Hoi An and Cua Dai, 2012–2014 (Authors' compilation from data provided by Hoi An Department of Trade and Tourism).

	Hoi An			Cua Dai		
Year	International	Domestic	Total	International	Domestic	Total
2012	521,647	139,537	661,184	88,599	31,034	119,633
2013	606,733	152,845	759,578	124,542	29,386	153,928
2014	636,456	150,522	786,978	112,784	29,251	142,035

resort area of Cua Dai. What tourists like the most is the peace and quiet of the old town of Hoi An, which reflects the highest level of visitor satisfaction (5.3) in comparison to other major destinations in Vietnam, placing the city among the top three destinations for future visits. However, international tourists are critical in their evaluation of Hoi An as an environment-friendly destination, providing ratings of 4.3 and 4.7 from international and domestic visitors, respectively, among the lowest scores in comparison to other destinations. Clearly the impacts of climate change on the destination have caused tourists concern with respect to Hoi An's environmental sustainability.

Beach erosion has a negative economic impact on resort attractiveness, room rates and property values (Houston, 2002; Cowell *et al.*, 2008). While structural protection makes sense for certain tourism sector assets, it poses great challenges to coastal resorts struggling to maintain sufficient beach area and aesthetics to attract tourism clientele. Typical structural coastal protection is not well suited to the key objectives of coastal resorts of providing unobstructed views of the sea, maintaining unhindered access to the beach and sea, and offering a pristine beach environment. Once this aesthetic value is lost owing to coastal protection measures (Fig. 16.3), a decline in the number of visitors is unavoidable. Booking cancellations as a result of a less aesthetic beach landscape affect the incomes and lives of those who work in the tourism industry (*Quang Nam Newspaper*, 2015).

Interviews in May 2015 with the officers of a resort affected by coastal erosion revealed that the number of visitors to the resort declined by 70% in 2014. The resort has invested billions to build a concrete dam to prevent further erosion approaching its property. Despite offering a room rate heavily discounted by 50%, the resort could not secure bookings after tour operators saw that coastal squeeze left the resort with no beach. In addition, many resort projects under construction along Cua Dai are on hold owing to coastal erosion (Fig. 16.4). The current strategy of the private sector is to protect the properties while waiting for reaction from the public sector (*Quang Nam Newspaper*, 2015).

Climate Change: Managing Solutions

A comprehensive strategy to effectively combat the impacts of climate change is not yet in place. While waiting for a long-term solution from provincial governments, private-sector businesses and communities have found their own ways to adapt. Private businesses and local residents assume full financial responsibility for protecting their properties since the provincial government provides

Fig. 16.3. Seaside resort without beach.

Fig. 16.4. Resort projects are on hold owing to coastal erosion.

financial support only for protection of public properties (Quang Nam Newspaper, 2015). One of the immediate public-sector solutions is a construction project for a 1.5-km embankment at a cost of VND115 billion (US$5.5 million). The dike helps protect the most heavily damaged section of the road from erosion (Fig. 16.5).

> The embankment is one of the city's seven projects in the fight against climate change. We will also dredge rivers and upgrade roads (interviews with officer of Hoi An City's People's Committee, May 2015).

To minimize damage to the buildings, the Hoi An People's Committee has called on people and businesses to implement emergency measures such as placing pillars and using sandbags for wave prevention. The majority of forest land in the area has been destroyed in recent years to make way for aquaculture farms. However, many of these farms have now been shut down so that replanting can commence. The city is also about to invest VND7.5 billion (US$357,000) to replant 140 ha of Nipa palm along the Thu Bon River. This ecological forest is expected to protect the city from sea erosion and ease the flow of the river as well as reduce sand drift on the beach (*Vietnam News,* 2014).

At the local community level, a survey by UN Habitat (2014) showed that the community is accustomed to flooding. Because of their experience with floods and storms every year, communities in Hoi An have particularly high rates of disaster preparedness, social cohesiveness and mutual support. Interviews with local communities showed that household to household support during the flood and storm season is very common. In many cases, households with multi-storey houses offer rooms to others living in temporary housing during flooding evacuation. Each hamlet has an assembled rescue team that gives priority to people living in temporary homes, the elderly, women and children. Another example of adaptation at the community level includes initiatives for forest plantation to prevent coastal and riverbank erosion and the application of crop rotation to cope with salinity (UN Habitat, 2014).

From a policy viewpoint, Hoi An still lacks an institutional body to coordinate the

Fig. 16.5. Sandbag emplacement on public beach.

preparation and implementation of climate change action plans, as well as to ensure the link between these action plans and existing interventions related to sustainable urban development. Three major administrative issues have been identified as limitations of Hoi An's adaptation to climate change (UN Habitat, 2014). First, there is a lack of consistency and coordination between the Climate Change Response Action Plan and the City Master Plan as these plans are presently being prepared separately by different teams (Division of Natural Resource and Environment and Division of Urban Management, respectively) with different objectives. Second, significant knowledge and skill gaps exist with respect to strategic urban planning among local urban planners. In light of decentralization these weaknesses are a concern, because if urban planners are not capable of preparing strategic and participatory urban development plans, their ability to contribute to the city-level action plan to respond to climate change is also questionable. Third, local governments have no enabling mechanism for increasing their budget for climate change response activities. The master plan does not include participation from either the private sector or the local community, which is also a constraint for cities in mobilizing various sources of funding for urban development and climate change response. In fact, this approach has largely been contested in the current situation of Hoi An, where preventive actions to remedy the impacts of climate change have largely been carried out by the private sector and local communities to protect their properties. Therefore, the gaps between policy proposal and implementation exist not only among different departments of the public sector, but also between the public and private sectors and communities.

Conclusions: Future Expectations

This chapter analyses multiple effects of climate change on the World Cultural Heritage Hoi An Ancient Town. The results suggest a number of conclusions about the current impact of climate change on this tourist destination. First, the heritage city has been affected in three ways: directly from sea-level rise causing coastal erosion along the seaside resort area of Cua Dai, indirectly from increased severity of floods and storms causing negative impacts on wooden houses of the heritage asset, and induced impacts resulting from climate change mitigation, which has affected the attractiveness of the destination.

Second, long-term solutions for mitigating the impacts of climate change have not yet been developed. Ability for adaptation to climate change is apparent at the community level, which is accustomed to the threat of floods. The private sector has reacted independently to protect properties from the surge of coastal erosion without much support from the public sector. In contrast, the public sector has been unable to combat climate change owing to a lack of coordination between planning sectors along with limitations of authority and capability. Further decentralization of governance and promotion of more active roles for local communities and the private sector would address these limitations.

Finally, the city of Hoi An has largely depended on tourism for its economic development in terms of job creation and attracting investment. On the one hand, development of seaside resorts interferes with the ground structure and worsens coastal erosion. On the other hand, the resort sector is itself the victim of erosion, floods and storms that lead to degradation of commercial and cultural properties and a declining number of visitors. In other words, excessive tourism in Hoi An both adds to the causes and suffers the consequences of climate change. This circle of development demonstrates the need for more attention to a sustainable planning approach that should be taken by many destinations in the age of climate change.

References

Asian Development Bank (2014) VIE: *Urban Environment and Climate Change Adaptation Project (Hoi An City, Quang Nam Province)*. Asian Development Bank, Hanoi.

Cowell, R.K., Brehm, G., Cardelus, C.L., Gilman, A.C. and Longino, J.T. (2008) Global warming elevational range shifts, and lowland biotic attrition in the wet tropics. *Science* 322, 258–261.

ESRT (2014) *Tourism Satisfaction Survey*. Environmentally and Socially Responsible Tourism Programme, Hanoi.

Hoi An Center for Monuments Management and Conservation (2008) *Cultural Tourism and Heritage Management in the World Heritage Site of Ancient Town of Hoi An, Vietnam*. United Nations Educational, Scientific and Cultural Organization, Bangkok.

Houston, J.R. (2002) The economic value of beaches: a 2002 update. *Shore and Beach* 70, 9–12.

McIntyre-Tamwoy, S. (2008) The impact of climate change and cultural heritage: grasping the issues and defining the problem. *Historic Environment* 21(1), 2–9.

Quang Nam Newspaper (2015) Beach Erosion is Putting Tourism Business in Difficulties. Available at: http://baoquangnam.com.vn/du-lich/201509/sat-lo-bien-hoi-an-doanh-nghiep-du-lich-gap-kho-634469 (accessed 6 November 2015) (text in Vietnamese).

Scott, D., Hall, M. and Gössling, S. (2012) *Tourism and Climate Change Impacts, Adaptation and Mitigation*. Routledge, London.

UNESCO (2007) *Climate Change and World Heritage*. United Nations Educational, Scientific and Cultural Organization, Paris. Available at: http://whc.unesco.org/en/activities/474 (accessed 10 November 2015).

UN Habitat (2014) *Hoi An – Vietnam: Climate Change Vulnerability Assessment*. United Nations Human Settlements Programme, Nairobi.

Vietnam News (2014) Resorts Suffer as Beach Disappears. Available at: http://vietnamnews.vn/environment/240524/resorts-suffer-as-beach-disappears.html (accessed 4 November 2015).

17 Case Study Sri Lanka: Climate Change Challenges for the Sri Lankan Tourism Industry

J. Buultjens,[1]* I. Ratnayake[2] and W.K. Athula Gnanapala[2]
[1]*Southern Cross University, Australia;* [2]*Sabaragamuwa University, Sri Lanka*

Introduction

Sri Lanka, a small island located in the Indian Ocean at the southern tip of India, has a coastline that stretches for 1600 km. The country is highly vulnerable to climate change (Sri Lanka Ministry of Environment (MOE), 2011) and the accompanying types of natural disasters such as floods, landslides, droughts, coastal erosion and sea surges (EML, 2012). In addition to the natural disasters, the predicted sea-level rise of 0.5 m over the next two decades will also have severe consequences for the country's coastal areas and the 25% of the population that reside in them (Ahmed and Suphachalasai, 2014). The consequences from sea-level rise include increased destruction of mangroves and coral reefs, a decline of ecosystems and marine habitats and damage to shelter, infrastructure and human safety. Areas adjacent to coastal regions will also experience gradual but intense salinization of inland freshwater sources (EML, 2012). Negative impacts on the coastal zone will have substantial effects on the country's economy since the zone accounts for approximately 43% of the nation's GDP (MOE, 2010a).

Tourism, an important sector of the national economy, is substantially located in the coastal zone. Historically, the tourism industry has been dependent upon the 'sea, sand and sun' segment and consequently has been concentrated in coastal areas (Gunawardana and Sanjeewani, 2009; Buultjens *et al.*, 2015a). This concentration and dependency on coastal resources makes the sector very vulnerable to climate change (EML, 2012). This vulnerability is unfortunate since the industry, which has experienced a rapid expansion since the end of the civil war in 2009, has been earmarked as an important contributor to the country's ongoing economic development (EML, 2012). The ability of the industry to contribute to future economic development is likely to be compromised by the impacts of climate change and its ability to respond to these impacts. Not only will climate change impact on the nation's natural and built tourism assets, it is also likely to disrupt the transport system, result in increased competition for reducing water resources as well as increase energy consumption through additional use of air conditioning as a response to increasing temperatures (EML, 2012).

This chapter examines the potential impacts from climate change on coastal zone tourism as well as government and private-sector attempts to mitigate and adapt to these

*Corresponding author, e-mail: jeremy.buultjens@scu.edu.au

changes. The chapter begins by providing an overview of the Sri Lankan tourism industry and is followed by the predicted impacts of climate change on the country and its coastal zone. The next section examines the projected climate change impacts on the tourism industry and the approaches that have been taken to address these impacts.

The Sri Lankan Tourism Industry

Prior to World War II Sri Lanka's location on the world's sea lanes enabled the country to attract many cruise ships; however this market collapsed after the start of the war. In 1948, after gaining independence from the British, the new government established a Tourist Bureau with a view to reviving the industry. The Tourist Bureau undertook, among other things, tourist promotional work internationally as well as developing tourist accommodation facilities. As part of the development process a number of colonial buildings were converted into a network of accommodation establishments, many of which continue to be popular attractions today (Sri Lanka Tourism Development Authority (SLTDA), n.d.b; UGC Marketing Group, 2011). In 1964 another important initiative, the establishment of the country's first hotel school, further facilitated the development of the industry. This was followed in 1966 by the government's establishment of the Ceylon Tourist Board to replace the Tourist Bureau. Under the Ceylon Tourist Board provisions for tourists were largely left to the private sector, although the Board did provide facilities in areas where the private-sector establishments were considered inadequate (SLTDA, n.d.b). In the following year, 1967, the country's first Tourism Management Plan was released (Jayawardena, 2013).

The introduction of the Tourism Management Plan underpinned an annual growth in international visitation in excess of 21% between the mid-1960s and the early 1980s (SLTDA, 2005). However, the expansion of the industry was substantially disrupted by the start of the long-running civil war between the government forces and the Liberation Tigers of Tamil Eelam (LTTE). During the 25-year civil war international visitation stagnated at between 400,000 and 500,000 visitors punctuated by some intermittent and sharp falls as a consequence of specific terrorist acts as well as the tsunami that struck in 2004 (SLTDA, n.d.a). The ending of the war in 2009 has underpinned the rapid expansion in international visitation from 447,890 tourists in 2009 to over 1.5 million in 2014, with the government planning for 2.5 million visitors by 2016 (SLTDA, n.d.a). The domestic industry, a relatively small segment of the industry, has also undergone an expansion since 2009 due to rising personal incomes.

The industry's dependency on the comparatively low-yield, mass international 'sea, sun and sand' market (Fernando and Shariff, 2013; Buultjens *et al.*, 2015a) is not surprising given its tropical weather and substantial coastline. There has also been a relatively popular cultural tourism product offered in the so-called 'cultural triangle' based around Kandy, Anuradhapura and Polonnaruwa (Ranasinghe, 2015). In recent times there have been attempts to diversify the industry further with the promotion of more 'niche' products such as health, nature-based and wildlife tourism, however, despite these efforts, the country continues to be predominantly a 'sea, sun and sand' destination driven by natural and built coastal attractions. For example, in 2008, 62% of visitors to Sri Lanka stated that the beach was the main reason for their visit to Sri Lanka (MOE, 2010b). In 2013, more than 55% departing foreign tourists indicated that they had had a 'sea' experience while visiting the country (SLTDA, 2013). Additionally, in 2014, approximately 76% of all foreign guest nights in graded accommodation were spent in the coastal regions of Colombo City, Greater Colombo, the South Coast and the East Coast (SLTDA, 2014).

The industry's focus on the coast is reflected by the fact that over 70% of the country's tourist infrastructure is located in the coastal zone (MOE, 2010a). Furthermore, in 2014, approximately 76% of the 334 tourist hotel units and 1265 supplementary accommodation units were located in the coastal zone (SLTDA, 2014). Much of the other major tourist infrastructure including the Bandaranaike International Airport in Colombo as well as a second, recently developed airport at Hambantota in the south, and the port facilities at Colombo, Galle and another at Hambantota are also located in the coastal zone. In addition, a large proportion of the road network servicing the tourism industry and which has undergone considerable development recently (UGC Marketing Group, 2011) is also located in the coastal areas. These developments have considerably improved road connectivity between various important tourist cities but are also vulnerable to the impacts from climate change. There is also a considerable railway network throughout the country including the Coastal Line that follows the edge of the Indian Ocean between Colombo and Matara in the south. A substantial proportion of the transport infrastructure is expected to come under severe threat due to climate change (MOE, 2010b).

These impacts from climate change have substantial economic implications. Since 2009 the industry has rapidly become an important component of the Sri Lankan economy. For example, in 2014, the travel and tourism sector contributed 4.8% directly to GDP, with this contribution forecast to increase further. The total contribution of travel and tourism was 11.1% of GDP. In 2014, travel and tourism directly supported 352,000 jobs or 4.3% of total employment. The total contribution by the industry was 819,500 jobs or 10.0% of total employment. In addition, international visitation generated 21.3% of total exports in 2014 (World Travel & Tourism Council, 2014). The foreign exchange earnings from the industry increased by 43.6% from US$1715.5 million in 2013 to US$2431.1 million in 2014. In 2014, tourism was the third largest earner of foreign exchange (SLTDA, 2014).

Notwithstanding its importance, the tourism industry faces a number of problems in addition to climate change. As in many other countries, the industry has grown in a rather haphazard fashion. This has resulted in many inappropriate developments taking place, particularly in the coastal zone, and many negative impacts on a range of natural resources. Instructively the last *Tourism Development Strategy: 2011 – 2016* (Sri Lanka Ministry of Economic Development, 2010) was almost exclusively focused on growth; there was little reference to sustainable development and none to climate change. Despite this 'Sri Lanka Tourism has recognised "Sustainable" development as a timely requirement and as a priority target. Sustainable Tourism will be the foundation to achieving [a] ... set [of] objectives which satisfy the definition of sustainability' (EML, 2012, p. 104).

The almost total focus on growth in combination with poorly planned tourism has contributed to substantial biodiversity loss especially in the coastal zone. The construction of beach resorts has resulted in considerable damage and/or destruction of coastal wetlands, particularly mangroves. There has also been considerable damage to coral from its use as a construction material for hotels, and sewage and sedimentation runoff from human developments (Buultjens *et al.*, 2015a).

Another consequence of the rapid growth in tourism and the type of development is an expected increase in the demand for water (Buultjens *et al.*, 2015a). The style of development being encouraged by the government is the larger, luxury, resort-style hotels with their golf courses, irrigated gardens, swimming pools, spas and wellness facilities. The increase in demand for water is likely to exacerbate concerns over its general availability, especially in the dry and arid zone.

Climate Change: Assessing Problems and Challenges

Sri Lanka's mainland area is 65,610 km^2 and the coastal region covers about 23% of this land area (Ahmed and Suphachalasai, 2014). There is no area that is more than approximately 100 km away from the sea. The country has a tropical climate that depends largely on monsoon wind patterns.

The monsoons impact significantly on the country's three climatic zones. The wet zone is located in the south-western quadrant of the country; the dry zone in the north, east and north-central provinces; and the intermediate area lies between the other two. Much of the coastal zone lies within the dry zone (Ahmed and Suphachalasai, 2014).

The rainfall in Sri Lanka is governed by the two monsoons, the south-west monsoon, which is active during May–September, and the north-east monsoon, which is active during December–February. The mean annual rainfall in the wet zone ranges between 2500 mm to over 5000 mm, while in the dry and intermediate zones the mean annual rainfall is below 1750 mm. The annual mean average temperature is 27°C in lowlands while the annual mean average in the highlands is 15°C (MOE, 2011).

Climate change is likely to bring rapid temperature increases in Sri Lanka and at a faster than the average global rate of warming. Winter temperatures are expected to increase more than summer temperatures (Practical Action Sri Lanka, 2015). The maximum and minimum temperatures in the country have shown upward trends between 1961 and 2000, with up to a maximum 0.46°C increase per decade in terms of maximum temperature and 0.27°C increase per decade in terms of minimum temperature (MOE, 2011). The projected temperature rise for 2100 is expected to be between 1.1°C and 2.4°C with a more prominent increase expected in the areas influenced by the south-west monsoon.

The projected change in rainfall is more complex. The level of winter rainfall is expected to decrease, whilst summer rainfall will increase (Practical Action Sri Lanka, 2015). Between 1961 and 2000 there have been decreasing trends with the largest falls occurring in the high rainfall areas. Rainfall projections suggest that in the wet zone rainfall is increasing while in the dry zone in the north-east there is less prominent change (MOE, 2011). Weather patterns have also become less predictable, causing floods in certain areas and droughts in others. For example, floods in January and February 2011 displaced over 1.2 million people from their homes in 14 of the 25 districts in Sri Lanka (United Nations Human Settlements Programme (UN-Habitat), n.d.). In addition, cyclones are anticipated to increase in strength (Practical Action Sri Lanka, 2015).

Sea level rise scenarios are also expected to be considerable with some suggestion of a shoreline retreat of 10 m by 2050. Not surprisingly the coastal and low-lying areas will face the most substantial impact. Sea level rise will contribute to further erosion, the loss of land and the loss of livelihoods resulting in substantial migration from these areas (Practical Action Sri Lanka, 2015).

In summary, the MOE (2010b) identify a number of climate-related impacts, including:

- An increased frequency and intensity of floods and landslides causing extensive asset loss/damage and displacement of people, particularly the poor;
- An increased frequency and intensity of droughts and a saline intrusion in the coastal areas. This will exert greater pressure on freshwater resources resulting in more frequent and severe shortages in drinking water. These impacts are expected to be especially strong in the dry and intermediate zones. Water shortages will increase the competition between different sectors including households, agriculture and tourism;
- There will be reduced water storage capacity in irrigation tanks (reservoirs) and canals as a result of increased droughts and soil erosion. These tanks provide much of the current water supply but are in poor condition;
- Increased ambient temperatures will result in greater energy (electricity) demand for cooling purposes;
- Increased vector densities due to climatic factors will result in the rapid spread of diseases such as dengue, particularly in urban areas;
- There will be changes in growth rates, reproduction and geographic ranges of species and phenology of plants; and
- There will be changes in coastal and marine systems, species and ecosystem services due to sea level rise, global

warming and ocean acidification, with particular impacts on coral reefs and associated species (see Figs 17.1, 17.2 and 17.3).

The climate change will have substantial economic consequences. For example, it is estimated that if no action is taken to adapt to and/or mitigate global climate change, the average total economic losses for Sri Lanka are estimated to be 6.5% of GDP (Ahmed and Suphachalasai, 2014).

The Sri Lankan tourism industry

Climate change impacts on tourism are likely to be largely negative (Munasinghe Institute for Development, 2003; EML, 2012; Ahmed and Suphachalasai, 2014). Tourism developers in coastal areas have exposed themselves to considerable problems by consistently attempting to locate or actually locating tourism projects almost at the shoreline. These developments also invade coastal reservations and buffer zones, causing environmental damage to the coast, coastal vegetation and coastal stability (EML, 2012) and further expose the industry to the impacts of climate change. The vulnerability of the coastline and the industry to climatic events were highlighted by the tsunami that struck the north and east of the country in December 2004 (Samath, 2007). It was estimated that tourism asset loss from the tsunami was US$250 million and output loss, based on projections for 2005 and 2006, was estimated at US$130 million (Weerakoon *et al.*, 2007). In addition to the tsunami, in recent years many coastal cities have experienced a range of devastating climate-related impacts (EML, 2012).

The range of tourism infrastructure and related businesses and services in the coastal belt will be affected by climate change, sea level rise, storm surges and coastal flooding (Athulathmudali *et al.*, 2011). There will be a loss of natural and built assets, tourism revenue and employment opportunities in the industry. The beaches in the country, an important tourism drawcard, will be degraded as a consequence of the loss of land

Fig. 17.1. Coastal road at Akural Beach, Ambalangoda.

Fig. 17.2. Beach protection at Unawatuna.

Fig. 17.3. Unawatuna shoreline.

and damage to the shore front. Rising ocean temperatures and changing rainfall patterns may also cause substantial damage to coastal wetlands, flora and fauna. For example, there is likely to be degradation of coral reefs, including coral bleaching. The damage to the ecosystems will diminish the country's appeal as a biodiversity hotspot.

Another consequence of climate change is the reduced availability of water. A reduction in the availability of water combined with an increase in consumption from tourism expansion means the tourism industry will compete against other sectors, including households and agriculture, for the increasingly scarce water resources (Buultjens *et al.*, 2015a). In addition to depletion of water resources, tourism and climate change will also directly impact on the quality of water, for example the discharge of untreated sewage and freshwater abstraction (United Nations Environment Programme (UNEP), 2011).

A further outcome from climate change for tourism is a likely increase in energy consumption. As ambient temperatures rise tourists will have increased cooling requirements (Athulathmudali *et al.*, 2011). Rising energy consumption is a major problem since '[T]he increase in fossil fuel based energy is one of the largest climate change and development related issues for Sri Lanka' (Global Environment Facility (GEF), 2014). Finally, climate change and the related occurrence of natural disasters such as floods and landslides will cause a disruption to the country's transportation networks. This will reduce tourists' mobility and have a considerable impact on the industry (Athulathmudali *et al.*, 2011). The high vulnerability of the industry to climate change and its impacts is likely to encourage tourists to seek less vulnerable destinations (Athulathmudali *et al.*, 2011).

Climate Change: Managing Solutions

Although still in the early stages the government and private sector are attempting to address climate change and the vulnerabilities that arise from it. Sri Lanka agreed to the Kyoto Protocol on 3 September 2002. The Ministry of Environment and Natural Resources was given the responsibility of implementing the provisions of the Protocol (UN-Habitat, 2013). The major document articulating the government's approach to sustainable development is the *Mahinda Chinthanaya* – a 10-year development plan for sustainable development. Tourism is highlighted as a high priority sector in the *Mahinda Chinthanaya*. As part of this development plan the Ministry of Environment and Natural Resources launched the Haritha (Green) Lanka Programme in 2009 with the aim of greening economic development. One of the ten missions of the Haritha Lanka Programme is '[M]eeting the Challenges of Climate Change' (EML, 2012, p. 8).

In response to meeting the challenges of climate change, the Climate Change Secretariat was established in 2008 by the Ministry of Environment and Natural Resources. In addition, a Climate Change Coordinating & Steering Committee (CCC&SC) was established to develop climate change policy that was to encompass adaptation, mitigation and clean development (UN-Habitat, 2013). A National Climate Change Adaptation Strategy: 2011–16 and a Public Information and Awareness Strategy were developed to address the challenges of climate change (Buultjens *et al.*, 2015a). This was followed in 2012 by the release of the National Climate Change Policy (NCCP) that provided the national vision and policy principles on climate change (Senaratne, 2014). In 2015, the National Adaptation Plan for Climate Change Impacts: 2016 to 2015 was released and addressed nine vulnerable sectors, including tourism. The other vulnerable sectors were food security, water, coastal sector, health, human settlements, biodiversity, export development and industry energy-transportation (Climate Change Secretariat Sri Lanka, 2015).

Since the early 2000s, Sri Lanka has introduced various macro-measures aimed at mitigation and adaptation. There have also been some sector-specific measures, especially in agriculture. An example of a mitigation activity was the carbon trading mechanism through the Clean Development Mechanism (Pathiraja *et al.*, 2014). Many of the various climate change adaptation measures have been introduced at a grass-roots level, especially in the agricultural sector. However, adaptation efforts have often been fragmented and there has been an absence of a strong link with national climate change strategies and plans (Pathiraja *et al.*, 2014).

Until very recently there have been few policies or strategies addressing climate change and tourism. For example, the Ministry of Tourism has been given no mandate to address climate change. Not surprisingly, as a consequence the tourism industry has implemented very few iniatives to directly address climate change. For example, as noted earlier, the government's latest tourism strategy gave very little attention to explicitly promoting environmental sustainability and no attention to addressing climate change, but rather focused on growth. In addition, the strategy failed to acknowledge the importance of the environment to the tourism industry (Athulathmudali *et al.*, 2011).

One area where government action may have implications for tourism and climate changes is through planning. In an attempt to overcome the haphazard nature of tourism planning a number of Tourism Development Zones were established throughout the country in the aftermath of the tsunami (Buultjens *et al.*, 2015b). The developments encouraged by the government in these zones are characterized as 'mega resorts' and have the intention to provide opportunities for direct foreign investment in the tourism industry and also attempt to capture the high-end market (Rice and Haynes, 2005). In addition to supplying upmarket hotels, it is anticipated that the zones will also supply golf courses, water parks, light aircraft services and shopping malls. To ensure environmental and social safeguards are in place around these developments, Strategic Environmental Assessments (SEA) are to be conducted for all sites prior to development (EML, 2012). Through the introduction of zones the government has attempted to implement more effective infrastructure development. They have also attempted to introduce coastal buffer zones and coastal management plans. Despite the potential benefits from improved planning there have been critics of the Tourism Development Zones (Tjolle, 2011; Johansson, 2013; Buultjens *et al.*, 2015b) and there are fears the development in these zones may aggravate the impacts from climate change by increasing energy and water demand (Miththapala, 2014). As noted previously, the increase in fossil fuel-based energy is one of the largest climate change and development-related issues for Sri Lanka (GEF, 2014).

Another initiative to facilitate a sustainable tourism industry has been the promotion of niche markets such as ecotourism, wildlife and nature-based tourism. Green buildings and effective waste management systems have also been encouraged. However, at this point, there has been little government action to climate proof the industry (Athulathmudali *et al.*, 2011; Buultjens *et al.*, 2015a).

The lack of a specific focus on tourism is also reflected in the National Climate Change Adaptation Strategy for Sri Lanka, 2011 to 2016 (Climate Change Secretariat Sri Lanka, 2010). Despite noting the importance of the tourism industry, as well as its vulnerability to climate change, the document provides only two 'Priority Adaptation Measures' that address tourism specifically. These are: (i) the need to include climate change adaptations in tourism planning guidelines; and (ii) raising awareness about climate vulnerability through a 'sustained and broad-based dialogue to improve awareness, and to mobilize stakeholders within vulnerable industries (such as tourism and plantations) for climate change adaptation' (Climate Change Secretariat Sri Lanka, 2010, p. 9). The more recently developed National Adaptation Plan for Climate Change Impacts: 2016 to 2025 addresses tourism and recreation specifically and in greater detail (Climate Change Secretariat Sri Lanka, 2015). The plan also directly addresses the coastal sector in general.

The major actions for tourism and recreation from the plan are:

- Increase the awareness of tour industry operators on climate change and its impacts;
- Establish emergency communication channels for tourists and operators;
- Identify tourism facilities in vulnerable areas and make arrangements to increase their climate resilience;
- Assess the current promotional strategies with connection to emerging scenarios of climate change and adjust them accordingly; and
- Conduct research studies on climate change impacts on tourism and recreation.

In addition, the plan also identified the following actions for the coastal and marine sector:

- Implement a continuous programme for monitoring shoreline changes;
- Develop shore shoreline management plans including M&E programmes;
- Study impacts of sea level rise on costal habitats over short-, medium- and long-term horizons;
- Identify, declare, collect information and prepare maps on vulnerable areas to extreme events and inundation; and
- Conduct awareness programmes on sea level rise and extreme events to coastal communities to empower them for facing the risks of climate change.

Hopefully the release of the National Adaptation Plan for Climate Change Impacts: 2016 to 2015 will result in some tangible actions that can help the industry to adapt effectively for the impacts of climate change. In addition to the central government, local authorities can also play a central role in achieving responsible tourism by committing to supportive policy frameworks and adequate funding (EML, 2012). At this stage local authorities have had little involvement in managing climate change specifically and tourism more generally.

The relative inaction by government is replicated by the private sector, which has only implemented a limited number of actions aimed at addressing the challenges of climate change. Athulathmudali *et al.* (2011) suggest that the industry has neglected climate change since they have not been impacted by climate change to any great extent. In contrast, they have been more focused on addressing the pressing environmental problem of land and water pollution since it is viewed as having a direct and rather substantial impact on the industry in the resort belt.

The limited private-sector climate change initiatives include the construction of beach erosion structures to protect larger resorts and the implementation of the EU originated SWITCH Asia Programme. This programme, introduced in mid-2009 by the Ceylon Chamber of Commerce, targeted 350 small and medium enterprises (SMEs) in the hotel sector with the main objective of lowering their carbon footprint (switchasia, n.d.). This programme encouraged the SMEs to adopt best practices of the Sustainable Cities Program, which has had a favourable impact on the environment (EML, 2012). Other private-sector organizations have introduced carbon neutral programmes for their operations to offset the damage they have done to the environment. For example, Walkers Tours have a carbon-neutral vehicle fleet. Other companies have introduced initiatives to protect the natural environment. These activities do not directly focus on climate change adaptation but may contribute indirectly. Despite these efforts and those by the government more action is needed to address climate change and its impacts on the tourism industry.

A number of factors are currently inhibiting climate change adaptation. The Sri Lanka National Adaptation Plan: 2016-2015 suggests information, technological, policy and governance, institutional and coordination and resource mobilization gaps explain the lack of successful adaptation (Sterrett, 2011; see also Senaratne, 2014; Climate Change Secretariat Sri Lanka, 2015). As Senaratne (2014) suggests, adaptation to climate change is essentially an information-driven process that is required for capacity building. A lack of knowledge stifles this process. The lack of knowledge is particularly relevant for the tourism sector in Sri Lanka. It is also a factor prevalent in other countries (Jones and Phillips, 2011). The lack of sufficient integration between a range of relevant institutions, including intergovernmental agencies, civil society and the funders is also a major problem. Climate-change activities in Sri Lanka are being undertaken in an ad hoc manner by various agencies without proper coordination (see Sterrett, 2011; Senaratne, 2014). This results in substantial inefficiencies.

The lack of integration and coordination as well as the range of other 'gaps' in Sri Lanka can be attributed to a lack of access to adequate funds (Senaratne, 2014). For example, it is estimated that up to 1.2% of GDP per year could be required for adaptation by 2050.

Hopefully, as the Sri Lankan economy and the tourism industry in particular expands, the government and industry will be able to raise the necessary funds to make a more concerted attempt to address climate change. Relative inaction by government and industry in addressing the impacts of climate change on tourism in the coastal region will impact the whole community but it will have the greatest impact on the poor. The industry in the coastal belt has a high concentration of informal businesses. It is these businesses that are owned and/or managed by the poorer sectors of the community (Athulathmudali *et al.*, 2011). As a consequence it is the poorer members of the community who are more vulnerable to environmental or economic shocks resulting from climate change.

Conclusion

Like most island nations Sri Lanka is highly vulnerable to the impacts of climate change. The tourism sector, predominantly located in the coastal region, is especially vulnerable. Despite the vulnerability of the country and the industry there have been relatively few macro government policies, strategies and/or actions targeting climate change and even fewer that have directly addressed the tourism industry. The recommendations from the National Capacity Self-Assessment which was designed by the GEF to identify capacity needs and recommendations for addressing biodiversity, climate change and land degradation have not been systematically incorporated into institutional programmes or project designs (GEF, 2014). A major explanation in addition to a lack of resources committed to climate change is the existence of a range of gaps including information, technological, policy and governance, institutional and coordination and resource mobilization gaps. Developing appropriate policy and governance responses to climate change issues needs to occur with the participation of public sector as well as non-state actors such as the private sector, community organizations and civil society organizations.

References

Ahmed, M. and Suphachalasai, S. (2014) *Assessing the Costs of Climate Change and Adaptation in South Asia*. Asian Development Bank, Mandaluyong City, Philippines.

Athulathmudali, S., Balasuriya, A. and Fernando, K. (2011) *An Exploratory Study on Adapting to Climate Change in Coastal Areas of Sri Lanka*. NTNU Globalization Research Programme, Faculty of Humanities, Norwegian University of Science and Technology N 7491, Trondheim, Norway.

Buultjens, J., Ratnayake, I. and Gnanapala, A. (2015a) Sri Lanka and the green economy. In: Reddy, V. and Wilkes, K. (eds) *Tourism in the Green Economy*. Earthscan with Routledge, London, pp. 128–145.

Buultjens, J., Ratnayake, K. and Gnanapala, W. (2015b) Post-Conflict tourism development in Sri Lanka: implications for building resilience. *Current Issues in Tourism*, DOI: 10.1080/13683500.2014.1002760.

Climate Change Secretariat Sri Lanka (2010) *National Climate Change Adaptation Strategy for Sri Lanka 2011 to 2016*. Climate Change Secretariat Sri Lanka, Ministry of Environment, Colombo.

Climate Change Secretariat Sri Lanka (2015) *National Adaptation Plan for Climate Change Impacts in Sri Lanka: 2016 to 2025*. Climate Change Secretariat Sri Lanka, Ministry of Mahaweli Development & Environment, Colombo.

EML (2012) *Sri Lanka's Middle Path to Sustainable Development Through 'MahindaChintana – Vision for the Future', Country Report of Sri Lanka*. United Nations Conference on Sustainable Development / (Rio +20), Ministry of Environment, Battaramulla, Sri Lanka.

Fernando, S. and Shariff, O. (2013) Ecotourism in Sri Lanka Faces Serious Challenges. Available at: www.tourism-review.com/travel-tourism-magazine-challenges-of-ecotourism-in-sri-lanka-article2113 (accessed 2 September 2015).

Global Environment Facility (GEF) (2014) *Joint GEF/Sri Lanka Country Portifilo Evaluation: 1991–2012, Main Report, Volume 1*. GEF, Cancun, Mexico.

Gunawardana, M.R. and Sanjeewani, H.L.G. (2009) Planning implications and sustainability of tourism: a comparative study of Hikkaduwa and Bentota, Sri Lanka. *Journal of Tourism, Hospitality and Culinary Arts* 1(2), 67–85.

Jayawardena, C. (2013) Innovative solutions for future tourism development in Sri Lanka: 2013–2026. *Worldwide Hospitality and Tourism Themes* 5(5), 512–531.

Johansson, M. (2013) Sri Lankan Government Uses 'War Tourism' to Commemorate 2009 Victory. Available at: www.ibtimes.com/sri-lankan-government-uses-war-tourism-commemorate-2009-victory-994582 (accessed 5 October 2014).

Jones, A. and Phillips, M. (2011) Disappearing destinations: recognizing problems – meeting expectations – delivering solutions. In: Jones, A. and Phillips, M. (eds) *Disappearing Destinations: Climate Change and Future Challenges for Coastal Tourism*. CAB International, Wallingford, UK.

Miththapala, S. (2014) A Look Back at Sri Lanka Tourism 2013: The Good, the Bad and the Ugly. DailyFT. Available at: http://www.ft.lk/columns/a-look-back-at-sri-lanka-tourism-2013-the-good-the-bad-and-the-ugly/4-246930 (accessed 15 February 2014).

Munasinghe Institute for Development (2003) *Integrating Sustainable Development and Climate Change in the IPCC Fourth Assessment Report*. Munasinghe Institute for Development (MIND), Colombo.

Pathiraja, K., Balaraman, M. and de Silva, S. (2014) *Study of Climate Change Adaptation Measures Lacking Funding in Sri Lanka*. International Centre for Ethnic Studies, Colombo.

Practical Action Sri Lanka (2015) Promoting Adaptation to Climate Change in Sri Lanka. Available at: http://practicalaction.org/energy/advocacy/docs/advocacy/Adaptation_to_climate_change_Sri_Lanka.pdf (accessed 20 August 2014).

Ranasinghe, R. (2015) Strategic myopia of tourism development in Sri Lanka: a critique. *International Journal of Multidisciplinary Research and Development* 2(2), 604–609.

Rice, A. and Haynes, K. (2005) *Post-tsunami Reconstruction and Tourism: A Second Disaster?* Tourism Concern, London.

Samath, F. (2007) Sri Lanka: Climate Change Worse Than Civil War – UN Expert. *Colombo: IPS (Inter Press Services)*. Available at: www.ipsnews.net/2007/04/sri-lanka-climate-change-worse-than-civil-war-un-expert (accessed 20 August 2010).

Senaratne, A. (2014) *Policy Challenges in Climate Adaptation in Sri Lanka: Identifying Major Gaps*. Institute of Policy Studies. Available at: http://climatenet.blogspot.com.au/2014/04/policy-challenges-in-climate-adaptation.html (accessed 10 June 2015).

Sri Lanka Ministry of Economic Development (2010) *Tourism Development Strategy 2011–2016*. Ministry of Economic Development, Colombo.

Sri Lanka Ministry of Environment (MOE) (2010a) *Strengthening Capacity for Climate Change Adaptation Sector Vulnerability Profile: Urban Development, Human Settlements and Economic Infrastructure*. Climate Change Secretariat Sri Lanka, Ministry of Environment, Colombo.

Sri Lanka Ministry of Environment (MOE) (2010b) *Strengthening Capacity for Climate Change Adaptation Sector Vulnerability Profile: Water*. Climate Change Secretariat Sri Lanka, Ministry of Environment, Colombo.

Sri Lanka Ministry of Environment (MOE) (2011) *Sri Lanka's Second National Communication on Climate Change*. Ministry of Economic Development, Colombo.

Sri Lanka Tourism Development Authority (SLTDA) (n.d.a) Annual Statistical Reports. Available at: www.sltda.lk/statistics (accessed 11 January 2015).

Sri Lanka Tourism Development Authority (SLTDA) (n.d.b) Overview of Tourism Industry in Sri Lanka. Available at: www.sltda.lk/overview_of_tourism_in_sri_lanka (accessed 11 January 2015).

Sri Lanka Tourism Development Authority (SLTDA) (2005) *2005 Annual Statistical Report*. Tourism Development Authority, Colombo.

Sri Lanka Tourism Development Authority (SLTDA) (2013) *Survey of Departing Foreign Tourists from Sri Lanka*. Sri Lanka Tourism Development Authority, Colombo.

Sri Lanka Tourism Development Authority (SLTDA) (2014) *2014 Annual Statistical Report*. Sri Lanka Tourism Development Authority, Colombo.

Sterrett, C. (2011) *Review of Climate Change Adaptation Practices in South Asia*. Oxfam Research Reports, Climate Concern, Melbourne, Australia.

switchasia (n.d.) Greening Sri Lankan Hotels. Available at: www.switch-asia.eu/projects/greening-sri-lankan-hotels (accessed 2 June 2017).

Tjolle, V. (2011) Sri Lanka told to stop tourism land grabs. *Travelmole*. Available at: www.travelmole.com/news_feature.php?id=1148667 (accessed 24 March 2013).

UGC Marketing Group (2011) Sri Lanka Tourism Industry. Available at: https://www.slideshare.net/nrnslideshare/sri-lankan-tourism-industry (accessed 18 June 2013).

United Nations Environment Programme (UNEP) (2011) *Towards a Green Economy: Pathways to Sustainable Development and Poverty Eradication*. United Nations Environment Programme, Geneva.

United Nations Human Settlements Programme (UN-Habitat) (n.d.) A National Climate Change Policy for Sri Lanka. Available at: www.fukuoka.unhabitat.org/projects/voices/sri_lanka/detail03_en.html (accessed 10 March 2014).

United Nations Human Settlements Programme (UN-Habitat) (2013) *Climate Change Vulnerability Assessment*. UN-Habitat, Batticaloa, Sri Lanka.

World Travel & Tourism Council (2014) *Travel & Tourism Economic Impact 2014: Sri Lanka*. World Travel and Tourism Council, London.

Weerakoon, D., Jayasuriya, S., Arunatilake, N. and Steele, P. (2007) *Economic Challenges of Post-Tsunami Reconstruction in Sri Lanka*. Asian Development Bank Institute, Tokyo. Available at: www.adb.org/sites/default/files/publication/156714/adbi-dp75.pdf (accessed 14 November 2012).

18 Case Study Bangladesh: Addressing Climate Change Effects on Coastal Tourism in St Martin's Island of Bangladesh

Azizul Hassan[1]* and Roya Rahimi[2]

[1]*Cardiff Metropolitan University, Cardiff, UK;* [2]*University of Wolverhampton, Wolverhampton, UK*

Introduction

St Martin's Island is classified as one of the most beautiful islands in Bangladesh. It is the only coral island in the country and in close proximity to Cox's Bazar (a town which is a fishing port and district headquarters in Bangladesh), and has one of the longest unbroken sea beaches in the world. Geographically, the island is at the southernmost point of Bangladesh and lies in the Bay of Bengal, approximately 9 km from the Cox's Bazar-Teknaf peninsula and almost 8 km away from the north-west coast of Myanmar (Hasan, 2009).

The island was first inhabited by Arabian sailors who named it '*Zajira*'. The locals call the island in Bengali '*Narical Ginjira*' or '*Narikel Jinjira*' or '*Narikel Jinjera*' meaning 'Coconut Island'. The island is also known as the '*Daruchini Dip*', indicative of the presence of cinnamon traders from the East or Far East countries. However, the name was changed to St Martin's Island during the British colonial regime (Hasan, 2009).

Coastal tourism in Bangladesh is largely indebted to the coral colonies that nurture this special tourism area in this relatively unknown part of the world (Thompson and Islam, 2010; Afrin *et al.*, 2013). Geographical exquisiteness and a favourable national tourism policy (Hassan and Burns, 2014) have turned tourism into a thriving industry in this island. Between 1989 and 2004 only foreigners and non-resident Bangladeshis were allowed to visit the island. Lifting this restriction witnessed the influx of domestic tourists.

The most fascinating feature that St Martin's possesses is the 'Six Month Economy', the economic cycle (November–May) as a by-product of tourism. During the specific 6 months and for 3 h per day, more than five to six steamers carrying domestic and overseas tourists descend on St Martin's Island. The mass flow of tourists makes it difficult to determine the exact number of tourists visiting the island during this time. The number has risen rapidly over the years and it is estimated that more than 5000 tourists visit the island every day during these 6 economy months. Apart from the steamers, at least 3000–4000 boats and trawlers are engaged to carry tourists from the mainland. These 6 months are the main residents' earning period and the season makes a significant contribution to the local economy when almost all the population residing on the island live above the poverty line as a

*Corresponding author, e-mail: m.hassan15@outlook.cardiffmet.ac.uk

spill-over effect of tourism (SWAMP, 2006; NACOM, 2009; MRDI, 2015).

Research Context

This chapter relies on the review of published and unpublished resources and a detailed review of existing literature related to St Martin's Island. However, tourism research on the island is limited. In relation to this, personal observations by the authors supported by informal discussions both with tourists and islanders provided the main data sources. Critical discussions in the chapter have been assimilated from such analysis, and explanations and arguments thus determined and drawn.

Climate Change: Assessing Problems and Challenges

Possible effects of climate change

St Martin's Island has been facing challenges in the areas of reliable and quality power sources, energy crises and issues with the traditional power system to meet increasing power-supply demand on the island (Amin *et al.*, 2014). Energy supply is a challenge to the island and a grid connection electricity system will not be possible even in the near future due to its offshore location. Wood, kerosene and diesel are the main fuel types to fulfil energy demands of the residents, whereas wind and solar power as hybrid options have possibilities. Power supply in the island mainly faces three challenges: (i) energy crisis; (ii) issues with the traditional power system's reliability; and (iii) increasing demand for a quality power supply (Ullah, 2013).

Illegal extraction of corals is one of the main issues and is largely attributed to local incoming immigrants and the existing landless local population. The removal of the island's coral base is the island's main critical survival factor. Removal of these coral bases brings risks of beach soil erosion and even the island as a whole. According to Gornitz *et al.* (1982), the island has lost almost 25% of its coral bases over the last decade. This practice also makes the island more prone to sea level rise due to global warming as the outcome of climate change, which will remain as a threat for many decades in the future.

According to Alam *et al.* (2015), almost 7000 people are settled on the island within a relatively small land area and over-extraction of natural resources in the island causes imbalance and scarcity. The peak tourist season of the island is between November and February, and each year on average brings more than 5000 visitors to the island, which is beyond the accommodating capacity.

Ferries, engine boats, trawlers and other carriers operate between the mainland and the island to transport tourists. Figure 18.1 shows passenger ships carrying tourists to the island and causing air pollution. This transportation also leads to other environmental pollution; plastic or related non-biodegradable elements, crude oil and other wastes are thrown overboard. These vessels also cause irreparable harm to the corals by causing direct damage to the boulder coral reef by irresponsible anchoring. An investigation outlines that two-thirds of the islanders believe that the island is at risk; the main causes of pollution are shown in Table 18.1.

Most hotels in the west beach area directly connect sewage pipes into the sea, which contaminates both ground- and adjacent sea-water, causing adverse effects on the biodiversity of the island (Ali and Ahmed, 2000). Untreated domestic and sewage waste materials from both the local population and tourists are also discharged to the sea without treatment. All such waste and other harmful materials degrade the quality of marine water adjacent to the island. Lights from these hotels disturb nearby sandy beaches, which are normally used as turtle nesting grounds. It is believed that, if they encounter disturbance, turtles go back to sea without laying eggs (Islam, 2002). The turtle-nesting and tourist seasons coincide. Tourists also make unexpected noises and disturb the island's ambience with camping or bonfires at night for fun or enjoyment.

Fig. 18.1. Passenger ships causing air pollution.

Table 18.1. Causes of pollution (MRDI, 2015).

Response	Prioritized
Tourist-thrown waste	1
Hotel waste and sewage	2
Extracting coral from water	3
Local inhabitants' waste and sewage	4
No man-made cause... water itself is polluted	5

The majority of the island's younger generation residents are engaged in livelihood-generating tourism activities, while the middle and older generations work in fishing or fishing-related businesses. However, a considerable number of them become involved in tourism-related activities during the off-fishing season. The island has at least 50 residential cottages or hotels, which are mostly owned by relatively affluent people residing in Dhaka and other cities.

This mass flow of tourists and its resulting impact is one of the main reasons that have forced the Government of Bangladesh to announce St Martin's as an environmentally endangered zone. Following the governmental guidelines, any constructions using non-traditional materials are banned, aiming to encourage positive impacts on the island's sustainability. However, it remains very unlikely that the governmental guidelines are strictly followed. Rather, residents and businesspersons continue building non-traditional structures and residential houses are turned into rest houses, hotels or cottages to accommodate increasing numbers of tourists to ensure extra local sources of income. The island currently has more than 52 structures without a proper sanitation system, which has contributed to the island's environmental deterioration over recent years. Flies and mosquitoes are increasing rapidly and breeding over almost all the island. Nesting turtles are very often treated poorly – their nests are destroyed and their hatching capabilities disturbed. A number of fish species are over-fished, making them endangered and in turn limiting the capacity of local fishermen to earn a sustainable income. Fishermen are forced to move further out to sea in search of fish, with increased environmental and social risks.

Thus in the last few years, St Martin's has been facing serious ecological and natural calamities mainly due to the excessive number of tourist visits. Consequently, this is increasingly affecting the environment and nature and the island's fragile ecosystem, which can now hardly cope with increasing tourism numbers.

Responsible behaviour from tourists is also uncommon. Waste and sewage disposal in the island is becoming a serious burden on the islanders. Shells and corals mainly collected by tourists as souvenirs are posing serious threats to the nature and coral colonies of the island (Fig. 18.2). In addition, polyethylene-type non-biodegradable items thrown away by tourists have serious effects on the normal existence of coral. Global warming is becoming a serious concern for coral reef islands across the world, including St Martin's. In practice, global climate change and in particular global warming now also threaten the biodiversity of the coral reefs of this island (IUCN, 2010).

Unexpected natural calamities such as cyclones, sedimentation, beach erosion and storm surges as results of climate change are threatening the island (Ahmed, 1995). Impacts of global warming on the island's flora and fauna are particularly visible. The effect of global warming is predicted to have major consequences for the island's habitats. St Martin's is mostly popular for mangrove forests. Sea level rise resulting from climate change or global warming leads to serious threats to the island's mangrove ecosystem by increased salinity in landward zones, inundation stress and erosion of sediment. The losses of mangrove stands across the island as a result of 'coastal squeeze' are directly affected by demand for human settlements. Mangrove growth is also weakened by reduced silts as they are washed away by heavy rain. These silts are vital for the endurance of the mangrove forests (*The Daily Star*, 2009).

According to IUCN (2010), global warming as the consequence of global climate change leaves high risks to the biodiversity of St Martin's coral reefs. Increasing numbers of cyclones, storm surges followed by high levels of beach erosion and sedimentation threaten the island's coral reefs. Thus, global warming is treated as one of the serious threats to island coral reefs across the world. One of the commonly seen effects of global warming on coral reefs is coral bleaching, which results from high sea temperature (Fig. 18.3), following which they become colourless and unattractive (Fig. 18.4).

Hossain *et al.* (2009) argue that coral reefs in St Martin's Island have already started suffering from high sea temperatures and the effects are visible in their appearance. The coral reef itself depends on a particular algal species that symbiotically lives within its body and by photosynthesis. In cases when the sea temperature rises to above 28°C the algae are expelled by the coral; consequently, this leaves the coral to starve. As Hossain *et al.* (2005) and Brammer *et al.* (1993) mentioned in their studies, coconut trees are an asset of St Martin's. High temperature and water stress have sensitive effects on the reproductive development of these coconut trees. Coconut fruit is severely affected by the reduction of pollen viability leading to smaller, empty or elongated nut production. Pandanus or screw

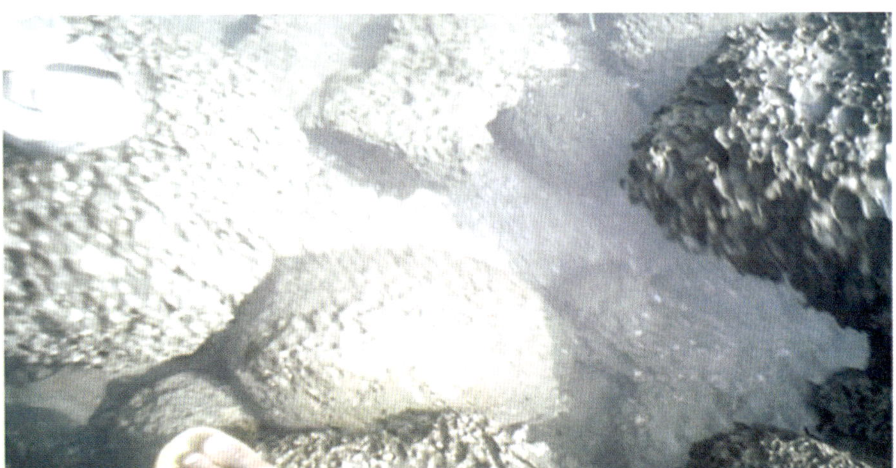

Fig. 18.2. Damage to the coral reef by tourists' footsteps.

Fig. 18.3. Example of coral bleaching.

Fig. 18.4. Decaying coral reef.

pine is iconic of St Martin's, and is affected by increasing salinity leading to untimely death. On the other hand, any removal of such species leads to enhanced beach and soil erosion. Ozone layer depletion as the outcome of global warming results in increased ultraviolet exposure. This can be harmful for both the existence and growth of seaweeds. Ultraviolet rays tend to affect biomolecules directly, reduce productivity and decrease photosynthesis of seaweeds. As a result of the negative effects of global warming, at least 40% of the algae population could experience an untimely death by the end of this century (Tomascik, 1997).

Climate Change: Managing Solutions for Tourism Practices

Relying on results from this research, Fig. 18.5 summarizes the relationship between climate change and coastal tourism development.

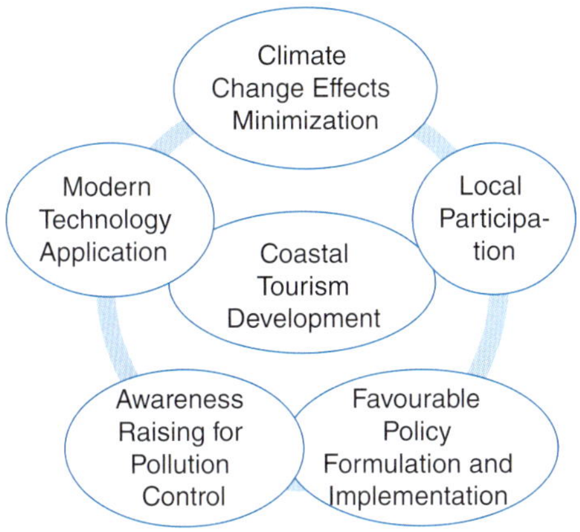

Fig. 18.5. Interrelated factors of climate change and coastal tourism.

According to Kamal (2015), in 2006 Bangladesh embarked into the global Convention of Biological Diversity (CBD) along with the declaration of her Ecologically Critical Areas (ECA). With two coastal and one inland wetland areas, St Martin's Island has been declared an ECA. This declaration encourages researchers to identify basic problems that the island faces and suggest solutions. ECA declaration by the Ministry of Environment and Forests of Bangladesh also encourages creating awareness within the local inhabitants for responsible use of resources. Innovative and sustainable technologies are used for diverse businesses (Azim and Hassan, 2013). Such technologies have acceptability in many areas, including destination image formation, wildlife conservation and visitor management, as relevant to the island (Hassan and Rahimi, 2016; Hassan and Ramkissoon, 2017; Rahimi *et al.*, 2017). St Martin's Island is separate from mainland Bangladesh. A considerable number of residents on the island rely mainly on fishing. A sufficient power supply for these residents along with many hotels and lodges becomes essential. A traditional power supply is not a possibility for this island mainly due to geographical distances and hence alternative sources become the only reliable option. A locally installed diesel generator, supplied by the Bangladesh Power Development Board, used to meet the demand but that becomes out of order from time to time. The islanders are getting used to wind turbine, solar PV, storage battery, diesel generator or biogas to meet their power demand.

St Martin's Island is facing serious consequences and the negative effects of pollution. There are a couple of suggestions to address this: first, St Martin's is desperately in need of cleaning followed by strong participation of the local residents. A similar approach experienced success in Cox's Bazar termed 'the Green World', and this can be replicated on St Martin's. Geographically, the island is far different from Cox's Bazar, meaning that effective initiatives need to be taken during October to March, which is during the busy tourist season, whereas during the other months workers can be given adequate training to raise the level of awareness of local residents and making handicrafts by using alternative resources. Increasing responsiveness of local residents is a core demand that can be achieved by training to protect the island from the negative effects of climate change. The cleaning workers in the off-season can be employed for environmental protection activities.

Second, St Martin's needs joint efforts to take responsibilities and duties to save the island from both natural and man-made disasters. A sound tourism management plan is a necessity for this environmentally fragile tourist attraction. Zoning, tourist access control and an innovative marketing initiative such as package tours can help (Hassan, 2012a, b, c). Trails with a visitor centre and tower need to be set up to reduce aimless movements and activities of the tourists. Corporate agencies through their corporate social responsibility programmes can come forward to incorporate the island's tourism management plan and to reduce negative effects of mass tourism. These activities should also develop branding, publicity and marketing of those corporate agencies. Local residents can also be involved with these activities, allowing them to learn their responsibilities in income-generating activities. Governmental initiatives should involve local residents. Planning and implementation of such activities is crucial to ensure responsibility to benefit all parties and beneficiaries. Local communities should play roles for making these activities successful. Community leaders have duties to initiate any programme in climate change effect reduction and responsible tourism promotion in the island. A considerable number of local residents become unemployed during the lean season forcing them to engage in the illegal coral extraction business to earn their livelihood. These people should be employed in the island's cleaning programme, resulting in both employment and income generation. This should reduce or even eliminate the illicit collection and destruction of coral. The island needs more plantations of different types; fruit trees should be an alternative source of income generation for the local inhabitants with fruits as a tourist commodity. This should rather help to navigate peoples' attention to search for alternative income sources as well as beautification of the island. The involvement of local residents with such initiatives is a demand for today to protect the future from adverse climate change effects. Effective and implementable ideas to support local indigenous technologies can be of particular help to tackle the ill effects of global climatic changes.

Conclusions: Future Expectations

St Martin's Island is a natural reservoir of land, marine and extra-terrestrial resources. The chapter explores awareness-raising against pollution, local participation and innovative technology application to help to minimize negative climate change effects, leading to coastal tourism development. The chapter also finds favourable strategic policy formulation and implementation as supportive for responsible tourism and to reduce negative effects of climate change. Responsible tourist behaviour and use of ecological resources are requirements. The chapter suggests zoning and restricted travel to support the ecology of St Martin's Island because the place itself is fragile and delicate. The chapter also suggests more use of appropriate technologies aiming to reduce natural resource exhaustion to tackle negative effects of global climate change on the island. In addition, the chapter proposes strategic policy formulation for ensuring greater restrictions on mass tourism and related activities with the support of local residents. The chapter thus concludes to ensure a balanced interrelationship between technology application and natural resource use to minimize possible negative effects of climate change on the environment and ecology of this pristine island.

Future expectations as implications are to:

- Strengthen the monitoring capacities of marine, biological and natural resources of St Martin's Island;
- Prepare and implement biodiversity management plans and conservation strategies to reduce negative effects of climate change and global warming;
- Develop *ex situ* conservation and protection for island-dependent species that are near extinction or endangered;
- Offer stimulus imbursements for island-dependent people aiming to find alternative livelihoods;
- Emphasize protecting turtle habitats, coral ecosystems, migratory birds and leaving them undisturbed;
- Establish an information centre to offer reliable and updated information helping to avoid threats to inhabitants, natural, marine and other resources, followed by awareness creation;
- Use print, electronic and other different media for creating public awareness;
- Create more opportunities for research on climate change, coral reef conservation and related areas;
- Introduce carbon reduction and carbon uptake measures;
- Involve both governmental and non-governmental agencies in the conservation and awareness programmes; and
- Introduce community-based conservation programmes attaching local populations to implement an ecosystem approach and fund creation for effective operation of such programmes.

References

Afrin, S., Sultana, N. and Islam, M.T. (2013) Exploring the potentials of 'Culture-friendly' tourism: an empirical study in the Saint Martin's Island. *American International Journal of Research in Humanities, Arts and Social Sciences* 3(1–2), 200–204.

Ahmed, M. (1995) *Biodiversity Conservation of Bangladesh*. Compilation of Theme Papers on Environmental Studies. Bangladesh Technical Education Board, Dhaka, pp. 136–150.

Alam, O., Deng, T.L., Uddin, M.N. and Amin, M. (2015) Application of environmental ethics for sustainable development and conservation of Saint Martin's Island in Bangladesh. *Journal of Environmental Science & Natural Resources* 8(1), 19–27.

Ali, M.O. and Ahmed, M. (2000) Bio-diversity conservation: vision of Bangladesh. *Bangladesh State of Environment Report (Dhaka: FEJB)* 208–223.

Amin, R., Roy, R.B. and Hasan, M. (2014) Modeling and optimization of decentralized microgrid system for St Martin's Island in Bangladesh. *International Journal of Energy, Information and Communications* 5(5), 1–12.

Azim, R. and Hassan, A. (2013) Impact analysis of wireless and mobile technology on business management strategies. *Journal of Information and Knowledge Management* 2(2), 141–150.

Brammer, H., Asaduzzaman, M. and Sultana, P. (1993) Effects of climate and sea-level changes on the natural resources of Bangladesh. *Briefing Document No. 3*. Bangladesh Unnayan Parishad (BUP), Dhaka.

ESCAP (The United Nations Economic and Social Commission for Asia and the Pacific) (1989) *Coastal Environmental Management Plan for Bangladesh Final Report*. ESCAP, Bangkok.

Gornitz, V., Lebedeff, S. and Hansen, J. (1982) Global sea level trends in the past century. *Science* 215, 1611–1614.

Hasan, M.M. (2009) Tourism and Conservation of Biodiversity: A Case Study of St Martins Island, Bangladesh. Available at: www2.warwick.ac.uk/fac/soc/law/elj/lgd/2009_1/hasan (accessed 26 January 2015).

Hassan, A. (2012a) 'Package eco-tour' as special interest tourism product-Bangladesh perspective. *Developing Country Studies* 2(1), 1–8.

Hassan, A. (2012b) Key components for an effective marketing planning: a conceptual analysis. *International Journal of Management & Development Studies* 2(1), 68–70.

Hassan, A. (2012c) Rationalization of business planning through the current dynamics of tourism. *International Journal of Management & Development Studies* 2(1), 61–63.

Hassan, A. and Burns, P. (2014) Tourism policies of Bangladesh – a contextual analysis. *Tourism Planning & Development* 11(4), 463–466.

Hassan, A. and Rahimi, R. (2016) Consuming 'innovation': augmented reality as an innovation tool in digital tourism marketing. In: Nikolaos, P. and Bregoli, I. (eds) *Global Dynamics in Travel, Tourism, and Hospitality*. IGI Global, Hershey, Pennsylvania, pp. 130–147.

Hassan, A. and Ramkissoon, H. (2017) Augmented reality for visitor experiences. In: Albrecht, J.N. (ed.) *Visitor Management*. CAB International, Wallingford, UK, pp. 117–130.

Hossain, M.M., Islam, M.H., Haque, A.H.S. and Mahmud, M.S. (2005) Threats to the Sunderbon mangroves biodiversity and the St Martin's Island of the Bay of Bengal, Bangladesh: need potential intervention for conservation. *Journal of Subtropical Agriculture Research & Development* 3(1), 92–99.

Hossain, M.M., Sultana, R., Ormond, R. and Siddiqui, P.J.A. (2009) Diversity of coral associated fishes in the St Martin's Islands, Bangladesh. *Journal of Taxonomy and Biodiversity Research* 3(4), 21–28.

Islam, M.Z. (2002) Marine turtle nesting at St Martin's Island. *Marine Turtles Newsletter* 96, 19–21.

IUCN (International Union for Conservation of Nature) (2010) St Martin's Island and its unique biodiversity face serious threats. Available at: http://bit.ly/1kld55v (accessed 20 October 2015).

Kamal, A.H.M. (2015) Biodiversity and Ecological Aspects of Saint Martin's Coral Island. Available at: http://bit.ly/1IenVPJ (accessed 26 January 2015).

MRDI (Management and Resources Development Initiative) (2015) Cleaning up of Saint Martin Coral Island. Available at: http://bit.ly/1SQWdxr (accessed 26 January 2015).

NACOM (Nature Conservation Management) (2009) Environmental education for conservation of threatened coral resources in the St Martin's Island Project. *Final Report submitted to South Asia Co-operative Environment Programme (SACEP) and Department of Environment, Government of Bangladesh*. Government of Bangladesh, Dhaka.

Rahimi, R., Hassan, A. and Tekin, O. (2017) Augmented reality apps for tourism destination promotion. In: Rezaei, S. (ed.) *Apps management and e-commerce transactions in real-time*. IGI Global, Hershey, Pennsylvania, pp. 236–251.

SWAMP (Coastal Wetland Biodiversity and Management Program) (2006) *St Martin's Island ECA Conservation Management Plan Draft Report*. United Nation Environmental Program (UNEP)–Global Environmental Facilities (GEF)/Government of Bangladesh (GoB), Dhaka.

The Daily Star (2009) Impact of climate change on St Martin's island. Available at: http://bit.ly/1NkWbh1 (accessed 20 October 2015).

Thompson, P.M. and Islam, M.A. (2010) *Environmental profile of St Martin's island*. United Nations Development Programme, Dhaka.

Tomascik, T. (1997) *Management Plan for Resources of Narikel Jinjira (St Martin's Island) Final Report*. Ministry of Environment and Forest, Government of the Peoples' Republic of Bangladesh, Dhaka.

Ullah, M.H. (2013) An efficient solar-wind diesel-battery hybrid power system for St Martin Island of Bangladesh. *International Journal of Renewable Energy Research* 3(3), 659–665.

19 Case Study Vietnam: Climate Change Vulnerability Assessment of Coastal Tourism in Cu Lao Cham Island

V. Dao Truong[1,2]* and Anh Le[2]
[1]North-West University, Potchefstroom, South Africa; [2]National Economics University, Hanoi, Vietnam

Introduction

Climate change is a major environmental concern that has attracted substantial media and policy attention on a global scale (IPCC, 2014). Within the tourism industry, climate change has emerged as one issue of growing importance in terms of both the contribution of tourism to climate change and the potential detrimental effects of climate change on tourism (Moreno and Becken, 2009; Scott *et al.*, 2012). Coastal and marine areas, which are arguably among the most important tourist destinations in many countries, seem particularly vulnerable to climate change impacts given that they are exposed to both extreme climate events and sea level rise (Moreno and Becken, 2009; Scott *et al.*, 2012; IPCC, 2014). The 2014 report of the IPCC suggests that coastal systems and low-lying areas will increasingly experience submergence, flooding and erosion in the 21st century and beyond due to sea level rise (IPCC, 2014). However, despite the significance of coastal tourism to many national economies, there is limited research into the coastal tourism and climate change interface, particularly in the Asian region.

Vietnam, a developing country where coastal tourism is an important segment of the tourism industry (Truong, 2013), is recognized as one of the world's five countries (the other four are Cambodia, Bangladesh, Senegal and Mozambique) most vulnerable to climate change due to its high population density, long coastline (3260 km) and geographic position in relation to cyclones (United States Agency for International Development (USAID), 2011). Globally, Vietnam is ranked seventh on economic risk, ninth on the percentage of land area and population exposed, and 22nd on mortality from climate change-induced hazards (International Fund for Agricultural Development, 2014). During the past 50 years, Vietnam's average temperature has risen by 0.5–0.7°C and sea level has risen by 20 cm. It is predicted that by the end of the 21st century Vietnam's annual average temperature will increase by 2–3°C. With the anticipated 1 m of sea level rise, about 40% of the Mekong River Delta and 11% of the Red River Delta (Vietnam's largest deltas) will submerge in the future (Government of Vietnam (GOV), 2012a).

Although Vietnam became a signatory to the United Nations Framework Convention on Climate Change (UNFCCC) in 1994 and the Kyoto Protocol in 2002, it was not until recently that climate change has become an

*Corresponding author, e-mail: vandao83@yahoo.co.uk

important item on the GOV's agenda (Ministry of Natural Resources and Environment (MONRE), 2008; GOV, 2012a, b). In 2008, the National Target Program to Respond to Climate Change was formulated, whose goal is to evaluate climate change impacts by economic sectors and geographic regions and develop feasible action plans to ensure sustainable development (MONRE, 2008). In 2012, the National Climate Change Strategy and the National Green Growth Strategy for the period 2011–2020 (Vision to 2050) were established. The former aims to increase overall forest coverage to 45% by 2020, reduce greenhouse gas (GHG) emissions by 20% annually, and recycle and reuse 85% of annual urban domestic solid waste (GOV, 2012a), while the latter seeks to promote green growth as an important component of sustainable development and a contributor to achieving the goals of the former (GOV, 2012b).

Tourism has become increasingly important to Vietnam's economy. In the period 2000–2013, the number of domestic tourists increased from 11.2 million to 35 million, while that of international tourists rose from 2.2 million to 7.5 million (Truong and Le, 2016). In terms of economic contribution, the tourism receipts generated were about Vietnamese Dong (VND) 200 trillion (US$9.5 billion) in 2013, an 11-fold increase from VND17.4 trillion (US$809 million) in 2000 (Truong and Le, 2016). Vietnam's Tourism Law affirms that tourism is a spearhead sector and an important contributor to poverty alleviation (Truong, 2013; Truong *et al.*, 2014; see also Truong and Hall, 2013). However, there are very few detailed assessments of the potential detrimental impacts of climate change on tourism or of the vulnerability of tourist destinations to climate change in Vietnam.

This chapter assesses the vulnerability of coastal tourism to climate change in Cu Lao Cham Island, which has a growing number of tourists but which is predicted to be among few destinations in Vietnam significantly vulnerable to climate change impacts (Truong, 2011; Chu, 2014). Such knowledge is important for the sustainability of the destination as it helps inform the development of potential strategies to anticipate and minimize the impacts of climate change. Vulnerability is broadly defined as the characteristics of a system that make it prone or unresilient to change (Adger, 2006). Climate change vulnerability can be examined through three main dimensions. The first is exposure, which is the nature or the extent to which the system is exposed to climate change. The second is sensitivity or the characteristics of the system that determine the way it may react to change. The third dimension is adaptive capacity, which is the capability of the system to respond successfully to climate variation (Polsky *et al.*, 2007; Ordóñez and Duinker, 2015). The next section provides background information about Cu Lao Cham

Cu Lao Cham Island

Cu Lao Cham Island consists of eight islets. It is located in the central province of Quang Nam and covers an area of 6710 ha (Fig. 19.1). The terrestrial area comprises 595 ha of protected forest and 790 ha of rehabilitation forest. The marine area includes 311 ha of coral reefs and 500 ha of underwater plant life (Nguyen, 2010). About 2500 people reside in Cu Lao Cham, 80% of whom earn their living by fishing (Chu, 2014). Due to its rich biodiversity value, Cu Lao Cham was recognized in 2009 as a UNESCO World Biosphere Reserve (Nguyen, 2010; Le, 2011).

Tourism only commenced in Cu Lao Cham in the early 2000s despite its growth potential. Visitor numbers have increased relatively quickly in the past few years. In 2005, there were less than 7000 tourists to Cu Lao Cham. However, this number rose to 195,000 in 2013 and 237,000 in 2014 (Table 19.1). Tourism is considered as an important sector in the local economy and a means to improve the livelihoods of local residents (Nguyen, 2011; Chu, 2014). In terms of economic contribution, in 2013 tourism reportedly contributed over 30% to Cu Lao Cham's total revenues (Chu, 2014). However, it needs to be noted that more

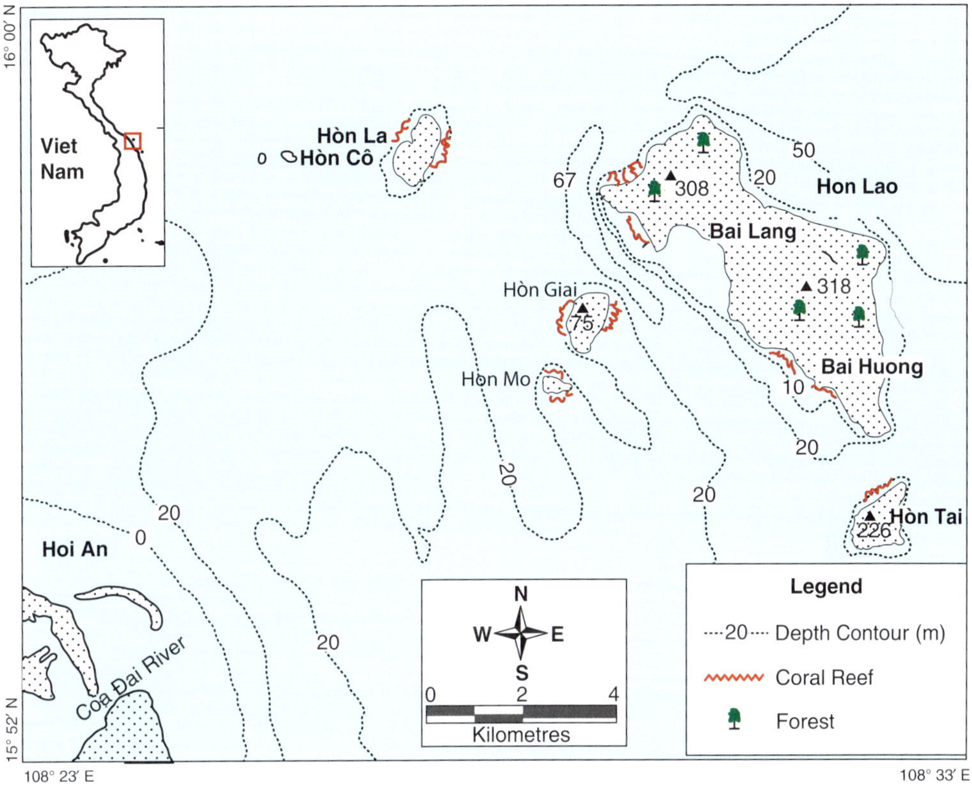

Fig. 19.1. Cu Lao Cham Island (Nguyen, 2010).

Table 19.1. Tourist arrivals to Cu Lao Cham 2005–2014 (Chu, 2014; Chu, personal communication, October 2015).

Year	Domestic	International	Total
2005	5,478	1,410	6,888
2006	4,849	2,361	7,210
2007	11,458	6,801	18,259
2008	12,376	7,784	20,160
2009	26,510	8,123	34,633
2010	30,948	12,165	43,113
2011	56,000	14,000	70,000
2012	77,989	27,085	105,074
2013	156,578	38,422	195,000
2014	160,800	76,200	237,000

than 70% of Cu Lao Cham's tourism products are based on local forest and marine resources (Chu, 2014). Given the continued increase in visitor numbers, this situation may pose a major threat to sustainable development in the island.

The issue of climate change has become a growing concern in Cu Lao Cham, given that Quang Nam is identified as among the few provinces in Vietnam that would be most affected by climate change impacts (Truong, 2011; Chu, 2014). With coastal tourism being an important segment in the province and the island becoming an increasingly popular tourist attraction, there needs to be a systematic assessment of the vulnerability of tourism to climate change, which would help inform policy making with respect to identification of anticipatory and mitigatory measures.

Methods

Our assessment of the climate change vulnerability of coastal tourism in Cu Lao Cham draws upon Moreno and Becken's (2009) five-step methodology (Fig. 19.2),

Step 1. System analysis

- The economic, environmental and social context of the destination is analysed.
- Tourism activities are identified and characterized.
- The importance of these activities is determined.

Step 2. Climate

- Climate conditions are characterized and key hazards identified.
- Activity-hazard subsystems are created.
- Subsystems are chosen for analysis.

Step 3. Vulnerability

- Vulnerability components and indicators are identified.
- Identified components and indicators are fine-tuned.
- Vulnerability is operationalized.
- Steps 1–3 are validated.

Step 4. Integration of individual vulnerability assessment

- Scenarios are established.
- Non-linearities, interdependencies and feedback loops are analysed.
- Results and overall vulnerability assessment are discussed.
- Assessment of validity, scenarios and uncertainties.

Step 5. Communication of results

- Vulnerability assessment results are disseminated to broader audience.

Fig. 19.2. Five-step tourism vulnerability assessment methodology (Adapted from Moreno and Becken, 2009).

which recognizes that vulnerability can only have meaning in a social system and that a tourism destination is a complex system consisting of multiple vulnerability situations. The first step involves analysis of the context of the destination of interest and identification of tourism activities. In the second step, climate conditions and potential climate hazards are identified and a link established between tourism activities and climate hazards. The third step identifies key vulnerability components and their measurement indicators. In step four, the overall climate change impact on the destination is assessed based on individual vulnerability assessments in order to construct climate change scenarios. In the final step, vulnerability assessment outcomes are communicated to relevant stakeholders. Moreno and Becken (2009) suggest that although these steps are sequential in principle, they can be circular depending on the context where they are applied.

We combined both qualitative and quantitative methods. While quantitative methods are useful for collecting numerical data (e.g. data on tourist population), qualitative methods are capable of capturing information that cannot be quantified such as the perceptions of local people and tourism businesses regarding the potential effects that climate change may have on the island (Ordóñez and Duinker, 2015). Focus groups and in-depth interviews were conducted with the island management board (n = 8), local residents (n = 43), tourism businesses (n = 16) and tourists (n = 35). Research data were collected in 3 months, from mid-August to mid-November 2014. Main findings are presented in the following section, which are supplemented by secondary evidence (i.e. the authors' previous work experience in Cu Lao Cham) where relevant.

Results

The five-step methodology, discussed above, provided a useful framework for integrating and structuring qualitative and quantitative information. It also allowed for combination of primary and secondary data in the analysis of the interaction between tourism and climate change.

Step 1

As mentioned above, Cu Lao Cham consists of eight islets, of which the largest is Hon Lao (Fig. 19.1). Its particular biodiversity marine species include corals (277 species), molluscs (97 species), crustaceans (25 species) and seaweed (76 species) (Chu, 2014; see also Nguyen, 2011). The island is home to about 2500 residents, most of whom are fishermen and generally do not obtain formal schooling (Chu, 2014). The local fishing community reportedly produce about 1500 t of sea products annually (Chu, 2014). However, coastal tourism has been an additional income source for the locals since the island became a Marine Protected Area (MPA) in 2005 (Le, 2011; Truong, 2011; Chu, 2014). Key activities undertaken by tourists include snorkelling, scuba diving, mountain climbing, swimming, and relaxing on the beach. Organized tours to Cu Lam Cham are offered daily by most local tour operators, which typically depart from Dang Nang/Hoi An and last 2 days and 1 night. Daytrips are also available. Domestic tourists far outweigh international tourists as shown in Table 19.1. As of 2014, there were 40 local households providing homestay service and 16 households operating restaurants in Cu Lao Cham, alongside providers of other tourist services such as souvenirs and transport services (Chu, 2014).

To ensure effective and efficient operations of the island, a management board was formulated in 2005 as soon as the MPA was established. Management regulations were also issued in the same year, which divided Cu Lao Cham MPA into three main zones: core zone (which is strictly protected), ecological restoration zone and development zone (e.g. ecotourism development). Since 2006, a number of management policies and development plans have been made, including the formulation of the Marine Protection Club, the Tourism Management Unit, and the Marine Development and Protection Plan (Chu, 2014).

Step 2

Vietnam has a tropical monsoon climate, with a considerable amount of sun, a high rate of rainfall and high humidity. Its annual average temperature ranges from 22°C to 27°C. While the northern region can be quite cold in winter and hot in summer, there may be only slight differences in temperature in the southern provinces that are relatively hot throughout the year.

Cu Lao Cham's climate is oceanic tropical. Its annual average temperature is estimated at 26°C. May and July are arguably the hottest months in the island, when the average temperature may reach up to 37–38°C. Between January and March, the average temperature may fall below 14°C and these are generally the coldest months in Cu Lao Cham (Truong, 2011). The dry season lasts from January to August and accounts for only 26% of annual rainfall. The rainy season, which lasts from September to December, produces about 74% of rainfall (Truong, 2011). It is in the rainy season that Cu Lao Cham is highly exposed to hazardous climate events, including storms, tropical depressions, the north-east monsoon and sea level rise. Our interviews and group discussions with the MPA management board and local tourism businesses identified storms and tropical depressions as the most critical impacts that climate change has on tourism operations. Storms and tropical depressions destroy natural (e.g. beaches, coral reefs) and built resources (e.g. infrastructure) and therefore significantly threaten the development of tourism in the island. As such, beach-storm is identified as the most important activity-hazard subsystem. In addition, our interviews with local fishermen and divers suggest that since 2010 coral bleaching has taken place, particularly in

Hon La, Hon Cu, Hon Dai and Hon Tai. With respect to tourism development, coral bleaching potentially reduces the attractiveness of the island given that diving/snorkelling is a major tourist activity as noted earlier. Therefore, snorkelling-coral bleaching is identified as the second main activity-hazard subsystem of concern.

Step 3

First, the vulnerability components of the beach-storm subsystem are presented using Polsky *et al.*'s (2007) Vulnerability Scoping Diagram (VSD) (Fig. 19.3).

Figure 19.3 shows that the exposure elements of the beach-storm subsystem include the statistical risk of storms and those that may be affected by a storm such as local residents, tourists, business owners and infrastructure. Statistical data suggest that on average Quang Nam is directly affected by a storm every 2 years although tropical depressions occur more often. Storms typically hit Cu Lao Cham in September and October, sometimes in May and June (Truong, 2011). The impacts of storms on tourism infrastructure can be measured in terms of monetary value. The exposure of tourists and local residents could be assessed in the form of the number of tourists on the island and population number during a storm, respectively.

The components of the sensitivity dimension identified include the condition of local infrastructure (can be measured by the number of storm-proof buildings), the characteristics of the beach and/or shoreline (can be assessed by the present erosion rates

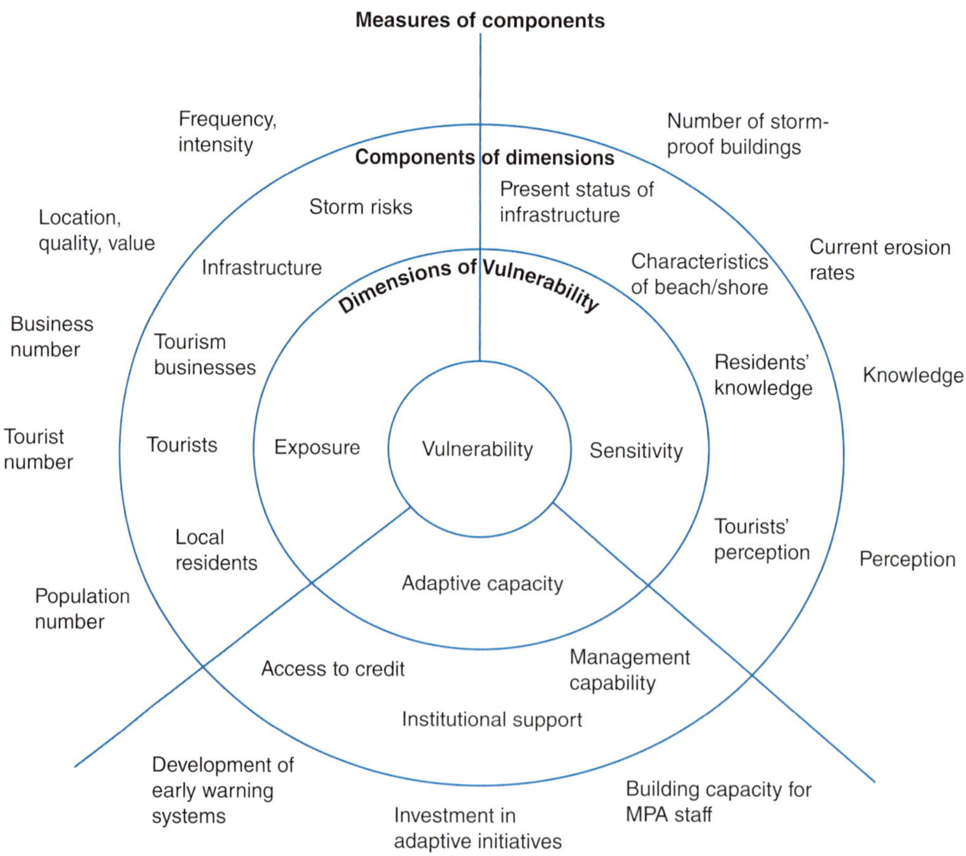

Fig. 19.3. Vulnerability scoping diagram for the beach-storm subsystem.

as well as the materials that are used for the construction of shoreline), the knowledge/experience of local people with respect to storms, and the perceptions of tourists (can be captured by surveys or depth interviews). Our interviews suggest that a large majority of visitors to Cu Lao Cham appear to be concerned about the risks of catching a storm during their visit, particularly in the rainy season. Our discussions with local tourism business owners also indicate that they are well aware of their operations being significantly affected by storms.

Adaptive capacities can be assessed by such components as the management capability of the MPA, access to credit for local residents and businesses, and institutional support. In the last few years, local residents and businesses have enjoyed favourable conditions with respect to bank credit, which allows them to participate meaningfully in tourism-related activities, for example as restaurant owners. Between 2006 and 2013, substantial training and capacity building have also been provided for the MPA staff and local residents, which focuses on tourist guiding, homestay operations and handicraft making. Yet limited attention has been paid to the development of an early warning system. Many interview respondents suggest that hazardous climate events should be notified as early as possible and that preventive instructions should be provided.

The vulnerability dimensions of the snorkelling-coral bleaching subsystem are shown in Fig. 19.4. The main components of the exposure dimension identified include the location and coverage of the coral

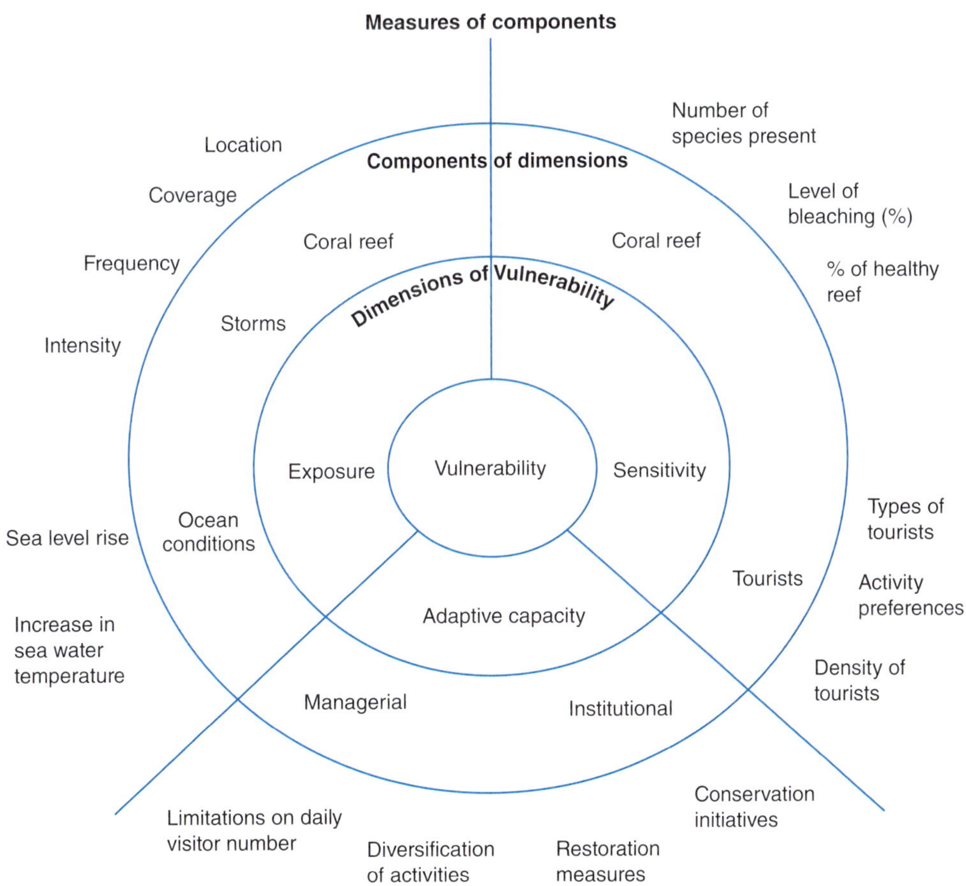

Fig. 19.4. Vulnerability scoping diagram for the snorkelling-coral bleaching subsystem.

reefs, and storms and ocean conditions that are key drivers of coral bleaching. Cu Lao Cham's coral reefs cover about 311.2 ha in total and are mainly found in the western and south-western ends of the island. A large majority of these reefs are located in shallow waters that are less than 14 m deep. Storms, sea level rise and increases in seawater temperature have resulted in serious bleaching of coral reefs, particularly in Bai Bac and Bai Huong. The extent to which the reefs are sensitive to changing conditions is influenced by reef characteristics such as the quantity and diversity of coral species as well as by the types and density of visitors to the island and their preferred activities. A number of adaptive measures have recently been proposed, including restriction of visitor number to less than 3000 arrivals between 7 am and 2 pm every day. It has also been proposed that vulnerable corals should be identified and taken to special zones (e.g. nursery ground) where they can be protected and nurtured (Nguyen, 2011).

Step 4

To assess the indicators that measure the vulnerability components of the two subsystems, a scoring system was developed where the MPA staff were invited to evaluate each component using a 5-point Likert scale. The degree of climate change vulnerability of each activity-hazard subsystem is presented in Table 19.2.

Stakeholder knowledge and existing literature suggest that under the intermediate emission scenario Cu Lao Cham's average annual temperature will likely increase by 1.3°C by 2050, 1.9°C by 2070 and 2.5°C by 2100. It is predicted that the island's annual precipitation will increase by 1.6% by 2050, 2.3% by 2070 and 3.0% by 2100. Sea level will rise by 30 cm by 2050, 75 cm by 2070 and 100 cm by 2100 (Chu, 2011). These changing climatic conditions will likely result in significant reductions in the area of the beach and shoreline, higher rates of shore erosion and decreases/losses of coral reefs due to bleaching and mortality. In 2008, the island's coral coverage reduced by 6.5% as compared to 2004 due to climate change wherein storms and tropical depressions were identified as one of the most critical causes (Nguyen, 2011). Unless adaptation strategies are put in place, the coverage of coral reefs will be significantly reduced in the future due to climate change impacts (Nguyen, 2011; Truong, 2011; Chu, 2014).

The identification of potential adaptation measures is an integral part of any vulnerability assessment (Adger, 2006; Moreno and Becken, 2009; Ordóñez and Duinker, 2015). The shoreline may be protected by seawalls, as in the case of the Mamanuca Islands in Fiji (Moreno and Becken, 2009). The impacts of storms and tropical depressions may potentially be mitigated with the development of an early warning system, which, however, is lacking. As discussed above, interview respondents have stated that they need to be notified of hazardous climate events as early as possible and that preventive instructions are required so that they can better prepare for those events. Such preventive instructions should be designed in consultation with community members given their local knowledge and experience (Truong, 2011).

It may be necessary to limit the number of daily visitors to the island. As discussed earlier, it has recently been proposed that the island should not accommodate more than 3000 visitors each day. While this proposal appears appropriate from a climate change perspective, it is not supported by tourism businesses whose revenues will be significantly affected. In addition, it contrasts starkly with the local government's

Table 19.2. Summary of vulnerability components of two activity-hazard subsystems.

Subsystem	Exposure	Sensitivity	Adaptive capacity	Overall vulnerability
Beach-storm	Very high	Very high	Very low	High
Snorkelling-coral bleaching	High	High	Low	Moderate

policy of developing coastal tourism as a means of improving local people's living standards and of protecting local forest and marine resources (i.e. as local people can participate in and earn income from tourism-related activities, fewer of them will likely rely on local forest and marine resources for food and income). Improving livelihood-seeking capability for island residents, for example through formal education and capacity building, may be a potential solution. It may also be important to raise the awareness of the public and tourism businesses regarding the issue of climate change so that they support the proposal of hosting less than 3000 visitors to the island each day.

Furthermore, conservation initiatives should be encouraged and undertaken where appropriate. For example, it may be necessary to develop nursery grounds where corals are nurtured and used to replace those that are bleached or destroyed due to climate change impacts. Where corals are seriously affected by climate change events, such as in Bai Bac and Bai Huong, they should be relocated to areas of strict conservation so that they can be restored properly. For these adaptation measures to succeed, the MPA staff need to improve their knowledge and capability and therefore training and capacity building need to be provided.

Step 5

The outcomes of this climate change vulnerability assessment could be distributed to interested parties in Cu Lao Cham (e.g. the MPA management board, tourism businesses), Quang Nam province (e.g. the Department of Tourism), or other island destinations in Vietnam (e.g. Nha Trang, Phu Quoc).

Conclusion

Tourism in coastal and marine areas has an established legacy (Moreno and Becken, 2009). It is still the most important form of tourism and is the largest segment of the tourism industry in many countries (Scott *et al.*, 2012). However, coastal destinations are also highly vulnerable to climate change, which potentially destroys natural and built resources, reduces the attractiveness of destinations, threatens the existence of businesses and puts community livelihoods at risk. Despite the significance of coastal tourism to many national economies, there is a dearth of research into the tourism and climate change interface particularly in the Asian region.

This chapter builds on a five-step methodology (Moreno and Becken, 2009) to assess the vulnerability of coastal tourism to climate change in Cu Lao Cham Island, Vietnam, where tourism has been considered a mechanism to improve the living standards of local residents (Chu, 2014). It has identified beach-storm and snorkelling-coral bleaching as the two most important activity-hazard subsystems of concern. A number of adaptation solutions have been suggested to mitigate climate change impacts. Although there has been a substantial body of research on vulnerability to climate change, detailed vulnerability assessments of tourism in coastal environments are relatively limited. This chapter, therefore, contributes to enriching the extant literature on the interaction between coastal tourism and climate change. It suggests that the five-step vulnerability assessment methodology developed by Moreno and Becken (2009) provides a useful framework for detailed analysis of different vulnerability situations in different contexts and environments. It may allow researchers to make systematic comparisons of results across independent vulnerability assessments, thereby developing a greater understanding of vulnerability hotspots on the world tourism map (Polsky *et al.*, 2007; Moreno and Becken, 2009).

Some limitations in this research need to be acknowledged. Although this study has analysed the two most important activity-hazard subsystems in relation to climate change vulnerability, other subsystems should also be examined. For example, water supply may present a challenge to homestay/restaurant owners in the dry season in the island.

In addition, given that this is arguably the first research on climate change vulnerability and coastal tourism in Cu Lao Cham, further studies are required to expand and/or validate the findings of this study. Further research may also examine the climate change vulnerability of tourism in other coastal destinations in Vietnam, such as Nha Trang and Phu Quoc, to assess the methodology's validity, compare vulnerability across destinations and develop potential adaptation strategies.

References

Adger, W.N. (2006) Vulnerability. *Global Environmental Change* 16, 268–281.

Chu, M.T. (2011) Waste Management in Cu Lao Cham Marine Protected Area and the Potential Effects of Climate Change. Available at: www.culaochammpa.com.vn (accessed April 2015).

Chu, M.T. (2014) Opportunities and Challenges in the Management and Preservation of Biodiversity Value of Cu Lao Cham Biosphere Reserve – Hoi An. Available at: www.culaochammpa.com.vn (accessed April 2015).

Government of Vietnam (GOV) (2012a) *National Climate Change Strategy*. GOV, Hanoi.

Government of Vietnam (GOV) (2012b) *National Green Growth Strategy*. GOV, Hanoi.

Intergovernmental Panel on Climate Change (IPCC) (2014) *Climate Change Report 2014*. IPCC, Geneva.

International Fund for Agricultural Development (2014) Comprehensive Environment and Climate Change Assessment in Vietnam. Available at: https://operations.ifad.org/documents/654016/507936/Viet+Nam+-+Environmental+and+climate+change+assessment/da669eb7-f21b-441d-acdf-abad4a4ff5e9 (accessed April 2015).

Le, V.H. (2011) Research Report: The Impacts of Climate Change on Biodiversity in Cu Lao Cham Biosphere Reserve. Available at: www.culaochammpa.com.vn (accessed April 2015).

Ministry of Natural Resources and Environment (MONRE) (2008) *National Target Program to Respond to Climate Change*. MONRE, Hanoi.

Moreno, A. and Becken, S. (2009) A climate change vulnerability assessment methodology for coastal tourism. *Journal of Sustainable Tourism* 17(4), 473–488.

Nguyen, T.T.N. (2010) Effectiveness evaluation of a marine protected area in Vietnam – The Cu Lao Cham MPA case study. Master's Thesis, University of Tromso, Norway. Available at: http://munin.uit.no/bitstream/handle/10037/2539/thesis.pdf?sequence=2 (accessed April 2015).

Nguyen, V.L. (2011) The Impacts of Climate Change on Biodiversity in Cu Lao Cham Marine Protected Area. Available at: www.culaochammpa.com.vn (accessed April 2015).

Ordóñez, C. and Duinker, P.N. (2015) Climate change vulnerability assessment of the urban forest in three Canadian cities. *Climatic Change* 131, 531–543.

Polsky, C., Neff, R. and Yarnal, B. (2007) Building comparable global change vulnerability assessments: the vulnerability scoping diagram. *Global Environmental Change* 17, 472–485.

Scott, D., Gössling, S. and Hall, C.M. (2012) *Tourism and Climate Change*. Routledge, Abingdon, UK.

Truong, T. (2011) Research Report: The Impacts of Weather, Climate and Natural Disasters on Cu Lao Cham. Available at: www.culaochammpa.com.vn (accessed April 2015).

Truong, V.D. (2013) Tourism policy development in Vietnam: a pro-poor perspective. *Journal of Policy Research in Tourism, Leisure & Events* 5(1), 28–45.

Truong, V.D. and Hall, C.M. (2013) Social marketing and tourism: what is the evidence? *Social Marketing Quarterly* 19(2), 110–135.

Truong, V.D. and Le, A. (2016) The evolution of tourism policy in Vietnam, 1960–2015. In: Hall, C.M. and Page, S. (eds) *The Routledge Handbook of Tourism in Asia*. Routledge, London.

Truong, V.D., Hall, C.M. and Garry, T. (2014) Tourism and poverty alleviation: perceptions and experiences of poor people in Sapa, Vietnam. *Journal of Sustainable Tourism* 22(7), 1071–1089.

United States Agency for International Development (USAID) (2011) Climate Change in Vietnam: Assessment of Issues and Options for USAID Funding. Available at: https://www.usaid.gov/sites/default/files/documents/1861/vietnam_climate_change_final2011.pdf (accessed April 2014).

20 Case Study New Zealand: Planning Responses to Coastal Climate Change Risks: The Case of Christchurch and the Akaroa Harbour, New Zealand

C. Michael Hall*

University of Canterbury, Christchurch, New Zealand; University of Oulu, Finland; and Linneaus University, Kalmar, Sweden

Introduction

That climate change presents a major challenge to coastal areas is clear in the scientific literature. However, what remains unclear in policy and planning terms is how best to respond to it, especially in multi-level political systems. This chapter examines the planning response to climate change and sea level rise (SLR) in the Akaroa Harbour area of Banks Peninsula near Christchurch in the South Island of New Zealand. Land use planning in the area is primarily the responsibility of the Christchurch City Council (CCC). Until March 2006 the peninsula had its own district council, however it was amalgamated with the CCC in 2006.

The area is of historical significance as a second home and day-trip destination for Christchurch residents (Fountain, 2002; Fountain and Hall, 2002; Nuth, 2007) with second homes concentrated in the settlements of Akaroa, Duvauchelle, Takamatua and Wainui, as well as some smaller settlements. However, it has also increasingly become an important heritage and ecotourism destination for international visitors, with Akaroa 'reliant on tourism and holiday makers for its economic viability' (Christchurch City Council (CCC), 2014, p. 1). In addition, following the loss of cruise ship facilities in Lyttelton Harbour (Christchurch's port) after the 2010–2011 Christchurch earthquake sequence (Hall *et al.*, 2016), Akaroa Harbour has received a rapid increase in cruise arrivals (Wilson *et al.*, 2015), with 85 cruise ship visits between October and April 2014 (CCC, 2014).

Although the economic significance of second homes in the region is recognized (CCC, 2015a), there are no statistics kept specifically on them. Therefore a useful proxy for second homes in New Zealand is the proportion of unoccupied houses at the time of the national census (Keen and Hall, 2004). Table 20.1 illustrates how almost two-thirds of houses in the Akaroa area are potential second homes. Significantly, many of the holiday homes (or baches as they are referred to locally) are located in small coastal settlements on the harbour and some significant tourism infrastructure, such as caravan parks, hotels, motels, restaurants and retail shops are also located in waterfront locations.

The first part of the chapter discusses the nature of the location before identifying

*E-mail: michael.hall@canterbury.ac.nz

Table 20.1. Unoccupied dwellings in Akaroa and Akaroa Harbour census districts 2001–2013 (Derived from CCC, 2015a).

Area unit	2001	2006	2013
Akaroa			
Usually resident population	576	567	624
Unoccupied dwellings	516	564	567
Occupied dwellings	324	315	342
Unoccupied as percentage of total dwellings	61%	64%	62%
Akaroa Harbour			
Usually resident population	681	735	777
Unoccupied dwellings	465	459	531
Occupied dwellings	315	342	372
Unoccupied as percentage of total dwellings	60%	57%	59%

the potential impacts of climate change as a result of SLR. The chapter then discusses how the local planning system sought to signal the likely future effects of climate change before the approach has halted as a result of pressure from affected parties and intervention by the national government. The study therefore illustrates not only the difficulties of initiating planning measures that respond to coastal climate change risk but also how tourism planning is almost incidental to general planning and policy measures, and how different communities of interest may coalesce around planning issues.

The Location

Banks Peninsula is an area of approximately 1150 km^2 that lies immediately to the south of Christchurch, the South Island's largest city. The peninsula comprises the eroded and partially flooded cones of the extinct Akaroa and Lyttelton Volcanoes. These two basaltic shield volcanoes, 30 km in diameter, formed an island 6 to 11 million years ago. Subsequent sea level change and deposition from weathering of the Southern Alps has resulted in the island subsequently becoming a peninsula of the mainland.

Akaroa Harbour is around 17 km long, oriented north to south, with a relatively constant width of 2–3 km. The shoreline along the southern end of the harbour is dominated by steep cliffs of volcanic rock. The harbour entrance is approximately 25 m deep and decreases to approximately 10 m halfway up the harbour. The upper section of the harbour has a number of bays interspersed between rocky headlands and wave-shore platforms (Tonkin & Taylor, 2013).

During glacial periods the volcanic rocks of the peninsula have become overlain by thick deposits of loess in many areas. This is significant for sedimentation of the upper harbour as it has been gradually infilled from its surrounding catchments. The loss of native forest vegetation on the peninsula, with most of the forest being felled or burned for timber or grazing land during the first 50 years of European occupation (Department of Conservation, 2006), together with the development of pastoral farming and small urban settlements has further contributed to sedimentation. Sedimentation runoff and the loss of vegetation cover are significant climate change issues because of the potential increase in the intensity of rainfall (IPCC, 2013) and the role of coastal ecosystems in shoreline stability (McGlone and Walker, 2011).

Although there is significant awareness of the impacts of climate change at a global scale, it is difficult to translate many of the global forecasts to the local level because of different geographical conditions, including built environments. Nevertheless, there is significant interest in the impact of coastal natural hazards in the study area not only because of SLR but also because the geomorphology of the area makes it highly susceptible to the impacts of tsunamis generated in New Zealand and South America

(De Lange and Healy, 1986; Goff and Chagué-Goff, 2012) and because of the impacts of the 2010–2011 Christchurch earthquake sequence (Hall *et al.*, 2016). The large number of earthquakes following the initial Darfield Earthquake on 4 September 2010 caused extensive liquefaction, lateral spreading and vertical displacement resulting in substantial damage to housing and infrastructure as well as loss of life. The impact of the earthquakes also led to substantial changes in ground levels in some locations of the order of +/–0.5 m (Tonkin & Taylor, 2013). This has also meant that some parts of Christchurch and the region have increased vulnerability to flooding as a result of high magnitude storm events (Allen *et al.*, 2014). Significantly, with respect to consideration of SLR, both tsunami risk and earthquake impact, e.g. susceptibility to liquefaction, have become considerations in land-use planning. For example, following the earthquakes, land that could be rebuilt on ('green zone land') was divided by the Canterbury Earthquake Recovery Authority (CERA) into three technical categories (TC) (CERA, 2013) describing how the land is expected to perform in future quakes:

> TC1: Future land damage from liquefaction is unlikely.
>
> TC2: Minor to moderate land damage from liquefaction is possible in future significant earthquakes.
>
> TC3: Moderate to significant land damage from liquefaction is possible in future significant earthquakes. TC3 houses requiring foundation repairs or new foundation require site-specific geotechnical assessment and specific engineering foundation design to be compliant with the building code.

Assessment of Sea Level Rise

DTec (2008) undertook an assessment of coastal erosion and inundation for all Akaroa Harbour settlements including the potential impacts of sea level rise on shoreline stability. The DTec report was based on a future SLR projection of 0.2 m by 2050. Based on this scenario the beaches of Takamatua and Duvauchelle were calculated to retreat approximately 15 m by 2050. Akaroa and Wainui were predicted to experience retreat of 2–8 m by 2050. DTec (2008) calculated a 50-year inundation level of 1.7 m above mean sea level (MSL), including tides, storm surge and SLR. Even at this level of impact they suggested (DTec, 2008, p. ii):

> The coastal hinterland of Akaroa Harbour is dominated by roads immediately behind the shoreline, with the stability of large sections of the roading being dependant on the 8 km of coastal protection structures. In many places there is little opportunity for re-location of these road networks, therefore, any erosion and/or failure of the coastal protection structures will affect these transport corridors.

Approximately 50% of these structures were recognized as being at risk from overtopping and inundation during storm surges (DTec, 2008). SLR will increase the frequency of overtopping along these structures as well as increasing the potential for failure: 'Up to 15% (1250 m) of these structures are assessed as already being in a poor to very poor condition, and will likely to almost certainly require major maintenance or replacement if they are to be effective at providing a protection function over the next 30 to 50 years' (DTec, 2008, p. ii).

A more recent analysis of climate change and SLR in the Akaroa Harbour was conducted by Tonkin & Taylor (2013) for the CCC post the Christchurch earthquake sequence and served as the basis for their recommendations to the council on coastal inundation and erosion hazard zones (Tonkin & Taylor, 2015). Tonkin & Taylor (2013) took a different SLR (1 m) and timescale (100 years) to DTec and used different methods of assessing SLR depending on the nature of the shoreline. For sandy beaches they used the Bruun Rule, also used by DTec (2008), which predicts that as sea level rises against a shore profile in equilibrium, beach erosion takes place to provide sediments to the nearshore so that the seabed can elevate in direct proportion to the rate of sea level rise (Bruun, 1988).

The second method employed by Tonkin & Taylor was the passive inundation method, which is simply the position of the high tide level (Mean High Water Springs (MHWS)) in 2115 allowing for a sea level rise of 1.0 m. The landward extent of shoreline retreat through passive inundation is the intersect position between the sea level rise projection elevation above MHWS and the existing profile. Tonkin & Taylor (2013) used a 10% MHWS level, which is exceeded by 10% of the MHWS. Based on a MWHS level of 1.15 m above the regional (Lyttelton) Vertical Datum 1937 (LVD) and the sea level rise projection of 1.0 m, the passive inundation level is 2.15 m (LVD). Therefore, if the backshore profile remained constant over time the level in 2115 would be the existing 2.2 m contour (after rounding) (Tonkin & Taylor, 2013). A third element of assessing SLR is the effects of storm surge. Future changes to storm surge will depend on changes in the frequency and intensity of low pressure systems and although Tonkin & Taylor (2013, p. 30) note that 'there are significant uncertainties over changes to storm events in New Zealand due to climate change over the next 100 years', the IPCC (2013) suggest increased extreme precipitation associated with tropical and extra-tropical storms. Nevertheless, Tonkin & Taylor took a very conservative approach, stating 'Since tidal characteristics are expected to remain unchanged by future sea level rise, storm tide characteristics are expected to remain similar (MfE, 2008). Therefore, to predict future extreme inundation levels, sea level rise can simply be added to the present day storm tide levels' (Tonkin & Taylor, 2013, p. 30). This meant that they used the current 1% annual exceedance probability (AEP) storm tide level for the region, added the 1.0 m for SLR to 2115 and a further freeboard allowance of 0.4 m to allow for some localized wave effects and other uncertainties, which is the existing freeboard set out in the Christchurch City District Plan for minimum floor levels in areas vulnerable to flooding. Therefore, the 1% AEP inundation level including sea level rise used for their study is 3.3 m (Tonkin & Taylor, 2013). The results of the different analyses of SLR are shown in Table 20.2 together with details of the shoreline for the main resort townships in Akaroa harbour (Figs 20.1–20.3).

Impacts of sea level rise

Tonkin & Taylor (2013, 2015) concluded that coastal erosion and shoreline retreat is likely to be exacerbated by climate change in all of their study areas within Banks Peninsula due to the following effects:

- rise in mean sea level;
- possible increase in the frequency and intensity of coastal storms; and
- possible re-orientation of shorelines in response to changes in wave climate.

The impacts of SLR identified by Tonkin & Taylor (2013, 2015) extend those identified by DTec (2008). Significantly, the quality of many of the sea protection measures evaluated by DTec (2008) and found to be in poor condition has only been further lessened post the earthquake sequence. A key issue for any destination is access and the greatest impact of SLR is on road access to all four of the settlements discussed in this study.

All of the main coastal roads are directly affected by SLR. Some of the roads are already protected by vertical seawalls but the seawalls have relatively low crests and are likely to be overtopped more frequently during future storm events and surges than they are at present (Tonkin & Taylor, 2013). However, the height of the crests can only be raised to the existing backshore levels as otherwise it would create drainage issues. As sea level rises there will also be an increase in maintenance costs for seawall remedial works associated with overtopping scour. The foreshore levels fronting the seawalls are also expected to lower over time as a result of shoreline retreat associated with SLR, resulting in increased scouring at the base of seawalls, leading to undermining and slumping; 'Therefore, unless significant reconstruction or replacement works are undertaken the existing sea walls are likely to fail as sea level rises providing limited protection from future shoreline retreat' (Tonkin & Taylor, 2013, p. 29).

Table 20.2. Summary of the estimated shoreline retreat due to potential sea level rise in Akaroa Harbour (DTec, 2008; Tonkin & Taylor, 2013; author).

Settlement	Characteristics	Predicted shoreline retreat (m) by 2058 at 0.2 m SLR	Equilibrium profile shoreline retreat (m) at 1 m SLR by 2115	Passive inundation shoreline retreat (m) at 1 m SLR by 2115	Area inundated (ha) from 1% AEP inundation event including SLR by 2115
Akaroa (North)	Akaroa is located in French Bay and consists of two shoreline types, with a natural shoreline located in the north of the bay and a modified shoreline located adjacent to the main settlement which is mostly protected by a vertical seawall. The northern bay has a silty sand beach and intertidal flats that are approximately 230 m wide	8.5	70	20	13.3
Akaroa (South)		2.5	20	170	
Duvauchelle	Duvauchelle comprises two bays with silty sand and gravel beaches. The bays are fronted by intertidal flats that are approximately 350 m wide and are protected with a vertical sea wall along the majority of the length	6.0–14.5	100	60	10.5
Wainui	Wainui is located on the western shoreline of Akaroa Harbour and has a relatively steep gravel beach. The backshore elevation is approximately 3.5 m above MHWS	3.5–8.5	10	10	3.4
Takamatua	Takamatua Bay has a silty sand and gravel beach with intertidal flats that are approximately 730 m wide. The backshore elevation is approximately 1.0 m above MHWS	3.5–16	200	100	7

MHWS, mean high water springs; SLR: sea level rise; AEP, annual exceedance probability

A summary of the impacts on the four main settlements in Akaroa Harbour is provided in Table 20.3. The impacts in the table are derived from CCC documents that have utilized the Tonkin & Taylor (2013) report. Both of the main business districts in Akaroa, which include a number of restaurants, hotels and accommodation providers,

Fig. 20.1. Apartment complex, Duvauchelle. Note the existing normal high water mark on the foreshore.

Fig. 20.2. State Highway 1 at Duvauchelle is already affected by storm surges and king tides, the effects of which will be exacerbated by sea level rise.

Fig. 20.3. The sea wall at Duvauchelle is already showing the effects of erosion and decisions will need to made regarding the cost of maintenance.

are directly threatened by a 1.0 m sea rise, along with accommodation and a caravan park at Duvauchelle. In addition, holiday homes in all four settlements are threatened (Fig. 20.4).

How accurate is it?

In New Zealand sea level was rising slowly (0.3±0.3 mm/year) from AD 1500 to AD 1900, but during the 20th century the rate increased to 2.8±0.5 mm/year (Gehrels *et al.*, 2008). Using new datum and a new approach Hannah and Bell (2012) provide an average relative sea level rise since the early 20th century of 1.7±0.1 mm/year with the Lyttelton figure being 1.9±0.1 mm/year. Tonkin & Taylor (2013) used Ministry for the Environment (MfE) (2008) guidance to local government in New Zealand that considers a 0.5 m base value of sea level rise by 2090 relative to the 1980–1999 average sea level, with 0.1 m additional rise per decade thereafter. The MfE guidance also suggests that scenarios above 0.8 m should also be considered for coastal risk planning beyond 2100 as well as

Table 20.3. Impacts of the estimated shoreline retreat and potential sea level rise in Akaroa Harbour (CCC, 2014, pp. 11–12; 2015a, pp. 445, 460, 467, 469).

Settlement	Statements in Christchurch City Council documents
Akaroa	Coastal erosion and shoreline retreat is likely to be exacerbated by climate change in all areas of Banks Peninsula due to a rise in mean sea level, possible increase in frequency of coastal storms, and possible re-orientation of shorelines in response to changes in wave action (CCC, 2014, p. 11).
	Potential effects could involve Jubilee Park, the Recreation Ground, Beach Road, the Boat Compound, lower Pompallier Village, lower Rue Grehan and Rue Jolie North. Most public assets affected are low-lying roads providing access to the town and south of the town centre. Current sea walls are likely to be overtopped more frequently during storm events (CCC, 2014, p. 12).
	Future development of and in identified areas will need to take into account projected sea level rise and the effects of projected climate change (CCC, 2014, p. 11).
Duvauchelle	The coastline at Duvauchelle currently already experiences overtopping events during coastal storms. These instances will increase as sea levels continue to rise (CCC, 2015, p. 460).
Wainui	Predicted shoreline retreat rates and distances for sea level rise at Wainui south end (3.5 m in 50 years), middle (8.5 m in 50 years) and north end (6.0 m in 50 years)... Storm wave heights and run-up elevations (water level at 1.7 m about MSL) south end (seawall 2–3.5 m, road 2–3.5 m) and north end (seawall 3.3 m, road 2–3 m). North of Wainui Valley Road likely to erode, maximum amount less than 10 m. Wainui Main Road likely to be affected within 50-year time frame. No other land likely to be affected.
	Set backs for road relocation advisable over next 50 years. Concrete seawall at end of Wainui Valley Road in very poor condition and will require replacement or major maintenance. For 400 m section of coast south of boat ramp and rock groyne, erosion rate of 5 m predicted which will affect sustainability of Wainui Main Road over next 50 years. Limited ability to reposition road, therefore coastal protection works required (CCC, 2015, p. 445).
Takamatua	Sea level rise is... a concern as the settlement slopes downwards from the beach. As such it is recommended that expansion be kept to a minimum and that which does occur should be on the higher ground to the north of the settlement (subject to land suitability).
	The lower elevations of Takamatua Bay are currently at risk from contemporary storm surge events and higher elevated land will become exposed to this risk as sea levels rise particularly around the vicinity of the Takamatua Stream (CCC, 2015, p. 469).

MHWS, mean high water springs; SLR, sea level rise

for low probability/high consequence considerations. In addition to its 2013 report Tonkin & Taylor (2015) also considered the IPCC's AR5 results although noted 'these figures do not include for the collapse of the marine-based sectors of the Antarctic ice sheet. However, this contribution is not likely to exceed several tenths of a meter of sea level rise during the 21st century' (2015, p. 6).

At the global level the IPCC (2013) scenarios suggest that for the period 2081–2100, compared to 1986–2005, global mean SLR is likely (medium confidence) to be in the 5% to 95% range of projections from process-based models, which give 0.26 to 0.55 m for RCP2.6, 0.32 to 0.63 m for RCP4.5, 0.33 to 0.63 m for RCP6.0 and 0.45 to 0.82 m for RCP8.5. For RCP8.5 (the business as usual scenario), the rise by 2100 is 0.52 to 0.98 m with a rate during 2081–2100 of from 8 to 16 mm/year. The IPCC (2013) also discusses semi-empirical estimates of SLR, which provide 0.7–1.5 m for the RCP8.5 scenario (936 ppm CO_2), but preference is given to the model-based estimate of 0.52–0.98 m. The IPCC scenarios are close to linear responses to the assumed rising climate forcing. In contrast, Hansen *et al.* (2015, p. 20119) raise fundamental questions as to 'whether ice sheet melt in response to rapid global warming will be nonlinear and better characterized by a doubling time for its rate of change or whether more linear processes dominate'. Rather than use an ice sheet

Fig. 20.4. Properties at Akaroa that are exceptionally vulnerable to sea-level rise and storm surge according to consultants' reports.

model to estimate rates of freshwater release utilized by the IPCC, Hansen *et al.* (2015) used *observations* for the *present* ice melt rate and specified several alternative rates of increase of ice melt. They suggest that the IPCC forecasts are far too conservative, concluding that multi-level SLR is practically unavoidable unless there is a dramatic reduction in emissions even further than what is estimated for a supposedly 'safe' 2°C. 'Humanity faces near certainty of eventual sea level rise of at least Eemian proportions, 5–9 m, if fossil fuel emissions continue on a business-as-usual course, e.g. IPCC scenario A1B that has CO_2 700 ppm in 2100' (Hansen *et al.*, 2015, p. 20119).

A further issue with respect to the assessment is with respect to the accuracy of the Bruun rule. Cooper and Pilkey (2004) suggest that several assumptions behind the Bruun Rule are known to be false and that the rule has not been adequately proven. Instead they argue that no universally applicable model of shoreline retreat under sea-level rise has yet been developed. They observe that the persistence of the concept as a management tool can be attributed to the following:

1. Appeal of a simple, easy to use analytical model.
2. Difficulty of determining the relative validity of 'proofs' and 'disproofs'.
3. Ease of application.
4. Positive advocacy by some scientists.
5. Application by other scientists without critical appraisal.
6. The simple numerical expression of the model.
7. Lack of easy alternatives.

Cooper and Pilkey (2004, p. 157) conclude, 'the Bruun Rule has no power for predicting shoreline behaviour under rising sea level and should be abandoned. It is a concept whose time has passed. The belief by policy makers that it offers a prediction of future shoreline position may well have stifled much-needed research into the coastal response to sea-level rise'. Indeed, it has been suggested that coastal risk management planning frameworks now require a more probabilistic rather than deterministic approaches such as that of the Bruun Rule (Ranasinghe *et al.*, 2012). Indeed, Ranasinghe and Stive (2009, p. 468) conclude that 'any predictions obtained via the Bruun rule should be considered only as broadly indicative estimates that are not suitable for direct use in coastal planning and management'. This does not mean that SLR is not occurring but it does mean that its shoreline effects in many circumstances are not easily predictable. Yet such an insight should suggest that even more caution needs to be applied to coastal zone management and the impacts of climate change rather than less.

Planning for Sea Level Rise

In undertaking their study of coastal hazards in the Christchurch region Tonkin & Taylor (2013, p. 20) observed that 'Climate change is not expected to create any new coastal hazards that do not already exist, but at many locations it has the potential to increase the impact and consequence of existing hazards'. Indeed, as noted above, natural hazard risk is already a recognized aspect of coastal management. Therefore, this section discusses how climate change risk was incorporated into the most recent district plan for the Christchurch region, and how it was then taken out.

District Plans are the predominant planning mechanism by which to engage in land use and development planning in New Zealand and are part of a group of planning and policy documents from all levels of government (national, regional and local) that together are intended to provide an integrated natural resource management system. The lead national legislation for planning is the Resource Management Act 1991. The Act provides for a series of National Policy Statements that set out resource management objectives and policies for matters of national significance that are relevant to achieving the purpose of the Act and which therefore guide subsequent decision making under the Act at the national, regional and district levels. The District Plan must give effect to National Policy Statements.

Under the Act a National Coastal Policy Statement by the Minister of Conservation is mandatory (Hall, 2009), but other national policy statements, which must also be approved by the Minister for the Environment, are optional (for example the National Policy Statement on freshwater management). National Environmental Standards are regulations that apply to the use, development and protection of natural and physical resources and which prescribe technical standards, methods or other requirements for implementing the standards in a consistent manner. National standards generally override existing provisions in plans that have a lower standard. Conversely, if a District Plan has a standard that is stricter than a national standard then that plan standard prevails. As of September 2015 there was no National Policy Statement with respect to climate change and sea level rise. However, there was already well-established recognition of the significance of climate change within planning and policy circles. For example, the Mfe had released a guidance manual to local government on climate change in 2004 and revised in 2008. In introducing a report on the science of climate change and rising sea levels the Parliamentary Commissioner for the Environment (PCE) noted, 'During my seven years as Commissioner, I have consistently said that climate change is the biggest environmental issue we face' (PCE, 2014, p. 5) and her comments with respect to the difficulties of planning for climate change have proven remarkably prescient for the planning process in Christchurch.

> Some areas of Christchurch have experienced an effective sea level rise of half a metre or more due to land dropping after the Canterbury earthquakes. Flood insurance has become harder to get and more expensive for some homeowners in

> the city. On the day I am writing this, the Insurance Council of New Zealand has released a 15-point plan on dealing with natural hazards, with the chief executive making special mention of sea level rise.
>
> Councils are obliged under law to plan and prepare for the impacts of a rising sea. But it is far from easy to introduce changes that may lower the value of people's homes or restrict development along sought-after coastal areas. In August 2012, the Kapiti District Council put coastal erosion risk on the Land Information Memorandum (LIM) reports of 1,800 houses, which was challenged by those who were affected. Such conflicts are understandable and inevitable.
>
> The President of Local Government New Zealand, Lawrence Yule, has called for greater direction from central government. 'Without any central government directive it is quite difficult for councils to do what effectively might be the right thing for the future, but is seen as being too aggressive for the people of the present'. It is not just private property that will be affected. Councils and central government will need to prepare for increased costs because some public infrastructure such as roads, waste water systems, and buildings will be affected by rising seas.
>
> PCE, 2014, pp. 6–7

Planning in Christchurch

On 2 July 2015 the CCC released a report by Tonkin & Taylor (2015) on coastal hazard assessment to the public as part of its mandatory land use planning process (CCC, 2015a). The report was commissioned by the CCC to help guide planning rules in the Replacement Christchurch District Plan (RCDP) for protecting people and property from the effects of SLR as part of a wider concern with planning for natural hazards. The rules were originally due to be publicly notified at the end of July and would affect development policy in the hazard zones over the long-term. The Tonkin & Taylor (2013, 2015) reports suggested that 5971 properties in Christchurch and Banks Peninsula could be under threat from coastal erosion by 2115 and almost 18,000 properties could be susceptible to coastal flooding. The CCC natural environment and heritage unit manager Helen Beaumont said the council would be looking to discourage new subdivisions and new building in high-risk hazard zones.

> If an application for a development in those areas comes in and they can demonstrate that they can mitigate the hazard, it might be possible for that development to go ahead.
>
> But if it is considered it would place people or property at risk, then it is unlikely to go ahead (quoted in Cairns, 2015a).

The proposed rules would not have stopped people in the hazard zones from rebuilding earthquake-damaged homes as they had existing use rights, but they would have to apply for permission (a resource consent) from the CCC if they wanted to change the building's footprint. Significantly, the Land Information Memorandums (LIMs) for those properties were immediately changed to reflect the hazard identified in the Tonkin & Taylor report. The CCC is required by the Local Government Official Information and Meetings Act to record if a risk is known to the council. According to Cairns (2015a), Beaumont said people should not be alarmed if their homes were within either an inundation or erosion zone as the report projected decades ahead and there was time for authorities to respond to the risk. Nevertheless, the CCC website page which sought to explain 'how is my property affected?' (CCC, 2015b) stated with respect to insurance and property values:

> The Council is unable to comment on how the insurance council, or insurance providers, will respond to the coastal hazard information. We suggest you contact your insurance company for further information and advice.
>
> The Council is unable to consider or assess how coastal hazard information will impact on property values.

Therefore, potentially generating uncertainty. Despite CCC assurances coastal property owners, nearly all of whom were in the eastern coastal suburbs of Christchurch, challenged the risk assessments and pushed for more time to lodge submissions to the planning process. According to Cairns (2015b), Burwood-Pegasus Community Board member Tim Sintes said he was worried the council was putting too much faith in the computer modelling and overreacting to the predicted risks and that residents were worried the approach would make it difficult to get insurance and eventually force people out of the area. Christchurch Mayor Lianne Dalziel acknowledged there was uncertainty about how much sea levels would rise but defended the robustness of the science behind the council's coastal hazards assessment:

> There are real challenges around the how much and the when but there is no dispute around the science on sea level rise... You wouldn't want intensified residential developments in areas where you know there is going to be risk unless they are relocatable or unless they are built to a level to withstand the impacts of what may occur.
>
> If we don't address these issues in a very open way then the insurance industry will address these issues. The insurance sector has access to the same information that we do. In my view, it is better that a council takes a position rather than leaves it to the insurance industry because the insurance industry will take a position (Lianne Dalziel quoted in Cairns, 2015b).

In an interview Sintes stated:

> I know that beach better than anybody... The implications of this are really scary. It has huge insurance ramifications and could destroy the equity people have in their homes... There's no way anyone is going to buy a house where the insurance is iffy... In the 70s we had car-less days because we were going to run out of fuel. In the year 2000 we were told planes were going to fall out of the sky. Now we're told we're going into a mini Ice Age. It is all crystal-ball gazing. To say we're going to have an eroding coastline in 50 years is like picking Lotto winners (quoted in Cairns, 2015c).

Christchurch City councillor David East also questioned the veracity of the computer modelling and was concerned that the council was overreacting to a 'theoretical threat' rather than taking an evidence-based approach. According to East, public meetings organized by the CCC to discuss its proposed response to the coastal hazard risk became heated, with residents voicing concern that getting insurance on their properties will become harder, that they will not be able to develop their existing homes, nor build on bare land they hold within the hazard areas. 'One of the points that has come through quite strongly is the experience of people who have lived there for a number of years – some of them up to 70 years or more – who have seen the coast pushing further and further out. Now suddenly they are been told they're going to be subject to this, that and whatever else and like me they're questioning the computer modelling' (quoted in Cairns, 2015c). In response the Mayor reinforced her position that the proposal to formally recognize the risk posed by SLR in the district plan was not about a managed retreat: 'It is about adaptation. We're not asking anyone to leave; we are not asking anyone to remove themselves. What we're trying to do is create an environment where we can tell the insurance industry we've mitigated the risk. If we don't address these issues in a very open way, then the insurance industry will address these issues' (quoted in Cairns, 2015c). Indeed, in 2014 the Insurance Council of New Zealand (ICNZ) released a position paper on natural hazards that noted that as a result of climate change, 'By 2050 about one million older New Zealanders will be living in areas vulnerable to severe flooding, coastal storm surges, land slips and wind storms' (ICNZ, 2014, p. 5) and have suggested that high-quality risk information should be publically available for every property in New Zealand.

According to Sintes and Mark Munro, speaking for a newly formed group Christchurch Coastal Residents United (CCRU), they had gone to the council to ask for more time but were rebuffed. However, CCC natural environment and heritage unit manager Helen Beaumont said the timeframe for submissions was dictated by an Order in Council that allowed a maximum of 30 days between notification and the end of the submission period (although this was later extended to the end of October). According to Cairns

(2015d), Beaumont said that the CCC had given the maximum time possible and released the science behind the proposals earlier to give people the greatest opportunity to familiarize themselves with the issues. The CCRU put in its own submission and established a website where people could get a pre-prepared submission form and find more information (Cairns, 2015d). The CCRU website included advice (CCRU, 2015) for non-residents of Christchurch to also make submissions on the planning process:

> New Zoning Sabotage NZ Communities
>
> Without community collaboration or consultation, Christchurch City Council has released Stage 3 of the proposed Christchurch Replacement District Plan. This reviews the Coastal Hazard Policy and has identified (created) 4 NEW Coastal Hazard Zones. The Christchurch City Council are excited to quickly push through this plan because it will form a template for other coastal and waterway cities and communities. If this plan is adopted in Christchurch it will set a precedent. Councils in other areas are watching with interest to see the outcome. If the provisions in this plan ARE adopted, you're next.
>
> Why your submission is valuable
>
> If the proposed changes are accepted then:
>
> - New development in coastal areas would be severely restricted meaning new builds, units, infill housing and subdivisions will become Non Complying
>
> - Bare land will become worthless as it has no existing usage rights
>
> - There will be no provisions for protection or hazard mitigation
>
> - There will be no requirement to replace or provide new community facilities or developments
>
> - The economic and social cost to communities will be un-measurable
>
> - The Christchurch City Council will destroy the future for over 18,000 home owners. This doesn't just affect Christchurch. This affects you, too.

The CCRU also announced that it was working closely with Coastal Ratepayers United, a group from Kapiti in the North Island, which forced its council to remove hazard zonings placed on 1800 coastal properties in 2013 (Blundell, 2014). The group's chairman, Christopher Ruthe and secretary, Salima Padamsey, flew to Christchurch to meet Sintes, Munro and council representatives. According to Cairns (2015c), Ruthe said the lack of time Christchurch residents had been given to make submissions was 'outrageous' and claimed that it appeared that the CCC 'were relying on the same flawed science as the Kapiti Coast District Council and taking an "ultra-conservative approach" that could have devastating consequences for coastal property owners' (Cairns, 2015d).

At the end of September 2015 the national government intervened and announced that it would create a National Coastal Hazards strategy and that the coastal hazard zonings would be dropped from the RCDP. In a press conference on 29 September the Environment Minister Nick Smith and Christchurch Mayor Lianne Dalziel announced that plan changes affecting property owners as a consequence of future coastal hazards would be dealt with through normal planning processes and not through the District Plan review process that had been fast-tracked as a result of the post-earthquake rebuild (Cairns, 2015e). Sintes said the decision was 'fantastic news... To get a result like this, it's democracy at its best. It has to go this way, with a national standard, rather than ticking off one town after another' (quoted in Cairns, 2015e). Dalziel commented that although she 'would have preferred the intervention had happened last year', the 'announcement will be welcomed by local residents who have been extremely concerned at the proposed changes to the District Plan and their ability to have meaningful input on the issue of coastal hazards. The fast-tracking of the District Plan Review was always intended to be about earthquake recovery. We do not need to move with the same speed with respect to these longer term issues' (quoted in Cairns, 2015e). Councillor David East, the only city councillor to vote against the coastal hazard provisions in the RCDP, said he was 'absolutely ecstatic' at the news the provisions were being withdrawn. 'He had always been sceptical about the science used to justify them and believed the council was acting too hastily' (Cairns, 2015e). In a press release Associate Minister Nicky Wagner stated, 'We should not allow our thinking to become so cautious that we block development in areas on the premise of worst case scenarios of sea level rise. We need to take the time on such far-reaching rules to rigorously test restrictions from both an economic and a risk management perspective' (Brownlee *et al.*, 2015). Nevertheless, it must be noted that at the time of writing SLR risk information remains on LIMs.

The Christchurch case is informative with respect to community response to planning measures enacted because of SLR risk, in that while there is an awareness of such risks few people appear willing to accept measures designed to adapt to future circumstance that reduce the present-day value of their assets. However, there also appeared very significant differences between coastal urban Christchurch and the communities of the Akaroa Harbour with respect to their responses. For example, submissions to the process from the Akaroa Harbour area have been limited: in Stage 1 submissions there were only two of 1483 submissions (0.1%), of which one was the Akaroa Civic Trust who concentrated on the heritage characteristics of Akaroa; in stage 2 eight of 594 (1.3%) and Stage 3 eight of 979 (0.8%) (as of 17 September). No submissions from the Akaroa Harbour area focused on the coastal risk assessment although one statement of evidence before the hearings panel regarding rezoning in Stage 2 explicitly acknowledged in the Tonkin & Taylor (2013) report as part of the design and rezoning process (Statement of Evidence of Patricia Harte on Behalf of: Mark Shadbolt – Submitter No 2375, Sunny Point Estates Ltd – Submitter No 2406). One reason for the lack of submissions from the area could be the large number of holiday homes whose owners live outside the district, alternatively the flooding and tsunami history of the harbour may have made the zoning proposals relatively non-controversial. However, regardless of the reason the district has a significant problem in getting its issues with SLR climate change addressed given the almost overwhelming focus in the local media on the coastal suburbs of Christchurch. The district therefore faces a challenge common to other rural locations that have been integrated into essentially urban planning systems in which spatial concerns are essentially overridden by a focus on the main population centres.

Conclusions

According to the CCC (2015a, p. 33) the key issues for the future of the Akaroa Harbour area 'are about managing small incremental changes to settlement areas to support local communities and economic growth in the area, and enable communities to adapt to future expected environmental changes, such as sea level rise'. SLR is a major challenge because it will greatly affect the relative accessibility of locations and hence travel times, while existing coastal road and protection infrastructure will be expensive to maintain. Arguably this may be even more difficult given a small permanent population in a rural area whose planning powers were ceded to the metropolitan centre a decade previously. The challenges facing the tourism-dependent communities of Akaroa Harbour are therefore similar to many other small tourism-dependent communities throughout the world with respect to the issue of who will pay for the required adaptation response.

Even though the economy of the harbour area is primarily tourism based, tourism has not been a major consideration in the coastal risk assessment process. Far greater attention is given in the district plan to the heritage values of Akaroa than other tourist planning issues, despite SLR threatening those values in the longer term as well as long-term accessibility. The challenges facing tourism with respect to climate change in the area have therefore become subsumed in broader planning debates and processes at the local, regional and national level. The RCDP placed a significant premium on identifying long-term risks from SLR for coastal development and land use. However, this first step in local adaptation in the planning process did not receive support from either the national government or from many coastal residents in the city who perceived the issue as one of potentially increased insurance premiums and loss of value of their housing assets in the short-term rather than any future loss. Given the recent history of Christchurch with respect to natural hazards, such as earthquakes, this stance is intriguing. Whereas the risk assessments of the impact of future earthquakes are broadly accepted as part of land-use planning, the potential impacts of climate change have not received such acceptance. It would indeed be unfortunate if a future flood or surge event were to be what would be required to accept the value of such risk assessments, especially when those proposed in the RCDP were relatively conservative. The Christchurch experience therefore has significant implications for the understanding of planning adaptation in coastal areas.

Writing in 2007 of risk perception in Akaroa, including that of climate change, Nuth (2007, p. 129) concluded, 'people in Akaroa have a general tendency to ignore global risks in favour of local issues'. Some 8 years later there may still be some suggestions that this is still the case, even though the responsible planning authority appears highly aware of risk. However, with respect to the Christchurch situation and the response to rising sea levels it may perhaps now be suggested that some people in Christchurch have a general tendency to ignore global risks in favour of their own short-term interests.

References

Allen, J., Davis, C., Giovinazzi, S. and Hart, D. (2014) *Geotechnical & Flooding Reconnaissance of the 2014 March Flood Event Post 2010–2011 Canterbury Earthquake Sequence, New Zealand, Version* 1: 19 June 2014. GEER, Christchurch, New Zealand.

Blundell, K. (2014) Residents win hazard lines challenge. *Dominion Post*, 21 March.

Brownlee, G., Smith, N. and Wagner, N. (2015) *Coastal Hazard Issue to be Uncoupled from Fast-track Earthquake Recovery Plan Process*. Press Release, 29 September. Parliament of New Zealand, Wellington.

Bruun, P. (1988) The Bruun rule of erosion by sea-level rise: a discussion on large-scale two-and three-dimensional usages. *Journal of Coastal Research* 4, 627–648.

Cairns, L. (2015a) 6000 Christchurch properties at risk from coastal erosion. *The Press*, 3 July.

Cairns, L. (2015b) Residents challenge coastal hazards assessment. *The Press*, 18 July.

Cairns, L. (2015c) Christchurch coastline under threat? If scientists are to be believed parts of Christchurch's coastline will be gobbled up by a rising sea over the next 100 years. Not everyone is convinced by their predictions. *The Press*, 25 July.

Cairns, L. (2015d) Christchurch coastal property owners demand more time on hazard zoning. *The Press*, 21 August.

Cairns, L. (2015e) Controversial coastal hazards zonings dropped. *The Press*, 29 September.

Canterbury Earthquake Recovery Authority (CERA) (2013) Overview of TC1, TC2 and TC3 Technical Categories. Available at: http://cera.govt.nz/residential-green-zone-technical-categories/overview (accessed 1 April 2015).

Christchurch City Council (CCC) (2014) *Community Profile - November 2014: Akaroa*. CCC, Christchurch, New Zealand.

Christchurch City Council (2015a) *Draft Section 32 Residential Chapter 14 – Stage 2 Additions Highlighted*, April 25. CCC, Christchurch, New Zealand.

Christchurch City Council (2015b) How is my Property Affected? Available at: www.ccc.govt.nz/environment/land/coast/coastal-hazards/how-is-my-property-affected (accessed 1 October 2015).

Christchurch Coastal Residents United (2015) Non-residents of Christchurch. Available at: www.ccru.co.nz/non-residents-of-chc (accessed 1 October 2015).

Cooper, J. and Pilkey, O. (2004) Sea-level rise and shoreline retreat: time to abandon the Bruun Rule. *Global and Planetary Change* 43(3), 157–171.

De Lange, W. and Healy, T. (1986) New Zealand tsunamis 1840–1982. *New Zealand Journal of Geology and Geophysics* 29(1), 115–134.

Department of Conservation (DoC) (2006) *Native Plants Natural to Banks Peninsula*. DoC, Christchurch, New Zealand.

DTec Consulting Ltd (2008) *Akaroa Harbour Basin Settlements Study – Coastal Erosion and Inundation Project, Report for Christchurch City Council: Strategy and Planning Group*. Christchurch City Council, Christchurch, New Zealand.

Fountain, J. (2002) Behind the brochures: The (re)construction of touristic place images in Akaroa, New Zealand. Unpublished PhD Thesis, Murdoch University, Western Australia.

Fountain, J. and Hall, C.M. (2002) The impact of lifestyle migration on rural communities: a case study of Akaroa, New Zealand. In: Hall, C.M. and Williams, A. (eds) *Tourism and Migration*. Kluwer, Dordrecht, the Netherlands, pp. 153–168.

Gehrels, W., Hayward, B., Newnham, R. and Southall, K. (2008) A 20th century acceleration of sea-level rise in New Zealand. *Geophysical Research Letters* 35, L02717.

Goff, J. and Chagué-Goff, C. (2012) A review of palaeo-tsunamis for the Christchurch region, New Zealand. *Quaternary Science Reviews* 57, 136–156.

Hall, C.M. (2009) Coastal tourism planning and policy in New Zealand. In: Dowling, R. and Pforr, C. (eds) *Coastal Tourism Development*. Cognizant Communication, New York, pp. 120–134.

Hall, C.M., Malinen, S., Vosslamber R. and Wordsworth, R. (2016) Introduction: the business, organisational and destination impacts of natural disasters – The Christchurch Earthquakes 2010–2011. In: Hall, C.M., Malinen, S., Vosslamber, R. and Wordsworth, R. (eds) *Business and Post-Disaster Management: Business, Organisational and Consumer Resilience and the Christchurch Earthquakes*. Routledge, Abingdon, UK, pp. 3–20.

Hannah, J. and Bell, R. (2012) Regional sea level trends in New Zealand. *Journal of Geophysical Research: Oceans* 117, C01004.

Hansen, J., Sato, M., Hearty, P., Ruedy, R., Kelley, M. *et al.* (2015) Ice melt, sea level rise and superstorms: evidence from paleoclimate data, climate modeling, and modern observations that 2°C global warming is highly dangerous. *Atmospheric Chemistry and Physics Discussions* 15(14), 20059–20179.

Insurance Council of New Zealand (2014) *Protecting New Zealand from Natural Hazards*. ICNZ, Wellington.

IPCC (2013) *Climate Change 2013: The Physical Science Basis. Contribution of Working Group I to the Fifth Assessment Report of the Intergovernmental Panel on Climate Change*. Cambridge University Press, Cambridge.

Keen, D. and Hall, C.M. (2004) Second homes in New Zealand. In: Hall, C.M. and Müller, D. (eds) *Tourism, Mobility and Second Homes*. Channel View, Clevedon, UK, pp. 174–195.

McGlone, M. and Walker, S. (2011) *Potential Effects of Climate Change on New Zealand's Terrestrial Biodiversity and Policy Recommendations for Mitigation, Adaptation and Research*. Department of Conservation, Wellington.

Ministry for the Environment (MfE) (2008) *Climate Change Effects and Impacts Assessment. A Guidance Manual for Local Government in New Zealand*, 2nd edn. Ministry for the Environment, Wellington.

Nuth, M. (2007) Risk environment: A case study of risk and risk perception in the tourist-dependent township of Akaroa. MA thesis, University of Canterbury, Christchurch, New Zealand.

Parliamentary Commissioner for the Environment (PCE) (2014) *Changing Climate and Rising Seas: Understanding the Science*. Author, Wellington.

Ranasinghe, R. and Stive, M. (2009) Rising seas and retreating coastlines. *Climatic Change* 97, 465–468.

Ranasinghe, R., Callaghan, D. and Stive, M. (2012) Estimating coastal recession due to sea level rise: beyond the Bruun rule. *Climatic Change* 110, 561–574.

Tonkin & Taylor (2013) *Effects of Sea Level Rise for Christchurch City*. Tonkin & Taylor, Christchurch, New Zealand.

Tonkin & Taylor (2015) *Coastal Hazard Assessment Stage Two*. Prepared for the Christchurch City Council. Tonkin & Taylor, Christchurch, New Zealand.

Wilson, J., Shone, M., Simmons, D. and Stewart, E. (2015) Making waves: examining the interface between cruise tourism and destination community in Akaroa, New Zealand. *Tourism in Marine Environments* 10(3–4), 211–223.

21 Case Study Turkey: Climate Change and Coastal Tourism: Impacts of Climate Change on the Turquoise Coast

O. Cenk Demiroglu,[1*] A. Akbas,[2] M. Tufan Turp,[3] T. Ozturk,[4] N. An[3] and M. Levent Kurnaz[3]

[1]*Umeå University, Sweden;* [2]*Ardahan University, Turkey;* [3]*Boğaziçi University Turkey;* [4]*Isik University, Turkey*

Introduction: The Case of the Turquoise Coast

The coastal tourism industry is one of the most important sectors for the socio-economic development and stability of many countries and regions located in the Caribbean and the Mediterranean. Turkey is one of the leading countries in this sense, owing to the existence of its well-established, so-called 'Riviera', also known as the 'Turquoise Coast' in the south-western region where the Mediterranean and the Aegean meet. The region is well known for some of its destinations such as Ayvalık in Balıkesir, Çeşme in İzmir, Kuşadası in Aydın, Bodrum, Datça, Marmaris and Fethiye in Muğla, and Kaş, Kalkan, Kemer, Belek, Side and Alanya in Antalya. The main conventional product is sun-sea-sand-based mass tourism, diversified by the recently emerging alternative tourism types such as yachting, cruising, diving and golf, while the concentration of numerous cultural heritage sites (Fig. 21.1) also plays a vital role as touristic attractions. The region offers 764,701 tourist beds in the provinces of Antalya (473,223 beds), Muğla (155,152 beds), İzmir (53,329 beds), Aydın (48,881 beds) and Balıkesir (34,116 beds) (Fig. 21.2). Moreover, it is home to 166,906 second homes mostly frequented during summers (Gökdeniz *et al.*, 2010). Altogether, such supply generates a significant proportion of the US$42 billion worth of receipts from 112 million annual (2014) domestic and international visits (TURKSTAT, 2015). A future expansion of the bed capacity by 150,401 for the region is already approved by the Ministry of Culture and Tourism (Fig. 21.3). Furthermore, policymakers aim at improving the infrastructure and encouraging the private sector into more investments via an integral planning approach for the region, according to the 'Tourism Strategy of Turkey – 2023' (Turkish Ministry of Culture and Tourism, 2007), in line with the centennial goals of the republic.

Although local or regional policy coverage of climate change impact and adaptation for tourism is uncommon in Turkey, recent communications from the central government emphasize the need of impact assessments and adaptation formulations for the industry, especially where the economy is strongly dependent on sun-sea-sand (3S) tourism (Turkish Ministry of Environment and

*Corresponding author, e-mail: cenk.demiroglu@umu.se

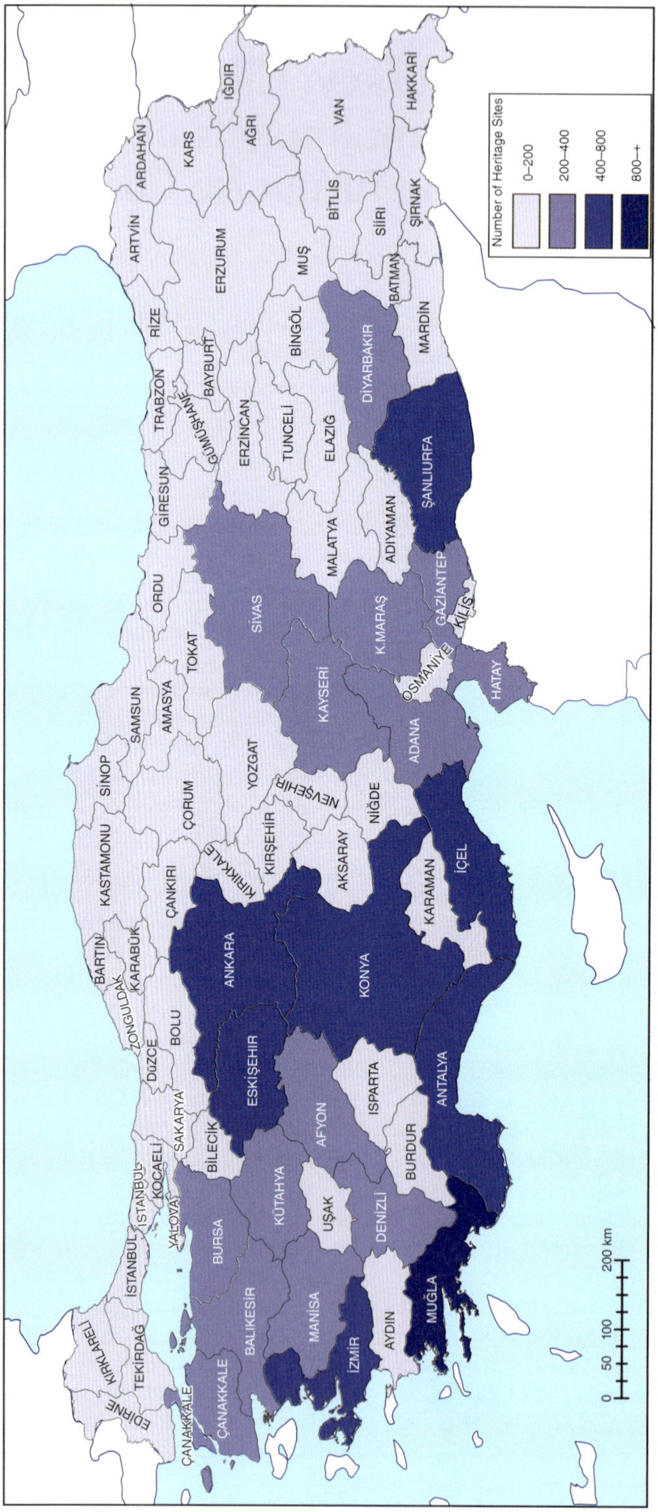

Fig. 21.1. Distribution of heritage visitor attraction sites in Turkey (Turkish Ministry of Culture and Tourism, 2015a).

Fig. 21.2. Distribution of operational tourist beds licensed by the Ministry of Culture and Tourism and the municipal governments in Turkey (Turkish Ministry of Culture and Tourism, 2015b).

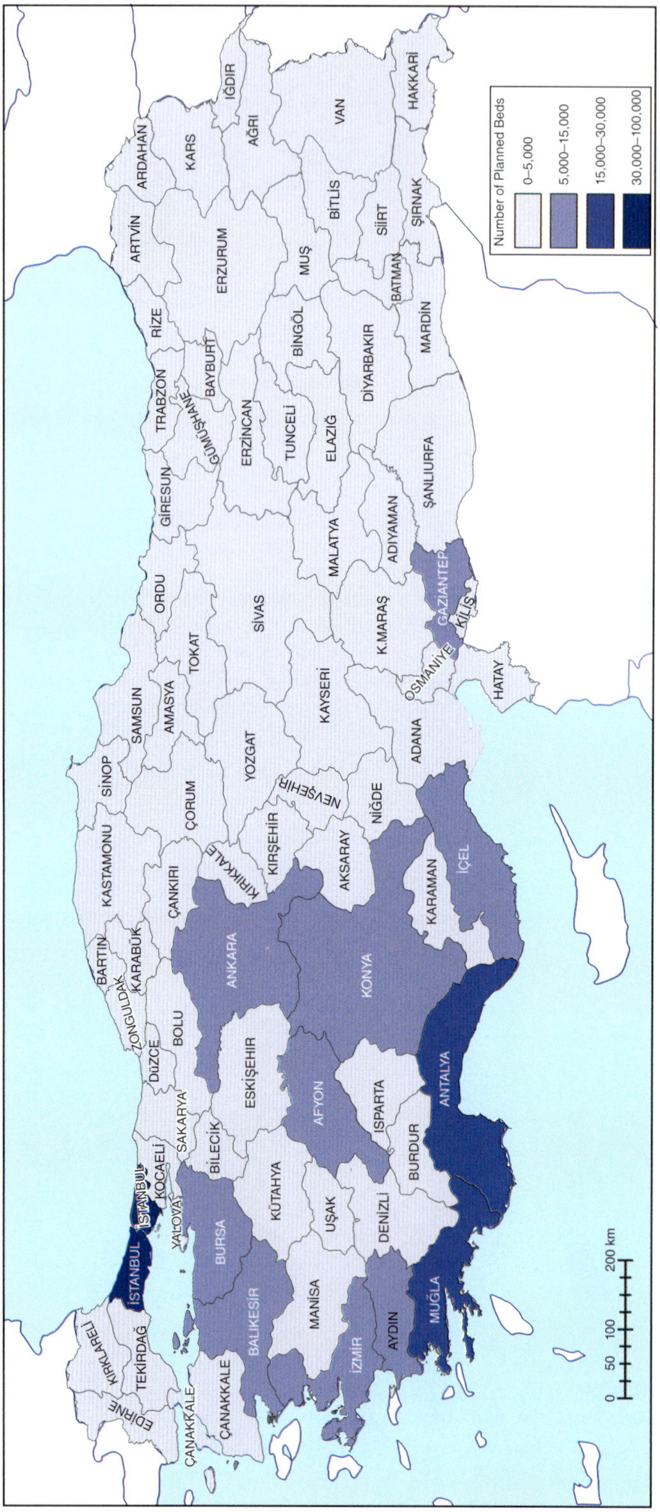

Fig. 21.3. Distribution of proposed tourist beds licensed by the Ministry of Culture and Tourism in Turkey (Turkish Ministry of Culture and Tourism, 2015b).

Urbanisation, 2013, p. 218). In that respect, various governmental, non-governmental and international parties are declared as either responsible or related for impact assessment and adaptation efforts according to the National Action Plan for Climate Change 2011–2023 (Turkish Ministry of Environment and Urbanisation, 2012). Such a political stance is critical as the anticipated impacts of climate change, such as rising sea levels and coastal erosion, changing touristic climate comfort, increasing frequencies and intensities of extreme weather events such as heatwaves, storms and floods, aridification and salinization and changing biodiversity, are expected to have severe consequences over the coastal tourism destinations (Mather *et al.*, 2005; Becken and Hay, 2007; Scott *et al.*, 2012a).

Research Context

This chapter tries to assess the severity of expected climate change impacts on tourism at the Turquoise Coast and build up an adaptation agenda for the concerned parties. First, sea level rise projections are implicated to understand the degree of exposure at the immediate coastal zone where a major proportion of the 3S infrastructure as well as many of the archaeological, ancient and historical heritage sites and golf courses lie. A storm and flood risk projection through a double-nested scaling of the MPI-ESM-MR global circulation model (Max Planck Institute for Meteorology, 2012) down to a 10 km grid resolution and according to the RCP8.5 scenario, where the radiative forcing is expected to reach 8.5 W/m^2 by 2100 with a CO_2-eq concentration of 700–1500 ppm on a business-as-usual trend (van Vuuren *et al.*, 2011), by the RegCM 4.4 regional climate model (The Abdus Salam International Center for Theoretical Physics, 2014) follows to visualize any additional risks on the coasts as well as for the nautical tourism routes. The climate model also provides us with outputs for determining the changes in the touristic climate comfort as well as spotting any increase of heatwave events. The changing windiness characteristics are also investigated by using RegCM 4.4 to reach implications for surf tourism. Last but not the least, we interpret an aridity index and water deficit projections to understand a conflicting future where water availability could decrease, whereas the growth of the tourism industry, particularly golf tourism, would imply more consumption of water resources. The assessments are finalized by implications on the future of a major component of diving tourism in the north Aegean – the red coral.

Climate Change: Assessing Problems and Challenges

Sea level rise

Warming over the past decades has resulted in more melting of the glaciers and thermal expansion of the oceans, leading to a global mean sea level rise (SLR) of 0.19 m during the period 1901–2010. RCP8.5 scenario projections pronounce a further rise of 0.52–0.98 m with an accelerated rate of 8–16 mm/year by the end of the 21st century, with respect to the pre-industrial level (Church *et al.*, 2013). Therefore the systems in and around the coastal zones will be ever more exposed to SLR-related risks of inundation, erosion and salinization.

In Turkey, SLR observations at two important touristic destinations, Antalya and Bodrum, have been measured as 0.17 m and 0.06 m, respectively, for the 1984–2002 period. Depending on the rate of SLR, total inundation of Turkish coastal land could total 107–416 km^2 by 2045 and 271–1057 km^2 by 2105 (Demirkesen *et al.*, 2008). Karaca and Nicholls (2008) suggest that Turkey is relatively less vulnerable than the southern Mediterranean countries to accelerated SLR, and no significant effect would be observed at least until the 2020s, yet in the long term, touristic regions could be threatened and seaside ancient heritage sites, such as Patara and Phaselis in Antalya, could particularly be subject to risk. Likewise, Kuleli (2010) has mapped Antalya's beach, heritage and golf attractions in Belek, Kemer, Phaselis, Olympos, Finike,

Patara and Beymelek as well as the Aegean coast destinations Ölüdeniz (Fig. 21.4), Dalaman, İstuzu (a major breeding ground for *Caretta caretta*), Dalyan, Gökova, the ancient city of Ephesus, Çandarlı, Dikili, Altınova, Ayvalık, Burhaniye, Edremit and Akçay as vulnerable to SLR in the long term. Such vulnerability is due to the relative socio-economic dependence of the region on tourism, besides the physical exposure, yet still favouring the Mediterranean over the Caribbean, where 60% of resorts are expected to be inundated or damaged by the consequent erosion of a 1 m SLR, and countries do not seem to have the adaptive capacity to cope with such impact on tourism – an industry highly vital for their survival (Scott *et al.*, 2012b).

Changes in touristic climate comfort and heatwaves

A direct impact of climate change on coastal tourism would be the changes in climate comfort. Touristic climate comfort, commonly assessed by the Tourism Climate Index (TCI; Mieczkowski, 1985) and its various modified versions, is considered as a major push-pull factor that determines travel decision. It is simply based on the assumption that people will desire trips to destinations where the climate conditions suitable for their chosen activity are the most ideal with respect to their residential origins.

In the worst case scenario, one could expect a double trouble for the Mediterranean destinations as the climate comfort conditions could improve at the tourist origins while deteriorating at the destinations. Such a path is exemplified in the PESETA Project, where the summer climate conditions are shaken from 'ideal' to 'marginal' along the Turquoise Coast but display an ideal picture for the coasts of the summer tourist-generating countries, such as the UK, France (north), Belgium, the Netherlands and Germany by the end of the 21st century, according to the RCAO model (Amelung and Moreno, 2009).

Fig. 21.4. Ölüdeniz – 'Blue Lagoon' – is a world renowned landmark of the Turquoise Coast (Ön, 2009).

Summer tourism climate comfort in the Mediterranean has lately been reassessed for Antalya, which has proven itself to be 'too hot' a beach spot to be avoided for the entire June–September season already, and including the month of October as well by the 2080s (Rutty and Scott, 2010). In addition, Kum (2011) has found out that the observed (1983–2006) TCI performance of major summer destinations such as Bodrum, Marmaris, Fethiye, Finike, Antalya (centrum) and Alanya was the lowest in the summer months and the highest in the shoulder season, especially in May in Marmaris and in April in Bodrum. Such findings stress the upper limit of touristic tolerance for hot weather, thus bringing in to the debate the implication that a warmer climate could bring in more tourists to the Turkish Mediterranean by extending its summer season (Şen, 2013). On the other hand, the aforementioned destinations have been and still are the most popular spots for summer tourism with the visitation figures ever increasing. This is probably due to the fact that the demand sensitivity to climate change is not as much as that of the supply side and/or the underestimation of indices for diurnal conditions. None the less, the changes must be assessed beforehand to be ready for the eventual end in the long-term range of the impacts, where a 0.45% loss of contribution to GDP from tourism revenues is expected in Mediterranean Europe (Barrios and Ibanez-Rivas, 2015).

In order to elaborate on the findings of Rutty and Scott (2010) and Kum (2011), we have assessed the changes in the annual frequencies of the 'too hot' days, as categorized to have a maximum temperature above 37°C (Rutty and Scott, 2010), and the 'tropical nights', defined as 24-h days with a minimum temperature of 20°C and above (Schulzweida, 2015). Moreover, heatwave events, assessed according to a threshold that looks for a minimum of 5 consecutive days with at least 5°C anomalies from their average maximum temperature in reference periods (Schulzweida, 2015), were also examined.

Regarding 'too hot' days, not much change is expected for the immediate coastlines, where the upper limit has already been fulfilled, whilst increases up to 20 days/year are projected for their relatively cooler hinterlands (Fig. 21.5). The touristic climate comfort will worsen even more with significant increases in the annual frequency of tropical nights in the near future, especially in Fethiye (Ölüdeniz), Antalya (centrum), Belek, Side and Alanya (Fig. 21.6), probably resulting in an increased need for air-conditioning and the consequent energy consumption, costs and emissions. Furthermore, 219 heatwave events, adding up to a total of 1703 days, are projected for western Turkey in the 2016–2050 period (Fig. 21.7), jeopardizing the collective health of the residents and the tourists, beyond just threatening the touristic climate comfort and season. It should be noted that these events are assessed as durations that deviate 5°C from their local, respective reference periods. Therefore the extremity of those in the 'too hot' zones (Fig. 21.5) will be more severe than those in the cooler zones, despite the lower frequency in the future period. Moreover, a consumer survey (Rutty and Scott, 2010) found out that the media stories on forthcoming heatwaves in the Mediterranean, let alone the events themselves, are likely to lead to more than half of the tourists changing their travel plans.

Changes in the frequencies of strong wind and heavy rainfall events

In addition to heatwaves, other extreme events such as floods and storms are major threats to coastal tourism as they impose direct threats over the infrastructure, whilst the latter is also a concern for nautical tourism.

Besides its coastal tourism market leadership, Turkey has recently emerged as a major player in cruise and yacht tourism by investing into its existing inventory of ports and marinas. The fact that Istanbul is nowadays one of the major embark/disembark ports has enhanced Çanakkale, İzmir, Kuşadası, Bodrum, Antalya and Alanya as frequent ports of call along the routes of the major cruise lines operating in the east Mediterranean. Moreover, provision of state-of-the-art marinas in Ayvalık, Izmir, Çeşme,

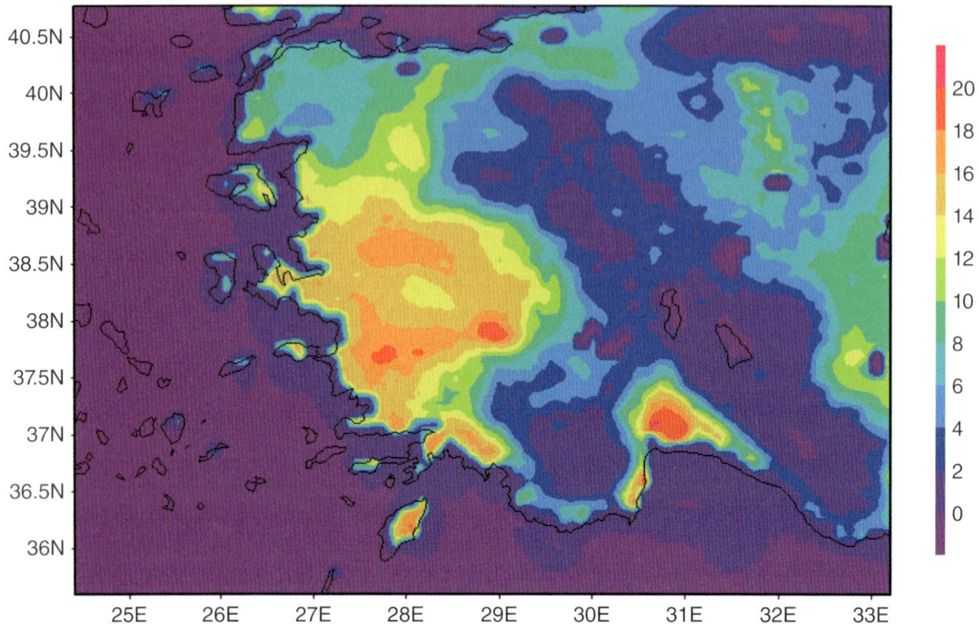

Fig. 21.5. Difference in the annual frequency of 'too hot' (>37°C) days from the 1970–2004 reference period to the 2016-2050 RCP8.5 scenario period.

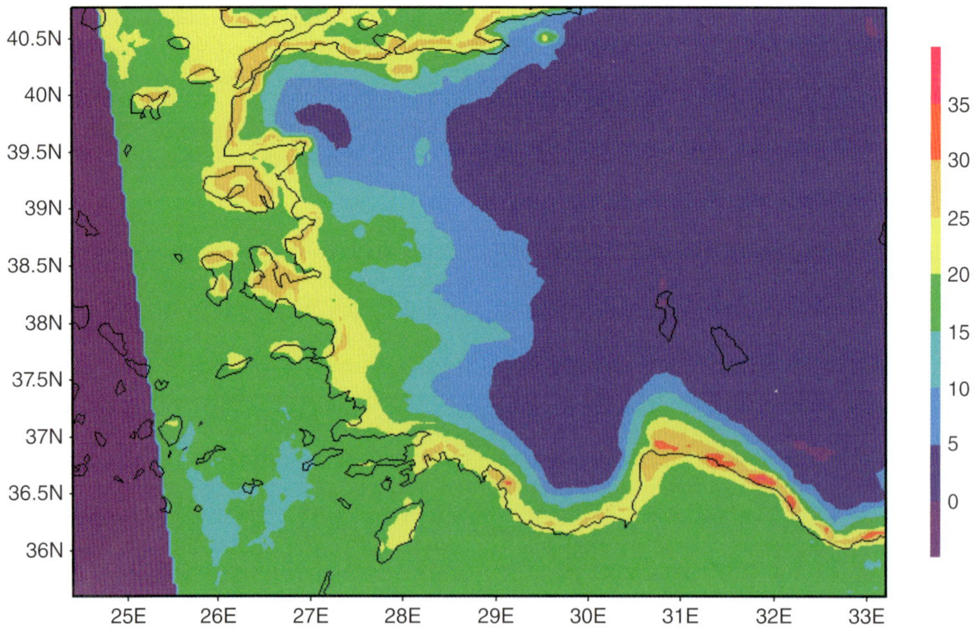

Fig. 21.6. Difference in the annual frequency of tropical nights from the 1970–2004 reference period to the 2016–2050 RCP8.5 scenario period.

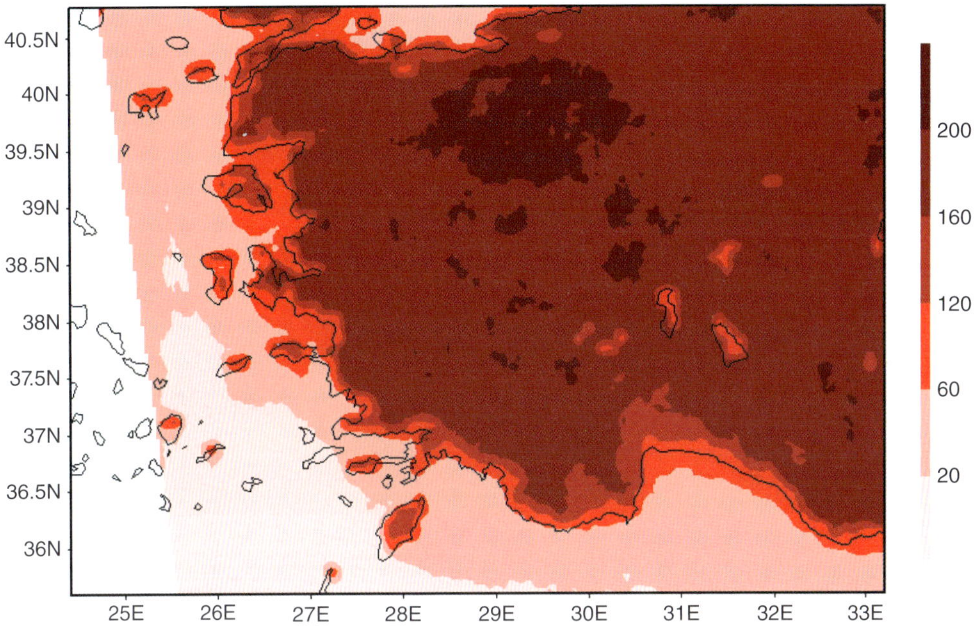

Fig. 21.7. Frequency of heat wave events in the 2016–2050 period (RCP8.5 scenario).

Kuşadası, Didim, Bodrum, Datça, Marmaris, Fethiye, Göcek, Kemer and Antalya together with the traditional 'Blue Voyage' tours on gulets (Fig. 21.8) has boosted yacht tourism, especially among the south Aegean parts of the country and the neighbouring Greek islands.

Despite the debates on the relationship of climate change and storm activity, increases in such activity could be expected for regions that lie along the poleward shift of the extra-tropical storm tracks (IPCC, 2012). Increases in the intensity and the frequency of these events could lead to consequences such as wind damage, coastal flooding and navigation risks. Sharp declines in tourism demand could be experienced during the event as well as due to the immediate negative post-event image. In order to see if and how such changes could affect our regions, we have looked into the future changes in the frequency of strong wind days, when the maximum wind speed reaches over 120 km/h, to be categorized as Force 12 on the Beaufort Scale, i.e. a 'hurricane' or a 'typhoon' (Schulzweida, 2015). The results (Fig. 21.9) indicate an increase in the frequency of such strong wind events around the touristic town and port of Alanya, especially towards the mountainous hinterland where a ski resort development has recently been approved by the Ministry of Culture and Tourism in line with the diversification goals of the Tourism Strategy 2023. Moreover, it seems that the Gulf of Çandarlı and Yeni Foça, home to many second homes and touristic facilities as well as maritime routes, would be experiencing a genesis of high frequency of storm events within the first half of the 21st century.

In addition to flooding in coastal regions due to storm surges, heavy rainfall, combined with slope aspects, would also lead to flood events, even for a wider area than the immediate coastline. The frequency of heavy rainfall is projected to increase globally (IPCC, 2012) and in the Mediterrenean Basin, especially in summer (Ozturk *et al.*, 2015). Indeed, the flood event that took place on 23 September 2015 in the touristic town of Bodrum was the result of such

Fig. 21.8. A Gulet on Blue Voyage along the Turquoise Coast (Dilmen, 2006).

precipitation (230 mm/m^2 – heaviest since 1988), destroying some parts of the urban areas as well as muddying the sea water during the extended Eid holiday, when many tourism businesses were hoping to generate revenues from a final peak inflow of visitors just before the end of the season.

Bearing in mind that local topography and soil and vegetation types also play vital roles in flood formation, we have simply taken a 240 mm daily rainfall as an analogue threshold to assess the future changes to the frequencies of heavy rainfall events that could lead to floods. The results (Fig. 21.10) display a slight decrease in the overall region, whereas parts of Antalya and Muğla seem to be vulnerable with increases up to 6 days a year. The only significant spot with deacreases of more than 18 days falls within the inner parts of Gökova Bay, yet still not beating the region's status of being exposed to the highest annual frequency of heavy rainfall in absolute terms (n=21) in the 2016–2050 period.

Changes in wind conditions suitable for surf tourism

Turkey is one of the leading windsurf countries, thanks to its established destinations with the sport's culture, facilities and the ideal climate conditions. Kiteboarding is also on the rise as a popular recreational and professional sport. Both sports are mostly practised at spots along the Aegean coast, aligning from south to north as Datça, Akyaka, Bodrum (Bitez and Akyarlar), Alaçatı and Gökçeada.

As with many aspects of climate, wind characteristics are also under (Hartmann *et al.*, 2013) and subject (Kirtman *et al.*, 2013) to change, as exemplified in the storm activity (Fig. 21.9). In order to assess the impacts of such changes on the windsurfing and kiteboarding potential along the Turquoise Coast, we, based on our personal communications and universal references (Hornik, 2001), have followed a simplified 'ideal surf day' threshold that takes account of the daily wind speed range as 10–25 knots/h, but ignores the wind directions and gusts. The projections imply no substantial changes (−2/+1%) for the high potential of 240–330 days/year observed across the destinations. However, higher resolution analyses that consider the microclimatology of aforementioned wind features, along with thermal comfort range as well as urban topography, need to be carried out before a final conclusion could be reached. Results of such analyses could be distinguished by different wind ranges suitable to various skill levels.

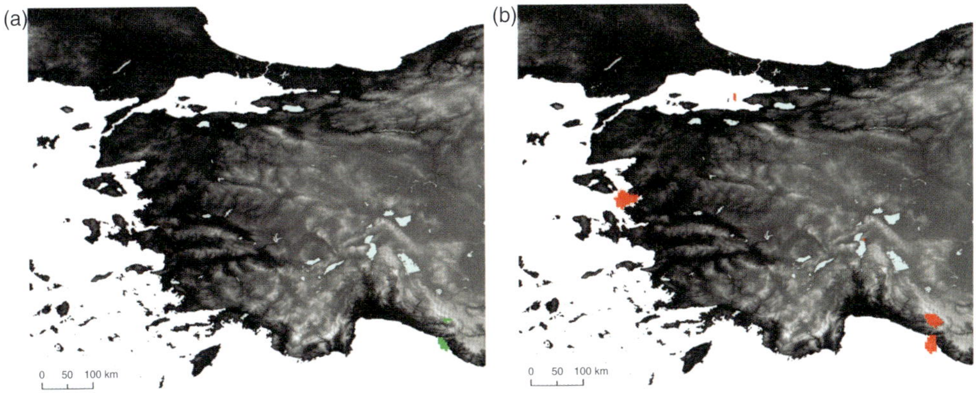

Fig. 21.9. Frequency of days with strong (>120 km/h) winds in the (a) 1970–2004 (n=23) and the (b) 2016–2050 (n=143) (RCP8.5 scenario) periods.

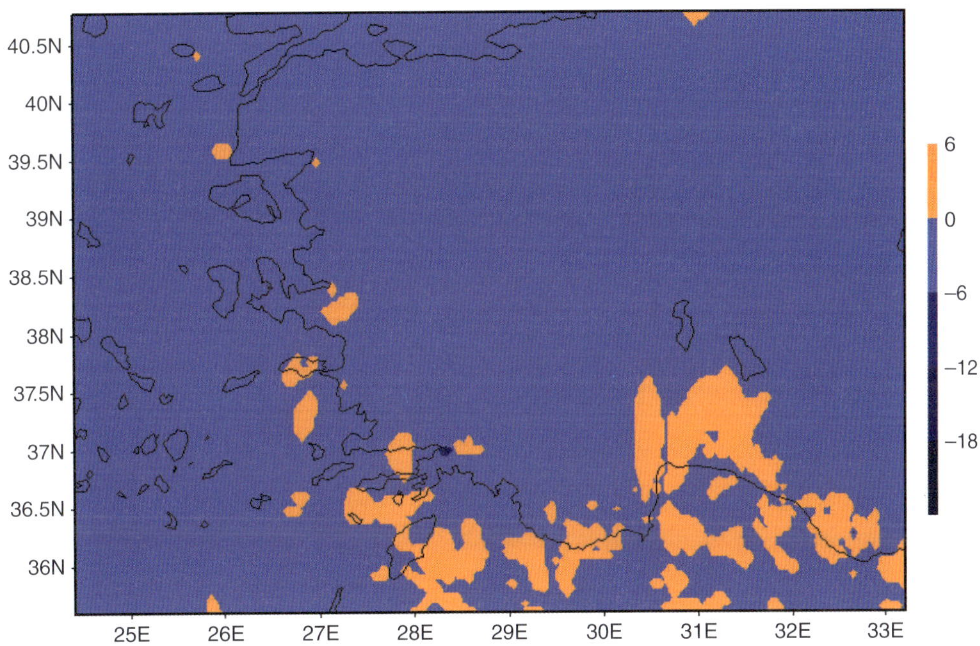

Fig. 21.10. Difference in the annual frequency of days with heavy rainfall from the 1970–2004 reference period to the 2016–2050 RCP8.5 scenario period.

Aridification versus golf tourism growth-led water consumption

On the one hand, golf tourism will need to adapt to climate change impacts (Scottish Golf Environment Group, 2004) or even benefit from them, depending on the region (Scott and Jones, 2007); on the other hand, the water-intensive industry, where a standard 18-hole course requires vast amounts of water (as much as 150,000–200,000 m^3/year; Gössling *et al.*, 2015), will more often be debated in a future where the water sources will be scarcer due to aridification (IPCC, 2014) as well as groundwater salinization from sea level rise (Jiménez-Cisneros *et al.*, 2014). Here, we examine the aridification–golf tourism growth relationship for the case of Belek in Antalya.

Belek was specially developed as a golf tourism destination in the 1990s and today is an economic success with 21 of the country's 29 golf courses that attract 2 million visitors in a year (Turkish Ministry of Culture and Tourism, 2015c). It is also one of the most favourable locations to be considered for new development projects (West Mediterranean Development Agency, 2011). However, Belek is also located in one of the most exposed regions to climate change impacts such as changing climate comfort and rising sea levels, as mentioned above, and one that is highly subject to dryer conditions and increased drought (Ozturk *et al.*, 2015).

In order to have an idea on how water availability could decrease due to aridification and growth of golf tourism in Belek, we have looked into the notion of 'socio-economic drought', where water supply trend is very crucial for a region in order to sustain some economic activities such as tourism, which in return could be a threat over the sustainability of water supply itself. As a practical tool, we have utilized the Aridity Index (AI) by Thornthwaite (1948), where total precipitation (P) over potential evapotranspiration (PE) yields the climate type in terms of aridity (Türkeş, 1999), and water deficit, defined as net precipitation (P–PE).

According to the results in Table 21.1, the most arid, thus critical, period in the Belek region is May–September. RegCM 4.4 projections according to RCP8.5 scenario indicated no change in the monthly climate type in terms of aridity, but an aridification trend for the wettest months. Projected values for both the P and the PE variables had very strong correlations with the AI values (Spearman's r: 0.95 and –0.97, respectively; $p<0.01$), indicating that both phenomena are the causes of this trend. The water deficit, on the other hand, is projected to increase by 44% in 2016–2050 with respect to the reference period of 1970–2004. Coupled with the possible (golf) tourism development in the region, these results imply a heated debate on water conflict for the first half of the 21st century.

Table 21.1. Changes in the aridity index in Belek, Antalya.

	Aridity index (AI)		
Months	1970–2004	2016–2050	Climate type[a]
January	6.29	5.20	Humid
February	3.42	3.25	Humid
March	2.10	1.97	Humid
April	0.66	0.53	Semi-arid
May	0.26	0.13	Arid
June	0.09	0.06	Arid
July	0.02	0.03	Arid
August	0.04	0.04	Arid
September	0.21	0.22	Arid
October	0.81	0.93	Sub-humid
November	4.34	3.18	Humid
December	5.47	4.67	Humid

[a]If AI<0.4, then 'Arid'; 0.4<AI<0.8, 'Semi-arid'; 0.8<AI<1.2, 'Sub-humid'; AI>1.2 'Humid'

Rise of the red coral?

The red coral, *Corallium rubrum*, is endemic to the Mediterranean Basin, at depths from 20 to 200 m, where rock formations and weak luminosity exist (Bann and Başak, 2013) (Fig. 21.11). It has a strong economic value in terms of jewellery and diving tourism industries. In Turkey, the coral is observed only around Ayvalık Adaları (Islands) Nature Park, making it a major attraction for diving tourism (Artüz *et al.*, 1990; Bann and Başak, 2013). However, as the red coral is highly sensitive to changes in the physical conditions such as temperature and luminosity (Artüz *et al.*, 1990), their existence in their original habitats is under threat by climate change, as the observed sea temperature increases are expected to continue into the future (IPCC, 2013). Yet, given the migratory skills of coral species, environmental changes could even lead up to an enriched population and/or diversity to be utilized as an underwater heritage asset for diving tourism in Ayvalık, or vice versa.

Conclusion

Coastal tourism will continue to be an important socio-economic sector in Turkey, despite all

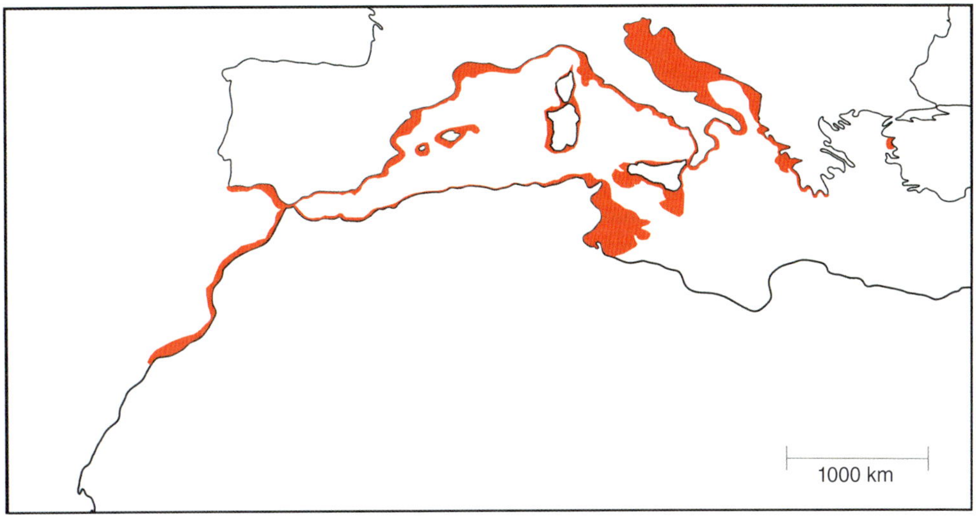

Fig. 21.11. Distribution of red coral (*Corallium rubrum*) (FAO, 2015).

the challenges ahead. Such a future is already evident in the approved investment expansion as well as the political will stated in the Tourism Strategy 2023. Therefore, it is imperative to identify the direct and indirect climate change impacts on this development goal, and to take the necessary precautions for action.

This chapter aimed at moving beyond doom saying and speculation on the climate change impacts for the Turquoise Coast, the lifeblood of Turkish tourism, and coming up with an agenda of the present and future problems based on scientific projections and implications. The assessment results on different impacts signalled the following conclusions and suggestions:

- Significant impacts of sea level rise would be felt in the very long term, by or even beyond the end of the 21st century. Such temporal range would not imply much threat for the relatively short-term, e.g. 30 years, lives of the tourism properties. In the longer term, new investments could be located at retreated zones. However, the same approach would not hold for the natural and cultural heritage assets that need to be sustained forever. Therefore, the options of beach nourishment and coastal protection should be set on the agenda.
- Studies show that the touristic climate comfort has already deteriorated on the coast, and our projections prove worse to come with warmer hinterlands and nights within the first half of the 21st century. As the demand response to such changes was not parallel, and even opposite, there need to more studies into finding the upper threshold of demand sensitivity to warmer seasons. Moreover, the institutional factors, such as holiday calendars in the market countries, should be examined thoroughly to come up with more realistic implications on the utilization of extended/shifted seasons.
- Increasing frequencies of extreme events such as heatwaves and hurricanes need to be distinguished intra-annually to identify any seasonal patterns. Such knowledge will be vital for the long-term development plans and the short-term business plans.
- Regarding flood risks and surf climate conditions, climate modelling needs to be coupled with changes in land use and topography at the highest spatio-temporal resolution possible. Otherwise, our limited findings come with a high spatial variance with respect to increase/decrease in flood risks, and

imply no major change for the ideal surf-tourism conditions.

- Today, water stress is a major issue in the wider region of the Turquoise Coast, even leading to wars and mass migration. Therefore, policies towards water-friendly tourism development is a hot topic. At the business scale, greening tourism establishments and adding retention ponds and improved irrigation systems to golf facilities would help for adaptation.
- The assessment on the impact of climate change on not only the red coral, but biodiversity in general, should be improved to see what other changes could have their implication for various tourism types, including the highly popular bird-watching industry.

The above conclusions and more should be communicated effectively by the central government to the local and regional actors, who should be directly acting on the relevant adaptation measures. Diversification of the tourism, or even the economic, mix in a way that minimizes vulnerability should be a priority. Last but not the least, all these conclusions should eventually lead the Turkish authorities and tourism industry to advocate not only adaptation to the impacts, but more importantly mitigation of the anthropogenic cause – greenhouse gas emissions – of the problem itself.

References

Amelung, B. and Moreno, A. (2009) *Impacts of Climate Change in Tourism in Europe*. PESETA-Tourism Study. European Communities, Spain.

Artüz, M.İ., Artüz, M. and Artüz, O.B. (1990) *Mercan Türlerine Getirilen Yasaklar İle İlgili Görüşler*. Turkish Ministry of Environment, Ankara.

Bann, C. and Başak, E. (2013) *Economic Analysis of Ayvalık Adaları Nature Park*. Turkish Ministry of Environment and Urbanization, Ankara.

Barrios, S. and Ibanez-Rivas, J. (2015) Time is of the essence: adaptation of tourism demand to climate change in Europe. *Climatic Change* 132(4), 645–660.

Becken, S. and Hay, J.E. (2007) *Tourism and Climate Change: Risks and Opportunities*. Channel View, Bristol, UK.

Church, J.A., Clark, P.U., Cazenave, A., Gregory, J.M., Jevrejeva, S. *et al.* (2013) Sea level change. In: Stocker, T.F., Qin, D., Plattner, G.-K., Tignor, M., Allen, S.K. *et al.* (eds) *Climate Change 2013: The Physical Science Basis*. Cambridge University Press, Cambridge, pp. 1137–1216.

Demirkesen, A.C., Evrendilek, F. and Berberoglu, S. (2008) Quantifying coastal inundation vulnerability of Turkey to sea-level rise. *Environmental Monitoring and Assessment* 138(1), 101–106.

Dilmen, N. (2006) İngiliz limanı. Available at: https://commons.wikimedia.org/wiki/File:Ingiliz_limani_04510_-_6115.jpg (accessed 22 November 2015).

FAO (2015) Species Fact Sheets: *Corallium Rubrum*. Available at: www.fao.org/fishery/species/3611/en (accessed 1 November 2015).

Gökdeniz, A., Dinçi Y. and Taşkır, H. (2010) Türkiye'de ikinci konut kavramı ve Ayvalık'ta ikinci konutlar üzerinde yapılan araştırma ışığında bir pazarlama model önerisi. In: Çolakoğlu, O.E. (ed.) *11. Ulusal Turizm Kongresi*. Detay Yayıncılık, Ankara, pp. 847–872.

Gössling, S., Hall, C.M. and Scott, D. (2015) *Tourism and Water*. Channel View, Bristol, UK.

Hartmann, D.L., Klein Tank, A.M.G., Rusticucci, M., Alexander, L.V., Brönnimann, S. *et al.* (2013) Observations: atmosphere and surface. In: Stocker, T.F., Qin, D., Plattner, G.-K., Tignor, M., Allen, S.K. *et al.* (eds) *Climate Change 2013: The Physical Science Basis*. Cambridge University Press, Cambridge, pp. 159–254.

Hornik, V. (2001) WindGURU: News. Available at: http://old.windguru.cz/int/news.php?oldnews=1 (accessed 17 November 2015).

IPCC (2012) *Managing the Risks of Extreme Events and Disasters to Advance Climate Change Adaptation. A Special Report of Working Groups I and II of the Intergovernmental Panel on Climate Change*. Cambridge University Press, Cambridge.

IPCC (2013) *Climate Change 2013: The Physical Science Basis*. Cambridge University Press, Cambridge.

IPCC (2014) *Climate Change 2014: Impacts, Adaptation, and Vulnerability*. Cambridge University Press, Cambridge.

Jiménez-Cisneros, B.E., Oki, T., Arnell, N.W., Benito, G., Cogley, J.G. *et al.* (2014) Freshwater resources. In: Field, C.B., Barros, V.R., Dokken, D.J., Mach, K.J., Mastrandrea, M.D. *et al.* (eds) *Climate Change 2014: Impacts, Adaptation, and Vulnerability*. Cambridge University Press, Cambridge, pp. 229–269.

Karaca, M. and Nicholls, R.J. (2008) Potential implications of accelerated sea-level rise for Turkey. *Journal of Coastal Research* 24(2), 288–298.

Kirtman, B., Power, S.B., Adedoyin, J.A., Boer, G.J., Bojariu, R. *et al.* (2013) Near-term climate change: projections and predictability. In: Stocker, T.F., Qin, D., Plattner, G.-K., Tignor, M., Allen, S.K. *et al.* (eds) *Climate Change 2013: The Physical Science Basis*. Cambridge University Press, Cambridge, pp. 953–1028.

Kuleli, T. (2010) City-based risk assessment of sea level rise using topographic and census data for the Turkish coastal zone. *Estuaries and Coasts* 33, 640–651.

Kum, G. (2011) İklim değişikliğinin Türkiye'nin güneybatı kıyılarında turizmin konfor şartlarına etkileri. Doctoral dissertation, Istanbul University, Istanbul, Turkey.

Mather, S., Viner, D. and Todd, G. (2005) Climate and policy changes: their implications for international tourism flows. In: Hall, C.M. and Higham, J.E.S. (eds) *Tourism, Recreation, and Climate Change*. Channel View, Bristol, UK, pp. 63–85.

Max Planck Institute for Meteorology (2012) New Earth System Model of Max Planck Institute for Meteorology. Available at: www.mpimet.mpg.de/en/science/models/mpi-esm.html (accessed 7 November 2015).

Mieczkowski, Z. (1985) The tourism climate index: a method of evaluating world climates for tourism. *The Canadian Geographer / Le Géographe canadien* 29(3), 220–233.

Ön, S. (2009) Ölüdeniz on the Turquoise Coast. Available at: https://commons.wikimedia.org/wiki/File:%C3%96l%C3%BCdeniz_on_the_Turquoise_Coast,_Turkey.jpg (accessed 22 November 2015).

Ozturk, T., Ceber, Z.P., Türkeş, M. and Kurnaz, M.L. (2015) Projections of climate change in the Mediterranean Basin by using downscaled global climate model outputs. *International Journal of Climatology* 35(14), 4276–4292.

Rutty, M. and Scott, D. (2010) Will the Mediterranean become 'too hot' for tourism? A reassessment. *Tourism and Hospitality Planning & Development* 7(3), 267–281.

Schulzweida, U. (2015) CDO User's Guide. Available at: http://ngweb.serc.iisc.in/facilities/wp-content/uploads/2015/08/cdo.pdf (accessed 27 November 2015).

Scott, D. and Jones, B. (2007) A regional comparison of the implications of climate change for the golf industry in Canada. *The Canadian Geographer / Le Géographe canadien* 51(2), 219–232.

Scott, D., Hall, C.M. and Gössling, S. (2012a) *Tourism and Climate Change: Impacts, Adaptation and Mitigation*. Routledge, London.

Scott, D., Simpson, M.C. and Sim, R. (2012b) The vulnerability of Caribbean coastal tourism to scenarios of climate change related sea level rise. *Journal of Sustainable Tourism* 20(6), 883–898.

Scottish Golf Environment Group (2004) *Climate Change and Scottish Golf Courses*. SGEG Ltd, Edinburgh.

Şen, Ö.L. (2013) *A Holistic View of Climate Change and Its Impacts in Turkey*. Istanbul Policy Center, Istanbul, Turkey.

The Abdus Salam International Center for Theoretical Physics (2014) Regional Model: REGCM4. Available at: www.ictp.it/research/esp/models/regcm4.aspx (accessed 12 November 2015).

Thornthwaite, C.W. (1948) An approach toward a rational classification of climate. *Geographical Review* 38(1), 55–94.

Türkeş, M. (1999) Vulnerability of Turkey to desertification with respect to precipitation and aridity conditions. *Turkish Journal of Engineering and Environmental Sciences* 23, 363–380.

Turkish Ministry of Culture and Tourism (2007) Tourism Strategy of Turkey – 2023. Available at: www.kulturturizm.gov.tr/genel/text/eng/TST2023.pdf (accessed 13 November 2015).

Turkish Ministry of Culture and Tourism (2015a) İllere Göre Sit Alanları İstatistiği. Available at: www.kulturvarliklari.gov.tr/TR,44974/illere-gore-sit-alanlari-istatistigi.html (accessed 13 November 2015).

Turkish Ministry of Culture and Tourism (2015b) Tesis İstatistikleri. Available at: http://yigm.kulturturizm.gov.tr/TR,9859/tesis-istatistikleri.html (accessed on 13 November 2015).

Turkish Ministry of Culture and Tourism (2015c) Golf Turizmi. Available at: http://yigm.kulturturizm.gov.tr/TR,10161/golf-turizmi.html (accessed 23 November 2015).

Turkish Ministry of Environment and Urbanisation (2012) *Türkiye Cumhuriyeti İklim Değişikliği Eylem Planı 2011–2023*. Ankara.

Turkish Ministry of Environment and Urbanisation (2013) *Türkiye İklim Değişikliği 5*. Bildirimi. Ankara.

TURKSTAT (2015) Tourism Statistics. Available at: www.turkstat.gov.tr/PreTablo.do?alt_id=1072 (accessed 13 November 2015).

van Vuuren, D.F., Edmonds, J., Kainuma, M., Raihi, K., Thomson, A. *et al.* (2011) The representative concentration pathways: an overview. *Climatic Change* 109(1), 5–31.

West Mediterranean Development Agency (2011) Golf Turizmi Sektör Raporu. Available at: http://baka.org.tr/uploads/1303486414GOLF-KATALOG--TURKCE-SON.pdf (accessed 5 June 2017).

22 Case Study Israel: Coastal Tourism, Coastal Planning and Climate Change in Israel

C. Michael Hall[1]* and Yael Ram[2]

[1]University of Canterbury, New Zealand; University of Oulu, Finland; and Linneaus University, Kalmar, Sweden; [2]Ashkelon Academic College, Israel

Introduction

Like many countries on the Mediterranean Sea Israel is facing major challenges from a changing climate. The country has over 270 km of coastline with 188 km located on the Mediterranean. There is also a small coastline on the Red Sea which is significant for tourism and is the location of Eilat, which has been a significant tourist destination for several decades (Knowles and Curtis, 1999; Azaryahu, 2005). In addition, Israel also has two internal seas, the hypersaline Dead Sea and the freshwater Sea of Galilee (Lake Kinneret). Israel lies in a transition zone between the hot and arid southern part of west Asia and the relatively cooler and wet northern Mediterranean region. As a result, there is a wide range of spatial and temporal variation in climatic conditions in the country (Ministry of Environmental Protection, 2010). In Israel most of the population lives north of the 200 mm isohyet and along the coast. Although the country is seeking to develop climate change mitigation strategies, carbon dioxide emissions in Israel increased over the period 2003 to 2010 from 64,233.63 Gg CO_2 to 66,083.57 Gg CO_2 (Israeli Central Bureau of Statistics, 2012). In September 2015 the Israeli government set a goal of decreasing emissions by 25% to meet a goal of 7.7 t of CO_2 eq per capita (Ministry of Environmental Protection, 2015).

As in other Mediterranean countries the coastal regions of Israel are important tourist destinations as well as sites of permanent settlement (Kutiel *et al.*, 1999, 2000; Israeli Central Bureau of Statistics, 2015). Approximately 70% of Israel's residents live within 15 km of its coastline while the coastal strip also provides the most fertile agricultural land (Coastal Area Management Programme (CAMP) Israel, 2000). Population and tourist pressure on the coastal zones has also increased. Markandya *et al.* (2009) report that there was a 92% increase in coastal residential density between 1970 and 2000 and a 72% increase in tourist density in the same period. A CAMP-Israel study by Tropp and Zaban (2000, cited in Markandya *et al.*, 2009) estimated that there were 13 million annual beachgoers in Israel generating NIS98 million on travel to sites, NIS25 million in entry fees and NIS8 million in parking. There were also another 18 million trips in areas close to the beaches resulting in a spend of NIS79 million. The CAMP-Israel study is also significant for understanding the long-term implications of climate change in the Israeli coastline as

*Corresponding author, e-mail: michael.hall@canterbury.ac.nz

they indicated a potential value for the willingness to pay for the 79 km of Israel's coastline that was undeveloped of €0.12 per household per kilometre per year. A more recent study of the Rosh Haniqra Marine Reserve in northern Israel, which is only 6 km long, by Becker and Lavee (2009) suggested that the total value of the site exceeds NIS50 million, only about half of this figure being associated with direct market values (primarily from tourism).

As of 2000, 50 km of the Mediterranean shoreline were used for national infrastructures and defence uses and are closed to the public. The remaining coastline was designated as follows: 59 km as urban shores, 43 km as nature reserves and national parks and 36 km for open space (free of all infrastructures and facilities) (CAMP Israel, 2000). Along the Mediterranean, 14 marine reserves, 20 coastal reserves (of which only 15 were statutory declared), two islet reserves and two protected natural asset belts have been either proposed or declared. The total area of declared, approved and proposed nature reserves along the coastal plain is approximately 48 km^2 which is 8% of the total area of coastal plain (Kutiel, 2001). The coastal sand dune region has a high degree of species endemism along with a number of rare and endangered species and comprises two-thirds of the area of coastal nature reserves. Significantly for managing the impacts of climate change along the coastline only about 17% of the Israeli coastal dunes are still ecologically intact, of which only half has protected status (Bar, 2013). The average area of conservation in each of the declared nature reserves is less than 0.5 km^2 (Kutiel, 2001), while the largest protected area along the Mediterranean coast is the Nizzanim LTER (Long Term Ecosystem Research) Nature Reserve of 2000 ha (Bar, 2013). There are also 16 beaches of high value for recreation in natural surroundings or sites of archaeological interest for visitors that have been designated as national parks (CAMP Israel, 2000). However, like many countries there are differences in the proportion of different ecosystems that are represented in the protected area system. Rotem and Weil (2014) estimate that in the natural ecosystem units of Israel and the Palestinian authority combined, 54.8% of rocky shores are protected but only 12.4% of sand dune soils along the coastline and 2.2% of shoreline salines (located at the estuaries of the Na'aman River and the Kishon at Haifa Bay). The Dead Sea and the Sea of Galilee are not represented satisfactorily in protected areas as well.

Despite growing concerns over the impacts of climate change on Israel and the coastal environment in particular, there is little direct research on the implications of climate change on coastal tourism in the country. As CAMP Israel (2000, p. 30) stated, 'assuming that worldwide climate change may result in a rise in global sea level, the implications of such a rise on the Mediterranean coastline and on different sectors of the economy which are dependent on the coast should be studied'. This chapter therefore highlights some of the problems of planning for tourism and climate change in the Israeli context. The chapter next discusses some of the implications of climate change for the coast before examining the way in which the Israeli planning system seeks to respond to coastal change.

The Coast and Climate Change

All of the coastal areas of Israel are under pressure from climate change in different ways. Sea level on the Israeli coast is expected to rise by 0.5 m by 2050 and 1 m by 2100. In comparison to the period 1961–1990, the A1B scenario of the IPCC predicts a rise of 1.5°C in Israel's average temperature by 2020. According to IPCC scenarios A2 and B2, average temperature in the years 2071–2100 is expected to increase further by 5°C and 3.5°C, respectively, with a 10% decrease in precipitation expected by 2020 and a 20% decrease by 2050, meaning that the climate overall will become more arid (Ministry of Environmental Protection, 2010).

Growth in the expected number and intensity of extreme weather events will not only have implications for flooding but also lead to an increase in the height and intensity of waves leading to the penetration of seawater inland in low-lying areas. Although not a major focus of the present chapter it should be noted that the changed climatic conditions will have significant implications for the internal seas, affecting the salinity of the Sea of Galilee as well as flows into the Dead Sea, although the national water desalinization plan may reduce this pressure in the future (Sternberg *et al.*, 2014).

Several significant consequences of sea level rise (SLR) are expected. A 10 cm increase in sea level (assuming a slope of 1–5% in the Israeli coast) is forecast to lead to a coastal retreat of 2–10 m, resulting in a loss of 0.4–2 km^2 every 10 years. A 1 m increase in sea level would flood a 50–100 m wide belt on the sandy beaches that constitutes more than half the length of the Israeli coastline. One scenario, in which such an increase will occur until 2060, predicts that 8.4 km^2 of beaches will have been lost by then, with an economic damage of NIS4–5 billion, much of which results from the impacts on coastal tourism and recreation (Ministry of Environmental Protection, 2010). However, other forms of beaches will also be affected by a 1 m SLR with the area affected by sea surges moving further inland, and changes in the wave regime and storm frequency affecting the strength of the sandstone (kurkar) ridges and cliffs along the coast. The three districts of Israel that are most at risk of SLR are Acre, Haifa and Tel Aviv (Lichter and Felsenstein, 2012). In terms of number of hotel rooms, person nights and employment generation, Tel Aviv is the third most important destination in the country (after Jerusalem and Eilat), while Haifa is also reasonably significant with the seventh largest number of hotel rooms (1338) by locality in 2014 (Israeli Central Bureau of Statistics, 2015), with both Tel Aviv and Haifa having a number of seafront hotels at risk of SLR. Although there are no direct assessments of the effects of SLR on tourist coastal visitation a de facto indicator may be the percentage of permanent population affected by SLR per region. Lichter and Felsenstein (2012) assessed the costs of SLR and flooding at a local level in Israel. Table 22.1 provides a snapshot of some of the potential impacts of different levels of SLR for Haifa and Tel Aviv, two of the most at-risk urban centres as well as nationally.

Given that 70 km out of Israel's 190 km coastline are characterized by a coastal cliff, measuring 30–40 m in height, SLR is expected to have a significant impact on the cliff retreat processes for which the annual average cliff retreat rates of the cliff crest was 20 cm/year until 2000 (Zviely and Klein, 2004). It is forecast that the area of risk for cliff destruction until 2100 is 20–30 m eastward from the current cliff line in 65–80% of the cliff length (Ministry of Environmental Protection, 2010). The acceleration of seawater acidification may also exacerbate sea cliff erosion (Ministry of Environmental Protection, 2014). Cliff retreat is a significant problem

Table 22.1. Risks from sea level rise: Haifa, Tel Aviv and national totals (Lichter and Felsenstein, 2012).

Spatial area	SLR permanent inundation (m)	Inhabitants at risk (%)	Total inundated area (thousand m^2)	Total built area (thousand m^2)
Haifa				
	0.5	0.06	4,164	284
	1.0	0.15	4,593	317
	2.0	0.48	11,355	497
Tel Aviv				
	0.5	0.11	822	39
	1.0	0.18	1,040	65
	2.0	0.62	1,622	124
National				
	0.5	0.03	19,295	380
	1.0	0.06	23,421	497
	2.0	0.18	41,735	1,234

in areas that also have a dense population base such as Ashkelon. The economic costs of damage to existing buildings as a result of a cliff retreat rate of an average of 0.5 m/year has been estimated at between NIS67 and NIS90 million; if the retreat rate reaches an average of 1 m/year the cost may reach up to NIS276 million (Ministry of Environmental Protection, 2010). In addition there are potentially significant effects on the market value of properties in areas affected by severe cliff erosion. These figures do not account for the impact of SLR on archaeological and heritage resources, which are often significant visitor attractions (Hall *et al.*, 2016). It is expected that underwater sites now covered with sand will be lost due to coastal erosion and the loss of sands from the sea floor, which will expose the sites to wave attack and oxidation with consequent rapid degradation, a phenomenon that has already started to occur at Caesarea (Pe'er and Safriel, 2000; Fig. 22.1).

The impacts of SLR on the Gulf of Eilat are not expected to be as severe as on the Mediterranean coastline because of the different coastal geomorphology although coral reef ecosystems there, which are a major tourist attraction, are expected to suffer from severe bleaching due to the increased water temperatures (Loya, 2004; Bahartan *et al.*, 2010; Furby *et al.*, 2013). It is expected that the degradation of the coral reef in the Gulf of Eilat by the synergistic effect of pollution and climate change will have a strong negative impact on tourism, as Eilat's international appeal is partly due to dive tourism on the coral reef (Pe'er and Safriel, 2000). The potential impacts of climate change also include the establishment of alien species in both the Mediterranean and Red Seas while seawater acidification will affect a wide range of species. One potential impact on coastal tourism in both marine and internal waters is an increase in algal blooms that may not only be unsightly but can also cause skin irritations. For example, since the mid-1990s, Lake Kinneret has been invaded by two species of nitrogen-fixing cyanobacteria from the Nostocales group that appear every summer and, in some years, form a massive summer bloom, the likes of which were never observed in Lake Kinneret in the past. One of these invasive species produces toxins that have a detrimental effect on water quality, including for drinking (Sternberg *et al.*, 2014). The opening of the bio-geographical barrier between the Mediterranean Sea and the Red Sea by the Suez Canal, combined with climate change and increased shipping mobility, led to proliferation of Red Sea species in the Mediterranean Sea together with shrinking and extinction of native species (Katsanevakis *et al.*, 2014). One of the worst introduced invasive species in the Mediterranean Sea is the nomad jellyfish (*Rhopilema nomadica*). Ghermandi *et al.* (2015) estimate that outbreak of the jellyfish causes an annual monetary loss of €1.8–6.2. The researchers also indicate that 41% of their respondents reported negative implications on their recreational activities due to the outbreak of jellyfish.

Adapting to Climate Change

Like most countries Israel does not have a tourism-specific or even tourism-focused climate change adaptation strategy (Scott *et al.*, 2012, 2016). Tourism industry adaptation is located at the level of individual firms. Environmental groups that have been at the forefront of coastal zone conservation in Israel have often used tourism and recreation values to restrict inappropriate coastal development. For example, in 1998 the Society for the Protection of Nature in Israel (SPNI) and the Israel Union for Environmental Defense (IUED) published a position paper on this subject and argued that areas with high landscape and environmental value, which are designated for tourism and recreation in the national tourism and other masterplans, should be preserved exclusively for such uses. They argued that there was a need to protect the coast from over-development with clear guidelines required to prevent the transformation of buildings constructed in areas allocated for tourism under the guise of 'apartment hotels' or 'resort apartments' to residential uses (CAMP Israel, 2000). Nevertheless, coastal zone planning

Fig. 22.1. The historical site of Caesarea is regarded as being at risk of sea level rise and wave attack.

in Israel remains relatively fragmented creating difficulties in connecting tourism and climate change responses within the overall planning framework.

The planning system of Israel is subject to a high degree of central planning. This has developed through the combination of high population densities and growth in urban areas, the small size of the country and the adoption of European urban and regional planning traditions at the time of the nation's creation (Shmueli *et al.*, 2015).

The Israeli planning system is determined by the Planning and Building Law of 1965. This law added a national planning authority over the two tiers that had existed since the British Mandate over Palestine – the local level and the district level. Lower-level plans must be strictly consistent with all higher-level plans. It means that land use nationally is determined by a binding masterplan for all actions of construction, demolition or change (Alterman, 2001). Additionally, the ownership of lands was defined by the Basic Law: Israel Lands of 1960, which stated that 93% of lands was to be publicly owned (Hananel, 2013). Some of the institutional arrangements for environmental planning in Israel was set even before the state was established by the central Jewish institution of the KKL-JNF (Keren Kayemeth LeIsrael – Jewish National Fund), which is still an active player under the Forests Act of 1926, and was later strengthened by the Planning and Building Law, the Land Law, the Water Act and the National Parks and Natural Reserves Law. In keeping with the then-dominant political statist worldview, these institutional settings were highly centralized (Shmueli *et al.*, 2015).

A series of amendments to the Planning and Building Law of 1965 (e.g. amendment number 43 of 1995, and 76 of 2006) and reforms were made in an effort to decentralize the planning process and broaden the powers of the local committees to improve economic efficiency and support the privatization process (Hananel, 2013). However, environmental goals were peripheral to this agenda. Instead, environmental plans remained subjects for central planning, national plans and laws, supported by policies of the Ministry of Environmental Protection. Although Israeli lawmakers passed the Law for the Protection of the Coastal Environment in 2004, the stated aims of which are:

- to protect the coastal environment and its natural treasures, to reduce and prevent coastal damage;
- to preserve the coastal environment and the coastal sand for the benefit and enjoyment of present and future generations; and
- to establish principles for the sustainable management, development and use of the coastline,

numerous coastal environmental conservation and management issues remain.

Like many jurisdictions Israel has a plethora of legislative items that relate to the coastal zone and that make integrated management extremely difficult. The situation is further complicated by several masterplans that apply to the coastal zone as well as numerous institutions. CAMP Israel (2000) identified seven masterplans that were relevant to coastal management: The National Masterplan for the Mediterranean Coast (NOS 13), The National Masterplan for Ports and Marinas (NOS 13B), The National Masterplan for the Resource Management of the Mediterranean Coastline for Tourism and Recreation (NOS 13C), The National Masterplan for Tourism (NOS 12), The National Masterplan for National Parks, Nature Reserves and Landscape Reserves (NOS 8), The National Masterplan for Forests and Afforestation (NOS 22) and The National Masterplan for Building and Development (NOS 35). Table 22.2 details some of the most significant institutional stakeholders in coastal tourism and climate change, although it needs to be emphasized that climate change considerations are not integrated into formal planning documents.

More recently the amendment 9/A to The National Masterplan for the Mediterranean Coast (NOS 13) provides a planning framework for protecting the limestone cliff along the Mediterranean coastline (see also Eidelman *et al.*, 2010). The introduction to the amendment explains the need for this updated plan in terms of natural and anthropogenic

Table 22.2. Institutional stakeholders in coastal tourism and capacities for climate change adaptation.

Institution	Responsibilities
Ministry of the Interior	National planning and building, declaration of new protected areas
Territorial Waters Committee	Approving offshore structures
National Planning and Building Board	National masterplans
District Planning & Building Commissions	Regional masterplans
Emergency Services, Special Duties & Bathing Beaches Administration	Supervision of bathing beaches
Ministry of Transport	Supervision of anchorages
Ministry of Environmental Protection	Marine pollution prevention, supervision of environmental legislation
Ministry of Health	Inspection of bathing beaches
Ministry of Tourism	Development of tourism and recreation
Ministry of National Infrastructures	Monitoring and research on coastal and marine resources
Ministry of Defence	Military areas on the coast
Ministry of Foreign Affairs	International climate change, environmental and marine conventions
Ministry of Justice	Marine legislation and conventions
Local Authorities	Coastal planning, licenses, maintenance of public areas
Israel Land Administration	Management of national land in Israel
Nature and National Parks Protection Authority	Nature protection and administration of coastal national parks and nature reserves (including marine reserves)
Antiquities Authority	Coastal and marine archaeology
Ports Authority	Commercial ports, loading and unloading facilities
Water Commission	Infiltration of seawater to groundwater
Artificial Islands Committee	Advancement of planning for islands
National Sewage Authority	Marine outlets

change without any direct connection to climate change or SLR. For planning purposes cliff locations were divided into different area 'cells'. The main objectives of the amendment were as follows:

1. To determine what are the allowed solutions for each area (cell).
2. To determine how to divide the coastline to different areas (cells).
3. To determine the conditions for building permits.
4. To instruct how to prepare an environmental document that has to be attached to each of the building permits requests.
5. To give instructions for maintenance works and prevention acts (Government of Israel, 2014).

The cliff coastline was divided into 40 cells and for each cell a group of management solutions to restrict undesirable change was offered. Each cell was analysed according to its physical characteristics and planning status to determine its urgency. In the context of tourism and heritage, built tourism facilities (hotels and resorts) and archaeological sites were identified as sensitive areas that require an immediate action. On the other hand, open spaces and nature reserves were defined as low priority areas.

A temporary solution that includes fences, signs and drainage of the cliff was offered to all 40 cells. Protecting the bottom of the cliff was recommended to 25 cells, most of them were also noted in the context of slope protection (18 cells). The least common management procedures include transfer of sand on the beach (11 cells), use of Geotubes (five cells), putting rocks in the shallow sea (seven cells) and beach nourishment (12 cells). None of the cells was suggested all the solutions, but three cells (from Netanya and Ashkelon) were offered five of the six solutions. In addition, in the longer term the evacuation of structures and the prevention of new building in areas of high risk will also be adopted. In August 2012 the Israeli cabinet approved the establishment of a public company to protect the Mediterranean cliffs. The Mediterranean Coastal Cliffs Preservation Government Company Ltd is responsible for all coastal cliff issues, including planning and implementation of recommendations to protect and strengthen the cliffs, as well as long-term maintenance and monitoring.

Ten municipalities are included in this plan, from which the public records contain the opinion of nine municipalities (Ministry of Interior, 2013). All municipalities, except one (Hof-Hasharon), were interested in long-term solutions (i.e. protection of the cliff and the slopes) and three municipalities (Tel Aviv, Netanya and Herzliya) even rejected the short-term solutions (signs and fences) that they had been offered and sought to promote their own long- and short-term plans. Other municipalities accepted the short-term solutions alongside the long-term plans they had been offered. However, not all management plans and strategies were well received.

Netanya promoted plans for building breakwaters in order to reduce the impact of waves on the beaches, even though breakwaters were not explicitly mentioned in the national plan as a solution for cliff protection (Rinat, 2014). However, the breakwater plans of Netanya, and the announcements of Herzliya and Ashkelon municipalities that they might consider building breakwaters, provoked a wave of resistance from academia, political stakeholders and the public. The Geological Institute indicated that breakwaters are not necessarily the ideal solution for cliff degradation (Rinat, 2014). Surfers, a very large recreational community in Israel, and in Herzlyia in particular, led a public protest against breakwaters because it would potentially destroy a number of beach breaks (Rinat, 2015).

The Green Party (Ha'Yerukim) also indicated the potentially harmful implications of breakwaters but, in addition, pointed to an interesting gap between adaptation and mitigation with regard to the protection of coastlines. They noted that the cliff protection plan will not stop the degradation process, since it protects assets and buildings but does not seek to eliminate the problem itself, which relates to an increase in storm and tidal surges (Ha'Yerokim, 2010). However, even this party did not relate the problem to climate change. The Israeli case therefore highlights a very interesting phenomenon in that when coastlines in Israel are discussed, climate change is still a 'hidden factor', leaving the focus on the development of adaptations but not on taking steps to mitigate emissions.

Conclusions

As the chapter has indicated, climate change and SLR is expected to lead to damage to coastal infrastructure (including marinas), ecosystems and archaeological sites, all of which will have implications for coastal tourism. In the Gulf of Eilat damage to coral reefs may also have significant implications for dive tourism. The implications of such change are substantial. SLR and increased frequency and intensity of winter storms will cause severe damage to the coastal cliff of Israel which, despite the development of specific environmental management strategies, is expected to continue to recede. The loss of coastal resources and damage to beaches is expected to have a significant impact on coastal tourism in Israel both directly as well as indirectly via the development of control measures that may restrict tourist activities or infrastructure. In addition, some proposed coastal defence measures such as breakwaters are problematic with some beach users because of their effects on wave profile and beach shape.

A key issue with respect to climate change and coastal tourism in Israel is the near absence of explicit connection between the two in government policy and planning processes. Government focus appears to be on adaptation to environmental change in coastal areas rather than mitigation (Ministry of Environmental Protection, 2015). Israel is better positioned than many countries in the eastern Mediterranean to respond to climate change. However, in the long run, some of the greatest threats to coastal tourism in the country may come from wider regional insecurities as a result of environmental change, particularly as climate change compounds the vulnerability of Palestinians in Gaza and the West Bank and also the broader security relationship between Israel and Palestine (Feitelson *et al.*, 2012; Mason *et al.*, 2012). Nevertheless, the lack of a clear integrated coastal resource management strategy to deal with climate change remains a significant challenge for tourism development on the coast. Although environmental change is recognized in planning strategies, considerations of the implications of climate change and SLR appear limited.

References

Alterman, R. (2001) National-level planning in Israel: walking the tightrope between government control and privatization. In: Alterman, R. (ed.) *National-Level Planning in Democratic Countries*. Liverpool University Press, Liverpool, UK, pp. 257–288.

Azaryahu, M. (2005) The beach at the end of the world: Eilat in Israeli popular culture. *Social & Cultural Geography* 6(1), 117–133.

Bahartan, K., Zibdah, M., Ahmed, Y., Israel, A., Brickner, I. and Abelson, A. (2010) Macroalgae in the coral reefs of Eilat (Gulf of Aqaba, Red Sea) as a possible indicator of reef degradation. *Marine Pollution Bulletin* 60(5), 759–764.

Bar, P. (2013) Restoration of coastal sand dunes for conservation of biodiversity: the Israeli experience. In: Martinez, L., Gallego-Fernández, J. and Hesp, P. (eds) *Restoration of Coastal Dunes*. Springer, Berlin, pp. 173–185.

Becker, N. and Lavee, D. (2009) Commercial development and conservation value the case of Rosh Haniqra Marine Reserve in Israel. *Journal of Infrastructure Development* 1(2), 193–217.

Coastal Area Management Programme (CAMP) Israel (2000) *Final Integrated Report*. United Nations Environment Programme, Mediterranean Action Plan, Coastal Area Management Programme, Ministry of the Environment, Jerusalem.

Eidelman, A., Bein, A. and Cohen, G. (eds) (2010) *Coping with Erosion along Israel's Cliff Shores: Economic, Public Policy and Environmental Implications*. Environmental Policy Center of the Jerusalem Institute for Israel Studies and the Ministry for Environmental Protection, Jerusalem.

Feitelson, E., Tamimi, A. and Rosenthal, G. (2012) Climate change and security in the Israeli–Palestinian context. *Journal of Peace Research* 49(1), 241–257.

Furby, K.A., Bouwmeester, J. and Berumen, M. (2013) Susceptibility of central Red Sea corals during a major bleaching event. *Coral Reefs* 32(2), 505–513.

Ghermandi, A., Galil, B., Gowdy, J. and Nunes, P.A.L.D. (2015) Jellyfish outbreak impacts on recreation in the Mediterranean Sea: welfare estimates from a socioeconomic pilot survey in Israel. *Ecosystem Services* 11, 140–147.

Government of Israel (2014) *Law of Planning and Construction: National Plan Number 13 – Mediterranean Coastline, Amendment 9/A* (in Hebrew). Government of Israel, Jerusalem.

Hall, C.M., Baird, T., James, M. and Ram, Y. (2016) Climate change and cultural heritage: conservation and heritage tourism in the Anthropocene. *Journal of Heritage Tourism* 11(1), 10–24.

Hananel, R. (2013) Planning discourse versus land discourse: the 2009–12 reforms in land-use planning policy and land policy in Israel. *International Journal of Urban and Regional Research* 37(5), 1611–1637.

Ha'Yerokim (2010) The Policy for Coastline Cliffs will Cause Damage for Beaches on Behalf of Real Estate [in Hebrew]. Available at: www.green-party.co.il/news/archives/3855 (accessed 16 September 2015).

Israeli Central Bureau of Statistics (2012) *Israel. Israel's GHG inventory*. UNFCCC ISR/INV/2 E. Ministry of Environmental Protection, Jerusalem.

Israeli Central Bureau of Statistics (2015) *Tourist Hotels, Rooms, Occupancy, Person-Nights, Revenue and Jobs in Tourist Hotels in Selected Localities 2014*. Available at: www.cbs.gov.il/shnaton66/st23_11.pdf (accessed 5 October 2015).

Katsanevakis, S., Coll, M., Piroddi, C., Steenbeek, J., Ben Rais Lasram, F. *et al.* (2014) Invading the Mediterranean Sea: biodiversity patterns shaped by human activities. *Frontiers in Marine Science* 1(32), DOI: 10.3389/fmars.2014.00032.

Knowles, T. and Curtis, S. (1999) The market viability of European mass tourist destinations. A post-stagnation life-cycle analysis. *The International Journal of Tourism Research* 1(2), 87–96.

Kutiel, P. (2001) Conservation and management of the Mediterranean coastal sand dunes in Israel. *Journal of Coastal Conservation* 7(2), 183–192.

Kutiel, P., Zhevelev, H. and Harrison, R. (1999) The effect of recreational impacts on soil and vegetation of stabilized coastal dunes in the Sharon Park, Israel. *Ocean and Coastal Management* 42, 1041–1060.

Kutiel, P., Zhevelev, H. and Eden, Z. (2000) Effect of experimental trampling and off-road motorcycle traffic on soil and vegetation of stabilized coastal dunes, Israel. *Environmental Conservation* 27, 14–23.

Lichter, M. and Felsenstein, D. (2012) Assessing the costs of sea-level rise and extreme flooding at the local level: a GIS-based approach. *Ocean & Coastal Management* 59, 47–62.

Loya, Y. (2004) The coral reefs of Eilat – past, present and future: three decades of coral community structure studies. In: Rosenberg, E. and Loya, Y. (eds) *Coral Health and Disease*. Springer, Berlin, pp. 1–34.

Markandya, A., Cassinelli, M., Arnold, S. and Taylor, T. (2009) *Coastal Zone Management in the Mediterranean: Legal and Economic Perspectives*. Fondazione Eni Enrico Mattei, Milan, Italy.

Mason, M., Zeitoun, M. and Mimi, Z. (2012) Compounding vulnerability: impacts of climate change on Palestinians in Gaza and the West Bank. *Journal of Palestine Studies* 41(3), 1–16.

Ministry of Environmental Protection (2010) *Israel's Second National Communication on Climate Change Submitted under the United Nations Framework Convention on Climate Change*. Ministry of Environmental Protection, Jerusalem.

Ministry of Environmental Protection (2014) *Adaptation to Climate Change in Israel Recommendations and Knowledge Gaps*. Ministry of Environmental Protection, Jerusalem.

Ministry of Environmental Protection (2015) *The National Agenda for Decreasing GHG Emissions Towards UN Agreements in Paris: A Presentation for Government Meeting 20 September 2015* [in Hebrew]. Ministry of Environmental Protection, Jerusalem.

Ministry of Interior (2013) Meeting Number 4 – A Master Presentation [in Hebrew]. Available at: http://mavat.moin.gov.il/MavatPS/Forms/SV4.aspx?tid=4 (accessed 6 September 2015).

Pe'er, G. and Safriel, U.N. (2000) *Climate Change: Israel National Report under the United Nations Framework Convention on Climate Change Impact, Vulnerability and Adaptation*. Ministry of Environmental Protection, Jerusalem.

Rinat, Z. (2014) The geological institute: breakwaters for cliffs protection will harm the beach [in Hebrew]. *Ha'aretz*, 9 September. Available at: www.haaretz.co.il/news/science/.premium-1.2428469 (accessed 16 September 2015).

Rinat, Z. (2015) Israel Turning to Breakwaters in Bid to Stem Cliff Erosion. Maritime defenses set to be built in Netanya, Herzliya and Ashkelon, but some are making waves about plan. *Ha'aretz*, 20 July. Available at: www.haaretz.com/israel-news/science/.premium-1.666796 (accessed 1 April 2016).

Rotem, D. and Weil, G. (2014) Natural ecosystem-units in Israel and the Palestinian authority-representativeness in protected areas and suggested solutions for biodiversity conservation. *Journal of Landscape Ecology* 7(1), 91–109.

Scott, D., Gössling, S. and Hall, C.M. (2012) *Tourism and Climate Change: Impacts, Adaptation and Mitigation*. Routledge, London.

Scott, D., Hall, C.M. and Gössling, S. (2016) A review of the IPCC 5th Assessment and implications for tourism sector climate resilience and decarbonization. *Journal of Sustainable Tourism* 24(1), 8–30, DOI:10.1080/09669582.2015.1062021.

Shmueli, D., Feitelson, E., Furst, B. and Hann, I. (2015) Scale and scope of environmental planning transformations: the Israeli case. *Planning Theory & Practice*, 16, 336–362, DOI:10.1080/14649357.2015.1054419.

Sternberg, M., Gabay, O., Angel, D., Barneah, O., Gafny, S. *et al.* (2014) Impacts of climate change on biodiversity in Israel: an expert assessment approach. *Regional Environmental Change* 15(5), 895–906.

Tropp, Z. and Zaban, S. (2000) *The Mediterranean Coast in Israel: Social and Economic Evaluation. CAMP Israel*. PAP/RAC, Split, Croatia.

Zviely, D. and Klein, M. (2004) Coastal cliff retreat rates at Beit-Yannay, Israel, in the 20th century. *Earth Surface Processes and Landforms* 29, 175–184.

23 Case Study Antarctica: Up Against the Ice Barrier: Antarctic Tourism Operators Prepare for the Polar Shipping Code

Julia Jabour*
Institute for Marine and Antarctic Studies, University of Tasmania, Hobart, Australia

Introduction

The 'cryosphere' is the world of ice. It includes Arctic and Antarctic sea and land ice, which scientists claim provides the most 'visible signatures of climate change' (Vaughan *et al.*, 2013, p. 319). When sea ice forms and expands, its extent can be seen and measured from space, and when it contracts the chlorophyll in ice-edge nutrient blooms that sustain cold-water life can also be seen from space. When land ice melts, that, too, is obvious – with massive icebergs calving off friable ice sheets and ice shelves at the water's edge and soaking the ocean with dense, cold, fresh water. These occurrences can be both sporadic and annual; they can be unpredictable, inconsistent and regionally asymmetrical; and occasionally they can defy developing trends. Understanding the ice regimes of the polar regions is a difficult task in these warming environments. This, together with the prospect of increased shipping activity in iceless polar waters, has prompted – in part – the development of the International Maritime Organization's (IMO) International Code for Ships Operating in Polar Waters (hereafter, the Polar Code), which came into effect in January 2017.

The Polar Code establishes a risk-based regime of ship classification, with corresponding operating requirements and limitations, determined by a vessel's ability to withstand contact with different classifications of ice. Vessels that operate in both polar regions will soon be required to comply with the Polar Code for the purposes of both safety of life and protection of the environment. In the Antarctic, the majority of these vessels will be passenger ships plying the small but lucrative Antarctic tourism trade.

The Antarctic tourism industry declined dramatically during the 2007–2008 global financial crisis, but tourist numbers have been steadily building back up to pre-crisis levels. The trajectory of tourist numbers (until that time) was exponentially upwards (Table 23.1), and there is little doubt that it will continue on this trend, however the changing operating conditions will provide a number of challenges and opportunities for existing or new tourism operators, potentially including route changes and increase in vessel capacity (Bender *et al.*, 2016).

Tourism to the Antarctic is inextricably linked to the cryosphere. The cryosphere plays a crucial role in the generation of the world's weather, and its importance attracts scientists from many disciplines to the polar

*Corresponding author, e-mail: julia.jabour@utas.edu.au

Table 23.1. IAATO Member Seaborne Tourism in the Decade 2005–06 to 2015–16 (IAATO (2015) Information Paper 53 to ATCM XXXVIII Sofia, 2015).

Years	No. operators	No. ships and yachts	No. voyages	No. landed passengers	No. cruise-only passengers
2005–06[a]	47	44	249	25,167	4,632
2006–07[a]	42	47	268	28,622	6,930
2007–08	48	55	308	32,637	13,015
2008–09	44	53	290	26,921	10,652
2009–10	44	51	239	21,622	15,020
2010–11	41	47	223	19,065	14,373
2011–12	36	41	234	21,131	4,872
2012–13	39	45	258	24,892	9,070
2013–14	42	51	283	27,374	9,670
2014–15	37	44	268	26,812	9,459
2015–16[b]	40	47	306	30,634	8,900

[a]Includes IAATO and non-IAATO member-operated yachts (sailing and motor) where the information was available.
[b]Based on pre-season estimates.

regions, particularly Antarctica, which, since it has no one sovereign, is preserved as a continent for peace and science through the Antarctic Treaty of 1959. The cryosphere is also exceptionally beautiful in ways that words cannot describe, and this is what attracts many tourists as well. Of course tourists also go to the Antarctic to view wildlife, witness historical artifacts, for adventure, and to check another destination off their 'bucket list'. It is fair to say, however, that the fate of the Antarctic tourism industry is – and will continue to be – shaped by ice.

Tourism to the Antarctic is small in global terms; only 0.003% of the total number of global tourists visits Antarctica (e.g. 36,702 Antarctic tourists in 2014–15 compared with 1.087 billion tourists worldwide; see World Tourism Organization, 2015; Table 23.1). Furthermore, it is almost exclusively ship-based, with smaller vessels carrying less than 500 passengers permitted to land tourists on the continent, and larger vessels over this limit permitted only to cruise Antarctic waterways without making landings (Fig. 23.1) (IAATO, 2016). There is a range of other tourism products, including yacht cruising, fly–cruise packages and air-supported land-based adventures for those who eschew the traditional small ship expedition-style cruising. One of the primary concerns of Antarctic commentators, operators and regulators is the ability of the ship-based industry to adapt to changing environmental conditions, and to withstand the increasingly more onerous regulatory requirements. Another concern is keeping Antarctic tourism sustainable. Less than 0.3% of the continent of Antarctica is ice-free, and these primarily coastal oases are creating conditions of overlap between scientific research facilities, safe tourism landing sites, and the natural wonders that both scientists and tourists come to study and see. Tourism, therefore, attracts the attention of the Antarctic Treaty Consultative Parties, who watch over but do not directly manage the day-to-day operations of the industry.

This update will describe the latest developments in preparation for the implementation of the Polar Code, and examine the state of readiness of the Antarctic tourism industry. While this study relates specifically to Antarctic ship-borne tourism, it is also acknowledged that many vessels operate in both polar waters during alternating tourist (summer) seasons.

Research Context

In my original chapter (Jabour, 2011) it was posited that the Antarctic Treaty Consultative Parties had a number of options open to them to take a firm hand in regulating

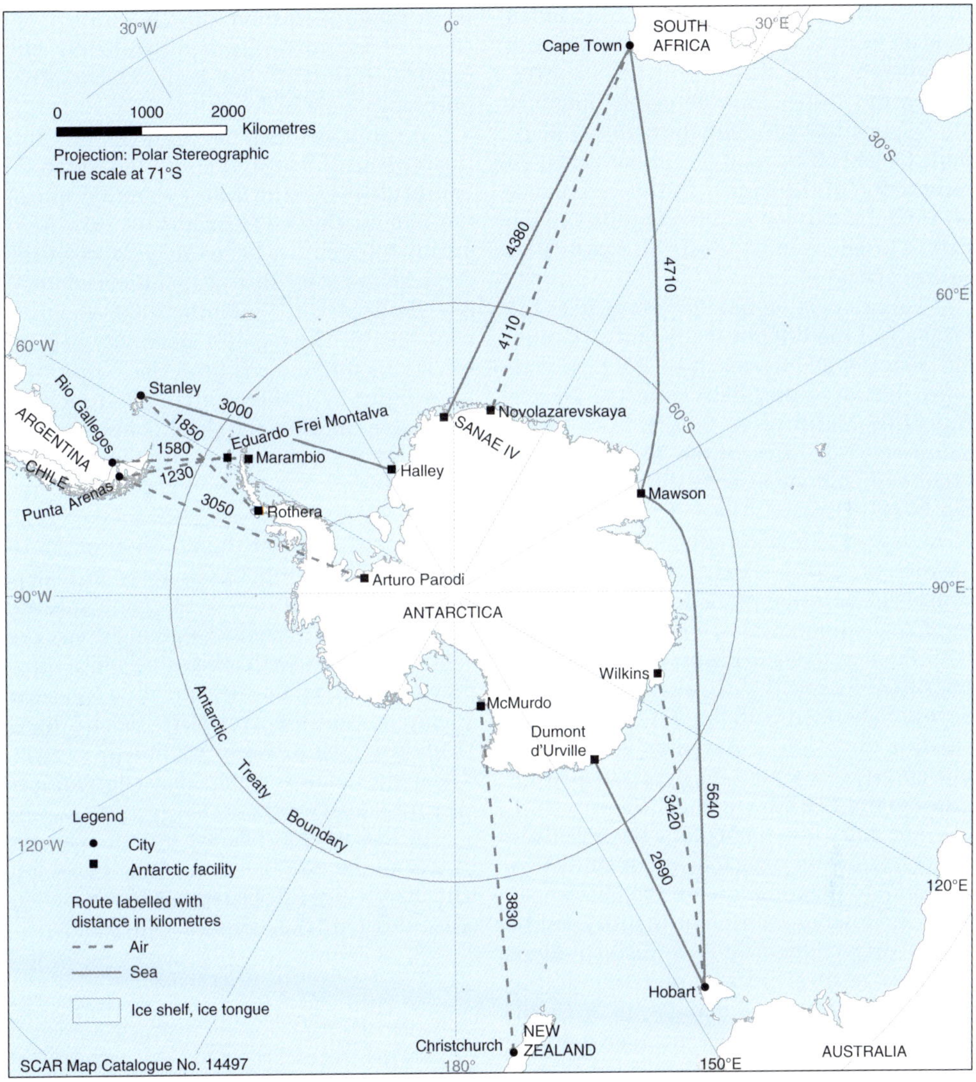

Fig. 23.1. Map of Antarctica (Antarctic Map – SCAR Map Catalogue No. 14497, © Commonwealth of Australia).

tourism, in the context of the dramatically changing environmental conditions being experienced in the Antarctic. It was proposed, for example, that the Parties could implement a sponsoring state provision, whereby an Antarctic Treaty Consultative Party that would accept strict liability for any environmental damage on their behalf must sponsor a tourism operator. Or that the area protection provisions in Annex V of the 1991 Protocol on Environmental Protection to the Antarctic Treaty (Madrid Protocol) could be used to designate specially protected areas with restrictions on the level of tourism (including, potentially, a prohibition on visits during certain times such as breeding). Or even, that the Parties might introduce industry-wide quotas or caps on tourism to specific locations in Antarctica – although the problems of implementing this kind of restriction, given the *sui generis* nature of Antarctic governance, made this choice virtually impossible. An expanded level of control over the industry has not eventuated since these thoughts in 2011. Rather, the Parties have abrogated much of the responsibility to the International Maritime Organization and its Polar Code, in recognition of the largely ship-based nature of Antarctic tourism and the inability of the Parties to agree unanimously on a particular course of action.

This abrogation not only recognizes the primacy of the IMO in matters of shipping, but also the greater reach of IMO conventions. For example, data collected by the Argentine Maritime Authority from ships' manifests reveal that of the 28 vessels that undertook 209 voyages to the Antarctic in the 2014/15 season, the majority – 63% – were flagged to non-Antarctic Treaty parties (Argentina, 2015). This statistic has been consistent for some years now, and because implementation of IMO conventions rests with the flag states of ships, it is logical for the Antarctic Treaty Consultative Parties to have collaborated with the IMO, thereby expanding the reach of Antarctic law beyond the current 53 Antarctic Treaty signatory states to the 171 IMO member states.

The entry into force of the Polar Code is imminent and the objective of this study was therefore to determine what the challenges and opportunities were for the industry and its state of preparedness for its implementation.

This research was undertaken during the austral spring and summer of 2015/16. Because the Polar Code was not then in force, much of what is contained in this study is based on published literature and official documents from the IMO, Parties to the IMO who have provided working documents and progress summaries, and discussions with key actors in the industry, insurance and classification services. No empirical data are yet available.

Climate Change: Assessing Problems and Challenges

The Antarctic Peninsula, bordered by the Weddell Sea to the east and the Bellingshausen/Amundsen seas to the west, is the most popular destination for tourists because of its proximity (approximately 800 nautical miles) to the main embarkation port of Ushuaia in Argentina (Fig. 23.1).

As well as being the prime tourist destination out of South America, the Antarctic Peninsula is also the fastest warming area of the planet. One consequence of this dubious distinction has been the collapse of the West Antarctic ice sheet at an unprecedented rate (Pollard and DeConto, 2009; Vaughan *et al.*, 2013). The region's instability has implications for both sea level rise – resulting from input by iceberg calving and melting ice shelves and the ice sheet – and, importantly, for the production, distribution and composition of sea ice, upon which the Polar Code bases its risk assessment. However, it is significant to note that unlike in the Arctic, where both the Russian Federation and Canada have designated specific maritime transit zones relative to sea ice concentrations, with corresponding shipping regulations, no such zonation has been attempted for the Antarctic (R. Hindley, Lead Specialist – Arctic Technology, Lloyd's Register, Vancouver, personal communication, 12 January 2016).

In the 5th Assessment Report released in 2013, the Intergovernmental Panel on Climate Change (IPCC) reported (in Vaughan *et al.*, 2013, p. 319, emphasis added) that:

> It is very likely that the annual Antarctic sea ice extent *increased* at a rate of between 1.2 and 1.8% per decade (0.13 to 0.20 million km^2 per decade) between 1979 and 2012. There was a greater increase in sea ice area, due to a decrease in the percentage of open water within the ice pack.

The albedo (reflectivity) of ice is greater than that of open water therefore ice- and snow-cover reflect the sun's heat more efficiently than darker surfaces such as soil or water. In a complex feedback mechanism, ice cover helps to keep temperature increases down, however once the melting or thaw begins and the albedo reduces, the temperature of the ocean or the land rises accordingly as the darker surfaces absorb more heat – becoming heat sinks. And thus the dynamic process begins. Vaughan *et al.*

(2013, p. 319, emphasis added) are careful to point out, however:

> There is high confidence that there are *strong regional differences* in this annual rate, with some regions increasing in extent/area and some decreasing.

They noted that 'the trends have been strongly positive in the Ross Sea and negative in the Bellingshausen/Amundsen seas' (Vaughan *et al.*, 2013, p. 330), meaning there is relatively more sea ice in the Ross Sea (off east Antarctica south of Australia and New Zealand) and relatively less off the west coast of the Antarctic Peninsula (south of South America). The Ross Sea region is an alternative Antarctic destination, but tourism to this area is a significantly smaller enterprise and an entirely different product because of the longer time spent at sea in comparison to the Ushuaia–Peninsula route. These regional differences will affect safe navigation for the variety of vessels that operate in Antarctic waters, as indeed such differences do now for purposes of emergency response and search and rescue.

Causality data supplied by Cruise Lines International Association (CLIA) to the IMO during the development of the Polar Code reported that only one Antarctic tourist vessel had a significant incident that involved contact with ice in the two decades of 1991–2011 (CLIA, 2011). The vessel was the MV *Explorer*, which sank in 2007 after entering an ice field inaccurately assessed by the Master as being safe to travel through at the ship's current speed at night. A number of other ships have grounded or been beset in heavy pack ice, but with no loss of life and insignificant pollution events.[1] Clearly, while ice is a major risk factor in polar waters, this CLIA report to the IMO concluded that often a number of factors are involved in an incident, including poor weather, poor charting, darkness, equipment failure and human error (CLIA, 2011). Indeed many of these factors existed prior to and during the sinking of the MV *Explorer* (Republic of Liberia, 2009).

Costs of compliance are already high for the polar tourism industry because of such characteristics as the vast distances

Fig. 23.2. MV *Orion* and Ross Ice Barrier (© Julia Jabour).

Fig. 23.3. Heavy pack ice off Terra Nova Bay (© Julia Jabour).

Fig. 23.4. Ice leaving Terra Nova Bay (© Julia Jabour).

involved, the relative costs of fuel and provisions, the requirement for trained crew and the retention of some waste materials onboard. These are likely to escalate in the future. The IMO is the platform for almost all of those increased costs. There is now a stronger Annex to MARPOL (the marine pollution convention) on controlling emissions from ships, as well as a change to MARPOL to ban the use or carriage of heavy or intermediate fuel oils south of 60° South (which coincides with the area of application of the Antarctic Treaty). Note, however, that this prohibition does not (yet) apply to the Arctic. There is a Convention on Ballast Water Management, which came into force in 2016 and requires vessels to have the capacity to comply with a strict routine of ballast water exchange and/or retention. The existing recommendatory guidelines for ships operating in polar waters became mandatory in 2017 as the Polar Code, and established new ice classifications for ships to help classification societies, insurers and flag states determine which ships can operate, when, where, and under what conditions.

Climate Change: Managing Solutions – The Polar Code

The International Code for Ships Operating in Polar Waters – the Polar Code – became mandatory through amendments to the International Convention for the Safety of Life at Sea (SOLAS) and the International Convention for the Prevention of Pollution from Ships, 1973, as modified by the 1978 Protocol relating thereto (MARPOL).[2]

The expected date of entry into force of the SOLAS and MARPOL amendments was 1 January 2017, under the IMO's tacit acceptance procedure. This procedure ensures swift entry into force of amendments because acceptance is inferred, i.e. it does not need to be formally expressed by a particular number of states representing a particular percentage of world shipping tonnage.

The Polar Code will initially apply to new ships constructed after 1 January 2017. Ships constructed before that date will be required to meet the relevant requirements of the Polar Code by the first intermediate or renewal survey, whichever occurs first, after 1 January 2018 (Z. Qamruzzaman, Australian Maritime Safety Authority, personal communication, 13 March 2015).

Vessels on government service (such as some research and resupply vessels) will generally be exempt, although it is anticipated that because of the difficult and dangerous operating conditions in the Antarctic, most will voluntarily comply. The provisions also do not apply to fishing vessels but it is known that very large fishing vessels operating in the Southern Ocean adopt SOLAS compliance regardless of the need for a SOLAS certificate. Furthermore, SOLAS only applies to vessels on international voyages and this has the potential to complicate matters in the sovereign-neutral Antarctic. There is insufficient scope to pursue this matter further here.[3]

A separate but important aspect of the Polar Code relates to the training of crew, which will be operationalized through the 1978 Standards of Training, Certification and Watchkeeping Convention (STCW). Amendments to Chapter V of STCW are required to establish a training and certification programme for masters and deck officers of polar vessels. Training to achieve a Polar Waters Certificate of Proficiency will be offered at both basic and advanced levels. There is some suggestion that there will be a transitional period for entry into force of the new crew credentials (Maritime Executive, 2015). A number of organizations are preparing for these requirements by providing a range of training options, including using a sea ice navigation simulator (AMC, 2016).

As far as it was possible to determine prior to the entry into force of the Polar Code, the administrative process will commence with the classification societies, after which time the flag states and operators will work together towards writing a set of operating instructions for each vessel. Insurers will play a pivotal role in ensuring that the operation of a vessel is commensurate with its proposed itinerary and that insuring the vessel under such conditions is reasonable.

CLIA have produced a flow-chart of the administrative procedures required for application of the Polar Code, reducing what is certain to be a very complex and comprehensive process to a more practical and comprehensible outline (CLIA, 2013). The information and discussion below incorporates and is based on their assessment of the probable administrative pathways and procedures.

Classification societies' involvement in the Polar Code

Classification societies, such as those members of the International Association of Classification Societies (IACS), will play a key role in the first part of the administrative process. The Polar Code regulates ship design and construction, on-board equipment, operational protocols, crew training, search and rescue, and the protection of the polar environments in the broadest sense through SOLAS and MARPOL, and to an extent, the STCW. The design and construction elements of the Polar Code will require qualitative and quantitative assessments by the societies of such features as a vessel's *intact* and *damaged* stability characteristics; its ice strengthening; and its essential machinery systems. Other aspects of the Code involve assessments of such features as internal and external communication systems; navigation systems; fire fighting, damage control and survival systems; and protocols for black- and grey-water systems (CLIA, 2013). This regime of assessments will determine the operational limitations of a vessel, and will help to inform the operator and the flag state about the vessel's suitability for operating in the polar regions.

Each SOLAS-certified vessel travelling into the Antarctic (i.e. south of 60° South) will be required to carry a Polar Ship Certificate. The certificate will indicate the ship Category assigned: either Category A, B or C, which principally relates to the ice strengthening (ice class) of the ship, corresponding to the existing IACS Polar Classes 1–5, 6–7, and everything else, respectively, but complicated somewhat by additional requirements for double hull construction[4] (R. Hindley, Lead Specialist – Arctic Technology, Lloyd's Register, Vancouver, personal communication, 12 January 2016).

Classification societies will carry out appraisal of plans and undertake surveys against the Polar Code requirements. At the end of this process, which involves the application of their own extensive codes and rules for the construction and equipping of ships, the society will either issue a 'Polar Ship Certificate' for a vessel if the flag state has delegated authority for it to do so, or the flag state will issue the Certificate. In Australia, for example, the Australian Maritime Safety Authority would issue the certificate (Z. Qamruzzaman, Australian Maritime Safety Authority, personal communication, 13 March 2015).

During the development of the draft code, and in response to a realization that a suitable means for evaluating the limitations of ships with various ice classes was not readily available to integrate into the code, an informal working group was established at MSC93 (the IMO Maritime Safety Committee's 93rd session in 2014) between the Arctic states with experience of setting operational limitations in their own coastal waters. The group was coordinated by IACS, in response to what were seen as weaknesses in the language of the code (for example, the situation where only one ice type is present is extremely rare, yet it would be this ice type that a ship would be limited to operate in). Through a series of joint workshops, gap analyses and troubleshooting exercises, IACS, the insurers, governments, the research community and industry (including ice experts) have created the Polar Operational Limit Assessment Risk Indexing System (POLARIS). POLARIS is 'a decision making system that can be used for voyage planning in advance and "on the bridge" that uses the actual ice conditions, ice class and operational mode' (i.e. whether a vessel is operating independently or is under escort) to calculate a level of risk and hence a safe operating regime (i.e. go—go cautiously at reduced speed—don't go or go only in exceptional circumstances).[5] Designed for use with ice charts, POLARIS is still in the development stage, but it, or something very like it, will clearly increase confidence in decision making by all agents. The final report of the IMO Correspondence Group, which was tasked with developing guidance for setting operational limitations in ice (of which POLARIS is a central component), resulted in IMO MSC.1/Circ.1519 (6 June 2016) 'Guidance on methodologies for assessing operational capabilities and limitations in ice', to be reviewed 4 years after the entry into force of the Polar Code.

While the classification societies will conduct the statutory assessments required for the issuance of a Polar Ship Certificate, it is the ship insurers that may have the ultimate power to impose limitations on when and where a ship can travel in the polar regions.

Ship insurers' involvement in the Polar Code

According to the Lloyd's (2012) report, *Arctic Opening: Opportunity and Risk in the High North*, the Arctic maritime environment today is characterized by uncertainty and significant knowledge gaps, and 'risk management is fundamental for companies to work safely, sustainably and successfully in the Arctic' (Lloyd's, 2012, p. 7). While this report was Arctic-specific, the same can be said for the Antarctic. Ship insurers have a huge stake in ensuring that the Polar Code is fit for purpose in both regions. According to M. Kingston (personal communication, 8 February 2016) the insurers have had a high profile role in the development of the code, and particularly in pushing for the inclusion of an 'ice regime limitation methodology for operation in ice' – risk assessment protocols applicable to all polar waters, i.e. POLARIS, described above. From the beginning of their involvement in the process of developing the code, the insurers saw an obvious need for proper planning, based on a safety case approach, for shipping in the polar regions (Kingston, 2015, p. 6).

The Arctic Council is being encouraged by the insurance industry to convene a 'best practice' forum to help states implement the Polar Code (M. Kingston, personal communication, 8 February 2016). This kind of forum would also be of great interest and benefit to Antarctic states. In fact, of the eight Arctic Council states, five – Finland, Sweden, Norway, Russian Federation and the USA – are also Antarctic Treaty Consultative Parties (decision-makers) and the other three – Iceland, Denmark and Canada – are Antarctic Treaty signatory states.

As a matter of current practice, insurers have difficulty insuring operations in polar regions. Hull and Machinery underwriters, who insure the property of the vessel, exclude operations above 70°N and certain areas below 50°S. Protection and Indemnity (P & I) Clubs, which cover liability to third parties for incidents such as to the passengers or the environment of the area in which the incident takes place, while they do not impose the same specific navigational limits as Hull and Machinery Underwriters, Club rules do require the operator to consult the P & I Club if they are operating 'outside normal trading patterns'. Of course polar waters are not a normal trading pattern for most operators.

This is obviously a key issue for large cruise ships not ordinarily operating in polar waters. Both Hull and Machinery and P & I insurers are most concerned about extreme cold, reduced coverage by navigation aids such as GPS, inaccurate charts, unreliable magnetic compasses, restricted visibility, inadequate weather reports, violent storms and virtually non-existent salvage facilities (Kingston, 2015, p. 27). For example, the closest heavy lifting gear able to effect a rescue or removal in Antarctica is probably located in the Gulf of Mexico in the case of the Antarctic Peninsula, and Singapore in the case of East Antarctica (Kingston, 2015, p. 29). This difficulty was demonstrated by the removal of the MV *Rena* in New Zealand where the equipment was sourced from Singapore.[6]

To reduce their own risks insurers could, therefore, place limitations on the location and/or timing of ship operations based on a range of anticipated polar hazards such as temperature, ice conditions and weather conditions. Or in extreme cases, where they have judged it is not possible to adequately mitigate risk, or the rules are too confusing, they could simply refuse to insure. If the insurers are to provide cover, they require clear regulation and a certain level of confidence that the flag states and operators will comply and best practice will be employed.

State responsibility for implementing the Polar Code

The IMO develops rules about shipping but plays no part in the actual implementation of those rules. This is the responsibility of the flag state of a vessel. In the case of the Polar Code, it is likely that the flag state will work with the ship operator to prepare the Polar Water Operational Manual (PWOM) for that vessel, as required under the Code.

Preparation of the PWOM will involve the identification of hazard and other historical data relevant to the region and the season in which ship operations will occur. According to CLIA (2013), these are split into two separate categories: (i) environmental, including such features as anticipated ice conditions, temperature, weather conditions, sea state, visibility, tides and currents; and (ii) operational, including such features as crew experience, light regime, charts, vessel traffic and tracking systems, and the availability of search and rescue assets. Each assessment will be relative to a planned voyage or series of voyages if the itineraries do not vary substantially, and the risk assessment will inform the content of a PWOM.

In summary, the PWOM outlines the vessel's operating capabilities and limitations, and will provide the owner, operator, master and crew with guidance to support their decision-making (Z. Qamruzzaman, Australian Maritime Safety Authority, personal communication, 13 March 2015). This might not be a great departure from what is already contained in the IMO's International

Fig. 23.5. Lemaire Channel, Antarctic Peninsula (© Julia Jabour).

Fig. 23.6. MV *Orion* in ice off East Antarctica (© Julia Jabour).

Safety Management (ISM) Code (IMO, 2017). This Code is designed to provide international standards for the safe management and operation of ships and for pollution prevention, and each ship operator is bound to retain a Safety Management Manual with practical guidance for how the vessel should be operated in order to comply with the safe management practices and procedures. The PWOM is essentially a polar-specific version of this Manual, in line with new requirements set out in the Polar Code.

Meeting the conditions of a PWOM is ultimately the responsibility of a ship's Master. At this stage the extent to which port state measures might be invoked to inspect tourist vessels to the Antarctic for certification and compliance with their PWOM is not known. It is likely, however, that the industry peak body will play a leading role in encouraging compliance, despite having no legal power to do anything other than foster best practice adherence to the Code.

Industry involvement in implementing the Polar Code

The International Association of Antarctica Tour Operators (IAATO) has taken a proactive role in the negotiations of the Polar Code, formally in the IMO through representation by CLIA, and in the Antarctic Treaty Consultative Meetings through being an invited expert.

IAATO has actively provided the Antarctic Treaty Consultative Parties with tourism statistics and trends, and information on industry developments, problems and potential solutions since its establishment in 1991. It has a reputation for being the competent day-to-day manager of Antarctic tourism, since most of its members are experienced operators in the region (in the Arctic also). Membership of IAATO is voluntary, but membership bears no relation to whether or not a vessel is permitted to travel to the Antarctic. Any regulation of this kind is a national and/or flag state duty. Most current Antarctic tour operators are members of IAATO.

IAATO is respected for its initiatives, e.g. for prohibiting the landing of passengers from member cruise ships carrying >500 passengers, for requiring one guide for every 20 passengers ashore, and for limiting site visits to one ship at a time and giving intervals of recovery between the number of ship visits per day (IAATO, 2015).

Consultation with IAATO has revealed that the industry peak body is once again playing a leading role in anticipating the challenges and opportunities its members will face with the entry into force of the Polar Code (M. Van der Hulst, International Association of Antarctica Tour Operators Marine Committee Chair, personal communications, 2016). IAATO has recognized that the diversity of its organization requires a standardized interpretation of risk assessment, and content and format of the PWOM. It is anticipated, for example, that IAATO will host a database containing ice and weather data from historical records and from such commercial providers as Polar View.[7]

Polar View staff are working with IAATO to help provide data and platforms for a database of information to assist in the operations of tourist vessels in both the Arctic and the Antarctic. It is expected the database will include sea ice forecasts (with detailed information on ice types, concentration, thickness, pressure and drift); detection of icebergs, including drift characteristics and historical trends; and meteorological information (D. Arthurs, Managing Director, Polar View Earth Observation Ltd, personal communication, 12 January 2016). It is anticipated that IAATO will host the database (M. Van der Hulst, International Association of Antarctica Tour Operators Marine Committee Chair, personal communications, 2016), which could also be made available to permitting authorities around the world, thereby potentially standardizing their administrative procedures, as well as becoming a reliable source of information for ship insurers to assist their decision-making.

Another potential, and considerably less expensive, source of information to add to the IAATO database, and to aid navigation in polar waters in real time, might come from an unmanned aerial vehicle (UAV). UAVs, or drones, can be deployed from a ship during sea ice operations to fly over a proposed sea lane and transmit pictures of ice coverage in the immediate vicinity. This technology is relatively new and is being tested for its utility through proof of concept flights during Antarctic shipping operations (see, for example, Australian Antarctic Division, 2015). The use of drones in Antarctica is currently the subject of discussion in the Antarctic Treaty Consultative Meetings, and it seems that their use is inevitable (see, for example, COMNAP, 2015).

Antarctic operators are familiar with the need to be flexible with itineraries; operators in East Antarctica, for example, have less continental sites available to them as landing options and often encounter extreme ice and weather conditions that might preclude any landings altogether. However, IAATO does not anticipate that its members will make radical changes to their itineraries once they receive their ship classification. Currently, IAATO is not aware of any operators who plan to leave the industry (M. Van der Hulst, International Association of Antarctica Tour Operators Marine Committee Chair, personal communications, 2016), but once the Polar Code has been in place for some years it will be worthwhile revisiting this. Bender *et al.* (2016, p. 9) predict that in the future:

> there may be reduced traffic at the margins of the Antarctic season (early spring and late autumn) when sea ice conditions and low temperatures may preclude the use of some vessels that have not met minimum standards for polar operations…the availability of smaller vessels, carrying<100 passengers is decreasing and there is a notable shift towards vessels carrying ≥150 passengers. This is largely due to the smaller vessels being no longer commercially viable.

It is now 26 years since the establishment of IAATO and the beginning of its involvement with the Antarctic Treaty Consultative Meetings. IAATO is in good standing within these meetings and open dialogue has characterized their relationship. In fact the Antarctic Treaty Consultative Parties tacitly admit that IAATO is the expert on the subject of sustainable, ship-borne tourism. In more than 50 years of meetings, the Parties have only made two legally binding regulations on the industry, one of which was the codification of existing IAATO by-laws.[8]

Conclusions: Future Expectations

Almost 40,000 tourists were expected to visit the Antarctic in the 2015–16 season, according to estimates from the peak body, IAATO. This number is almost back to pre-global financial crisis figures and is in line with the strong upward trajectory of Antarctic tourism historically (Table 23.1). The majority of tourists will travel by ship from ports in South America to the Antarctic Peninsula (with Australia, New Zealand and South Africa hosting just a few departing voyages), with operations managed on a day-to-day basis by IAATO. Membership of IAATO is voluntary and despite members being commercial rivals in many cases, there is a healthy and functional level of collaboration and communication among them, based on their shared respect for both the destination and the difficult and dangerous routes they must travel to get there.

Climate change, rather than disabling the Antarctic tourism industry, is actually helping to make tourism to the Antarctic safer. If the current industry perspectives and observations presented here prove to be accurate, this update indicates there is strong confidence that the industry is ready for the implementation of the Polar Code, designed to augment existing IMO safety and environmental conventions (Z. Qamruzzaman, Australian Maritime Safety Authority, personal communication, 13 March 2015; Hindley, 2016; M. Van der Hulst, International Association of Antarctica Tour Operators Marine Committee Chair, personal communications, 2016; R. Hindley, Lead Specialist – Arctic Technology, Lloyd's Register, Vancouver, personal communication, 12 January 2016). It has even been suggested that the code is welcomed by Antarctic tour operators, who see it as the codification of years of their collective polar experience and successful industry self-regulation. Further, they see the Polar Code as providing a barrier for new operators with little or no experience from entering the industry without restriction.

The Polar Code is not perfect and time will reveal the depth of any weaknesses, as well as its strengths. For example, it does not apply to vessels on government service, or fishing vessels, although there is some confidence that a subsequent round of negotiations will include a regime applying to fishing vessels, which statistics show are the most vulnerable category of vessels involved in accidents in the polar regions (CLIA, 2011). A follow-up study in 5 years' time will give a much more accurate picture of how the Polar Code operates.

The Antarctic is poorly charted, and the Polar Code does not alleviate this problem. It does, however, include known charting limitations as part of risk assessment. But since the surveying of waters is a coastal state responsibility, and arguably there are no coastal states in Antarctica, the problem will remain.

Weather, wind and currents together induce sea state and visibility, and these are some of the known hazards identified in the assessment of voyage risks. But Antarctic weather patterns are changing, as part of the overall warming trend. Providing synoptic forecasting and real-time local weather information is crucial to safety in Antarctica. Operators of assets in the Southern Ocean and facilities on the Antarctic continent will continue to rely heavily on scientific research to increase the accuracy and reliability of meteorological prediction.

The Polar Code does not satisfy the desire for sustainable tourism practices on the Antarctic continent and tourism will continue to be a topic of much debate within the Antarctic Treaty Consultative Meetings. But for the time being, at least, the Polar Code will meet long-standing concerns about the difficult and dangerous activity of shipping in the Antarctic. It was designed to enhance safety of life at sea and the prevention of marine pollution and it will do just that, but no more.

Acknowledgements

The author acknowledges with gratitude Rob Hindley of Lloyd's Register, Michael Kingston of DWF LLP, David Arthurs of Polar View, and Mark van der Hulst of IAATO for their assistance and comments on sections of this work.

Notes

[1] Significant loss of life has occurred in Antarctica, but this has been from fishing vessels and from the private Norwegian yacht, *Berserk*. For details of accidents and incidents in polar waters, see the comprehensive survey appended to IMO document DE 56/INF.9, 'IAATO Polar Risk Assessment', submitted to the Sub-Committee on Ship Design and Equipment by Cruise Lines International Association, dated 9 December 2011, referenced here as CLIA (2011).

[2] Two IMO Resolutions establish and adopt the Polar Code: MSC.385 (94) (London, 21 November 2014) and MEPC.264 (68) (London, 15 May 2015). The specific amendments to SOLAS are contained in IMO Resolution MSC.386 (94) (London, 21 November 2014), and to MARPOL Annexes I, II, IV, and V in Resolution MEPC.265 (68) (London, 15 May 2015).

[3] Under Chapter 1 of SOLAS, an 'international voyage' involves a ship travelling from the jurisdiction of one state, through international waters, to the jurisdiction of another state. In Antarctica, therefore, some would argue that ships travelling from a port in a claimant state, such as Australia, to an Australian base in Antarctica are *not* entering the jurisdiction of another state and thus that voyage would not be considered 'international' under SOLAS. This position is by no means universal.

[4] For a fuller understanding of the complicated definitions of a vessel's ice classification, refer to footnotes 6, 7 and 8 of Chapter 3 – Ship Construction, and footnotes 9, 10 and 11 of Chapter 6 – Machinery Installations, of the Polar Code (above, note 2).

[5] Bond, J. and Hindley, R. (2014) IMO Polar Code: Update on IACS Informal Working Group Operational Limitations. PDF of IACS PowerPoint Presentation, personal communication.

[6] For information on the MV *Rena* and its salvage, see www.renarecovery.org.nz (accessed 17 August 2017).

[7] www.polarview.org (accessed 10 January 2016).

[8] These two measures are related to emergency search and rescue insurance (Measure 4/2004, not yet in force) and site behaviour, e.g. landing of passengers, guide to passenger ratio, and no more than one tourist vessel at a site at any one time (Measure 15/2009, not yet in force). For further information, see Antarctic Treaty Secretariat (2016).

References

AMC (2016) Australian Maritime College, Icebreaker Crews Learn the Ropes without the Risks. Available at: www.amc.edu.au/about-amc/news-and-events/news-items/icebreaker-crews-learn-the-ropes-without-the-risks (accessed 28 May 2017).

Antarctic Treaty Secretariat (2016) Final Reports. Available at: www.ats.aq/devAS/info_finalrep.aspx?lang=e&menu=2 (accessed 10 January 2016).

Argentina (2015) *Report on Antarctic Tourist Flows and Cruise Ships Operating in Ushuaia during the 2014/2015 Austral Summer Season, Information Paper 126 to Antarctic Treaty Consultative Meeting XXXVIII, Sofia, 2015*. Antarctic Treaty Secretariat, Buenos Aires. Available at: www.ats.aq (accessed 10 December 2015).

Australian Antarctic Division (2015) Aurora Australis Uses Drone Technology to Navigate Sea Ice. Available at: www.antarctica.gov.au/news/2015/aurora-australis-uses-drone-technology-to-navigate-sea-ice (accessed 10 January 2016).

Bender, N.A., Crosbie, K. and Lynch, H.J. (2016) Patterns of tourism in the Antarctic Peninsula region: a 20-year analysis. *Antarctic Science*, DOI:10.1017/S0954102016000031.

COMNAP (2015) *Council of Managers of National Antarctic Programs, UAV Use in Antarctica – Risks and Benefits. Working Paper 22 to Antarctic Treaty Consultative Meeting XXXVIII, Sofia, 2015*. See Discussion as recorded in the Final Report. Antarctic Treaty Secretariat, Buenos Aires. Available at: www.ats.aq/devAS/info_finalrep.aspx?lang=e&menu=2 (accessed 10 January 2016).

Cruise Lines International Association (CLIA) (2011) IAATO Polar Risk Assessment, Document DE 56/INF.9 to IMO Sub-Committee on Ship Design and Equipment, 56th session. Available at: www.cruising.org/docs/default-source/imo/de-56-inf-9---iaato-polar-risk-assessment-(clia).pdf (accessed 9 December 2015).

Cruise Lines International Association (CLIA) (2013) Administrative Procedures for application of the Polar Code, Document DE 57/11/21 to IMO Sub-Committee on Ship Design and Equipment, 57th session. Available at: www.cruising.org/docs/default-source/imo/administrative-procedures-for-application-of-the-polar-code.pdf?sfvrsn=0 (accessed 10 January 2016).

International Association of Antarctica Tour Operators (IAATO) (2015) Information Paper IP53, IAATO Overview of Antarctic Tourism: 2013–14, 2014–15 Season and Preliminary Estimates for 2015–16 Season. Available at: http://iaato.org/current-iaato-information-papers (accessed 9 January 2016).

International Association of Antarctica Tour Operators (IAATO) (2016) Bylaws. Available at: http://iaato.org/bylaws (accessed 10 January 2016).

International Maritime Organization (2017) ISM Code and Guidelines on Implementation of the ISM Code. Available at: http://www.imo.org/en/OurWork/HumanElement/SafetyManagement/Pages/ISMCode.aspx (accessed 28 May 2017).

Jabour, J.A. (2011) Would you like ice with that? Antarctic tourism and climate change. In: Jones, A. and Phillips, M. (eds) *Disappearing Destinations: Climate Change and Future Challenges for Coastal Tourism*. CAB International, Wallingford, UK, pp. 177–190.

Kingston, M. (2015) The Finalisation of the Polar Code: The Concerns and Contribution of the Insurance Industry. Presentation to British Insurance Law Association, London, 10 April 2015. DWF LLP, London.

Lloyd's (2012) *Arctic Opening: Opportunity and Risk in the High North*. Chatham House, London.

Maritime Executive (2015) IMO Sets Polar Code Training Standards. Available at: www.maritime-executive.com/features/imo-sets-polar-code-training-standards (accessed 10 January 2016).

Pollard, D. and DeConto, R.M. (2009) Modeling West Antarctic ice sheet growth and collapse through the past five million years. *Nature* 458, 329–332, DOI:10.1038/nature07809.

Republic of Liberia (2009) *Decision of the Commissioner of Maritime Affairs, R.L., and the Report of Investigation in the Matter of Sinking of Passenger Vessel EXPLORER (O.N. 8495), 23 November 2007, in the Bransfield Strait near the South Shetland Islands. Submitted by Belgium on behalf of Liberia to Antarctic Treaty Consultative Meeting XXXII, Baltimore, 2009 as Information Paper 120*. Antarctic Treaty Secretariat, Buenos Aires. Available at: www.ats.aq/devAS/ats_meetings_doc_database.aspx?lang=e&menu=2 (accessed 19 January 2016).

Vaughan, D.G., Comiso, J.C., Allison, I., Carrasco, J., Kaser, G. *et al.* (2013) Observations: Cryosphere. In: Stocker, T.F., Qin, D., Plattner, G.-K., Tignor, M., Allen, S.K. *et al.*(eds) *Climate Change 2013: The Physical Science Basis. Contribution of Working Group I to the Fifth Assessment Report of the Intergovernmental Panel on Climate Change*. Cambridge University Press, Cambridge.

World Tourism Organization (2015) Facts and Figures Edition. Available at: www2.unwto.org/facts/eng/vision.htm (accessed 9 December 2015).

24 Case Study Morocco: Mediterranean Morocco, a Vulnerable Development Called into Question

Carlo Perelli*
Centre for North South Economic Research, University of Cagliari and University of Sassari, Cagliari, Italy

Introduction

The development of tourism in Mediterranean Morocco was a main source of interest to tourism geographers and planners in past years (Berriane, 1978; Hillali, 1994, 2007; Hazbun, 2004; Verdeguer, 2005; Araque Jiménez and Crespo Guerrero, 2010; Gauché, 2011), but these days concern is growing about the environmental impact of new resorts such as Saidia (Snoussi *et al.*, 2009; Boumeaza *et al.*, 2010) and the climate-change-related effects on the Saidia, Nador, Al Hoceima, Tetouan and Tangier coastlines (Anfuso and Nachite, 2011).

The Mediterranean coast attracted significant state investment during the first phase of development of mass tourism in Morocco (1956–1977) (Hillali, 2007). The government identified five priority areas in 1965, three of which (Tangier, Smir and Al Hoceima) were located on the Mediterranean coast. In Tangier, a state company known as the *Société Nationale d'Aménagement de la Baie de Tanger* (SNABT) was created to develop 30,000 hotel beds, but it turned out to be largely inefficient, even though it received generous state funding for two decades. Intense state involvement in the management of tourist resorts continued until the late 1970s, when international and internal economies began to suffer from a lengthy crisis. Tangier, the Tetouan coast and Al Hoceima only partially emerged as international destinations, and ended up by offering a low-budget product in order to compete with their Mediterranean competitors. Domestic demand was directed more towards second homes, camping and informal practices. As Berriane has observed (1990), in the 1980s, nearly one-third of Moroccans took their holidays in the summer and a large majority visited the Mediterranean coast, either renting or owning a holiday home (on this topic, see also Hazbun, 2010). During the 1990s, Mediterranean tourism proved unable to adapt its traditional low budget supply to the emerging demand for cultural tourism (Minca and Borghi, 2009), and this scenario still limits the opportunity to convert the supply offered by most destinations today.

In 2001, in accordance with the 'Vision 2010' National Tourism Plan, the Mediterranean coast was planned to be the site of the *Mediterranea Saidia*, one of the six national priority areas. In 2003, the Spanish developer Fadesa signed an agreement to develop a resort in Saidia on more than 700 ha, with

*E-mail: carlo.perelli@crenos.unica.it

a project that originally included hotels (16,000 beds), 1500 beds in villas and apartments, a marina and international golf courses. Despite the fact that the development faced budgetary and management problems and changes in ownership, the resort is now operating.

The aim of Vision 2010 was to increase the number of international tourists to over 10 million units by the year 2010. International tourism was targeted as a rapid growth sector, and tourist revenues were forecast to generate 20% of the country's GDP and to have created 600,000 new jobs by the end of the development plan. In line with Vision 2010's forecasts, visitor numbers reached around 9.2 million units in 2010.

In support of the Vision 2010 strategy, Morocco signed an 'Open Sky' agreement with the European Union in December 2005. EasyJet and Ryanair had already been flying to Morocco before the agreement was signed, but they increased their presence in the country and several other airlines followed them, including newly formed Moroccan low-cost companies. According to the government, the goal of 10 million tourists required growth from around 5 million (in 2001) to 15 million air passengers a year, and this target was reached in 2010; however, recent assessments of both the Open Sky and Vision 2010 strategies suggest that success in terms of tourist arrivals has been overestimated. Nevertheless, there can be no doubt that EU–Morocco connections increased, despite the fact that the number of charter flights decreased. It is likely that it was Moroccan immigrants rather than international tourists who caused this substantial expansion of airline traffic, as shown by the slower growth of nights spent compared with tourist arrivals (Dobruszkes and Mondou, 2013; Dobruszkes *et al.*, 2015).

In 2013, Morocco launched the Vision 2020 plan, which aimed to make Morocco a top-20 destination by 2020 (Kingdom of Morocco, 2011). The goal of Vision 2020 is to double the number of arrivals from 9.2 million in 2010 to 20 million in 2020. Accommodation capacity is also forecast to double from 175,000 in 2010 to 375,000 in 10 years. Tourism has once again become a driver of rapid economic growth, with 470,000 new direct jobs to be created across the country by 2020. Vision 2020 also plans to increase domestic tourism by tripling the number of trips taken by Moroccans.

Research Context

According to the IPCC (2014), coastal systems are forecast to be affected mainly by a rise in mean sea levels and surface temperatures due to climate change. Further research is needed on other drivers such as storm intensity or frequency. If we bear in mind that the scientific consensus on climate change and rises in sea level is now virtually unanimous (Hall *et al.*, 2015), the main issue for the Moroccan Mediterranean coastline would seem to be a mixture of human- and climate-induced changes:

> in Morocco, appropriate vulnerability assessment and adaptation strategies to SLR and climate change are required at a regional and national scale to improve the tools for proper coastal zone management. During recent decades, these have been greatly neglected because of more urgent requirements and problems or essentially the wish to create rapid economic growth linked to coastal urbanization and tourism.
>
> Anfuso and Nachite, 2011, p. 108

In recent years, on the other hand, Morocco has been targeting coastal areas as a laboratory in which to test new territorial policies based on mega-projects (Barthel and Planel, 2010; Ducruet *et al.*, 2011; Kanai and Kutz, 2011). Tangier has recently emerged as a 'globalising city-region' (Kanai and Kutz, 2011), but the Marchica and Nador West Med projects, which we will refer to later in this chapter, are also significant Mediterranean examples of this trend. Tourism is one of the pillars of these prestige projects, which are driven by both national and mobile international capital. As Barthel and Planel (2010) have shown, the peculiar power structure of the Moroccan monarchy seems in some way to offer the perfect framework for new forms of territorial governance based on a new partnership between the central power, local representatives and international capital, even though the level of local participation may be poor, and it may sometimes be contested by residents. The need for measures to adapt to climate change must now face the predominance of this type of development model. This chapter looks at this framework by focusing on the link between forecast climate change impacts, tourism development policies and adaptation strategies, and brief consideration is also given to the implementation of coastal planning tools in Mediterranean Morocco such as Integrated Coastal Zones Management (ICZM). Finally, some policy suggestions are offered.

Climate Change: Assessing the Problems and Challenges

In a previous contribution, Anfuso and Nachite (2011) provided a general overview of the debate on the impact of climate change on Mediterranean Morocco, and more recent studies have developed additional important scenarios: for instance, a rise in sea level of 0.44 m is forecast to be produced by an increase in temperatures of 2.6°C over the 1990 levels in the country, and if adaptation measures are not adopted, 1.82 million people could be affected in Morocco (Brown *et al.*, 2011). Another study has identified the Mediterranean coastline of Africa, including Morocco, as the area of the African continent that will be most affected by a rise in sea level in a scenario of rises of between 0.42 and 1.26 m by 2100 (Hinkel *et al.*, 2012).

Climate variations have been analysed not only by inter-annual variability, but also by extreme temperature events in recent years (Donat *et al.*, 2014; Filahi *et al.*, 2015; Khomsi *et al.*, 2015). There is general agreement that a rise in the maximum and minimum average temperatures on a seasonal basis, a rise in extreme temperatures and a decrease in the frequency of cold events can be observed in Morocco, as can a change in the frequency of both seasonal hot and cold extreme events. Significant differences have been reported regionally, although a lack of data to support regionally-situated analyses is currently limiting forecasting capabilities (Khomsi *et al.*, 2015). The same regional differences have also emerged with regard to precipitation data: despite an overall decrease in precipitation and drier conditions in southern Morocco, generally wetter conditions are being noted in Tetouan, and a possible increase in the frequency of heavy precipitation events was seen in Tetouan and Al Hoceima during the period between 1970 and 2012 (Filahi *et al.*, 2015). The effects of climate change on extreme daily precipitation in Mediterranean Morocco have been described by projections based on an analysis of daily data from ten stations between 1961 and 2007 (Tramblay *et al.*, 2012). The results show small differences for the period between 2020 and 2050 but a decrease (an average of −12%) for the period between 2070 and 2099 under the A1B scenario forecast in the IPCC Special Report on Emission Scenarios (SRES). Minor changes are also forecast in the Nador and Al Hoceima areas. Significant impacts on water and energy consumption patterns have also been forecast by UNEP/MAP in both 2030 and 2050 scenarios (El Badraoui and Berdai, 2011).

The Vision 2020 Plan identifies eight tourism development areas based on international demand segments. The Mediterranean coast is divided into two tourist areas: Northern Cape and Mediterranean Morocco (Fig. 24.1).

Northern Cape

The first area, known as Northern Cape, includes Tangier, Tetouan, Chefchaouen and the Atlantic destinations Asilah and Larache. Vision 2020 proposes a mixed offer of cultural and seaside destinations, and plans a 43,000 bed capacity, 2.8 million arrivals and the creation of 76,000 direct jobs by 2020. The Spanish Protectorate (1912–1956) developed the first tourism destination in Spanish Morocco in the Martil area (Araque Jiménez and Crespo Guerrero, 2010), and the urbanization of the coastline between

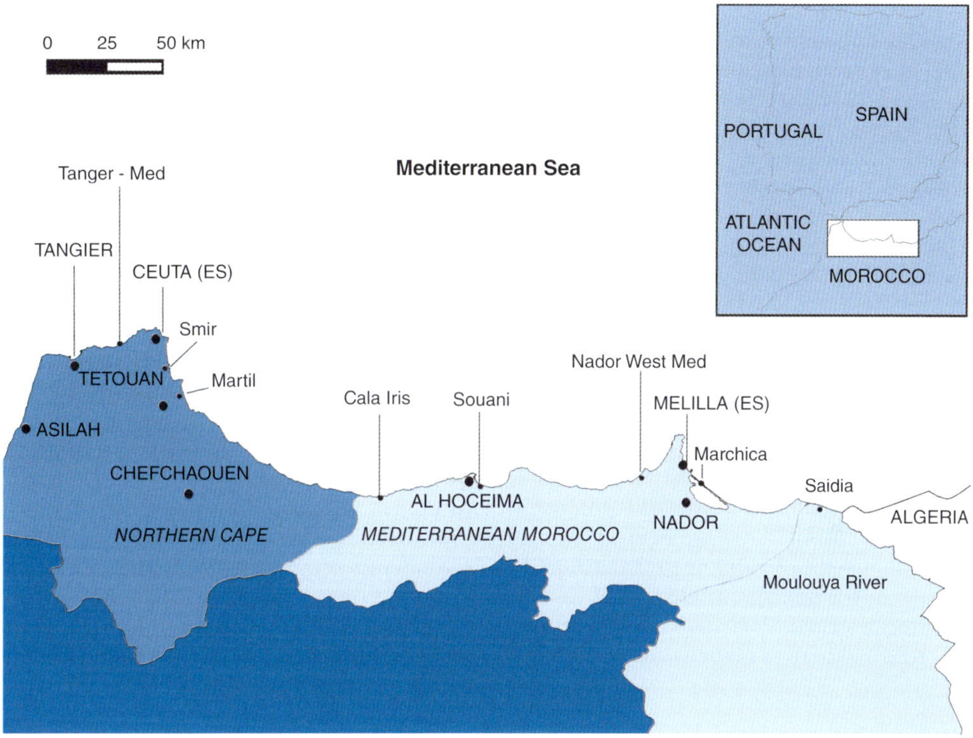

Fig. 24.1. Location map of Mediterranean Morocco and research area.

Ceuta and Tetouan has continued to expand. The coast already has well-established tourist zones, and is showing significant signs of beach vulnerability that are mainly related to second homes and hotels. Furthermore, the impact of the Tanger Med port on the Tangier area is still not clear. In 2015 port capacity is estimated to be over 8 million containers, 7 million passengers and 2.7 million trucks and vehicles (Mounime *et al.*, 2014). The effects on the coastline and on a whole range of port activities, including not only oil storage and refining but also related infrastructures such as free zones,[1] new roads and logistics facilities, raise the issue of risk and vulnerability assessment in the region (Pateau, 2014). Emerging erosion and flooding events are already observable, and also constitute a threat to the tourist industry (Anfuso and Nachite, 2011).

Flooding due to rises in sea level in the Tangier tourist area has been forecast by Snoussi *et al.* (2009). In the case of both the minimum (4 m) and maximum (11 m) hypothetical flood levels, beaches, tourist infrastructures and other coastal facilities will suffer significant damage. Projections to 2050 and 2100 describe a scenario in which Tangier's main beaches, such as Tarik, Marbel and Sanaa, will be seriously affected or completely disappear, and only a few beaches will survive in the area, even in the event of a moderate hypothesis of a sea level rise. Issues of tourist sustainability are emerging in the Tetouan area (Bourse, 2011), and an analysis of erosion dynamics over a 40 km-long coastline has shown the impact of human activities (dams, marinas and holiday homes) on dune morphology and the exchange of beach-dune materials (El Mrini *et al.*, 2012).

Mediterranean Morocco

The second tourist development area, known as Mediterranean Morocco, has been defined by the Moroccan Agency for Tourism Development (SMIT) as 'the new seaside leisure Mediterranean destination'. It includes the coastline from Al Hoceima to the Algerian border and the tourist sites of Cala Iris, Marchica and Saidia. By 2020, the Moroccan authorities plan to offer 27,000 beds for 0.9 million arrivals in this area, which will triple tourist revenues and create 23,000 direct jobs.

Al Hoceima Bay lies between the most important low-lying coastal areas of Mediterranean Morocco (Fig. 24.2). The area is very exposed to rises in sea level and extreme weather events, but is also the centre of ambitious development plans. In addition, over the last 20 years the Al Hoceima area has suffered the most powerful earthquakes to strike Morocco in the last century. In May 1994, a magnitude 5.5 to 6.0 earthquake caused major damage, injuries and deaths. A more destructive earthquake (magnitude 6.4) resulted in over 600 victims and several thousand buildings were demolished in February 2004. Risks related to coastline and sand dynamics and flood hazards are emerging (Khouakhi *et al.*, 2013). Pressures from human activities together with the physical vulnerability of the coastline must be taken into account. More than 40% of the Al Hoceima Bay coastline is highly vulnerable within 100 m of the shoreline to the impact of forecast climate change such as extreme weather events or sea level rise (Khouakhi *et al.*, 2013).

Nador Lagoon is included in the Ramsar Convention on Wetlands list. It has been preserved until now, and occupancy of land for residential use is still relatively limited. New settlements are planned by the Marchica project all along the barrier island, however.[2] The government is planning seven settlements on 20,000 ha, with a €4.1 billion investment between 2008 and 2025, mainly for residential and tourist use. Great emphasis is being placed on sustainability,

Fig. 24.2. New tourism developments in Quemado Beach, Al Hoceima city, May 2014.

notably with regard to energy production, emissions and nature conservation. The Nador area within the next 10 years will also host the Nador West Med port[3] on 850 ha. The port is intended to operate in collaboration with Tanger Med as a strategic player in the Mediterranean transport system.

As shown by Raji *et al.* (2013), the Marchica project will impact some of the most vulnerable areas of the barrier island directly and the whole lagoon system indirectly. In a sea level rise scenario, storm-generated waves are forecast to pass over the crest of the beach, with a significant impact on the lagoon and highly vulnerable parts of the Marchica project (Raji *et al.*, 2013). Furthermore, in Bou-Areg, the aquifer at the southern border of the lagoon, groundwater pollution and salinization have already reached critical levels. Human activities, together with a predicted reduction in water resources due to climate change, may impact planned tourism developments, leading to potential water-use conflicts in the future (Re *et al.*, 2014).

The Saidia tourist area is located at the mouth of Moulouya River. The vulnerability of the lower basin area near the coastline as a result of climate change and human activities has been variously analysed (Snoussi *et al.*, 2008; IUCN, 2010). All along the Moulouya Basin, the average annual precipitation is consistently below 400 mm, and agriculture consumes approximately 87% of the groundwater and surface-water resources (Tekken and Kropp, 2012). The tourism and agriculture development plans in the area are both highly water-consuming, and at present no direct measures are forecast to limit the pressure on water resources. Tekken *et al.* (2013) have estimated that the annual additional water needed by the Saidia tourist facilities will amount to 8 million m^3. Regional climate changes between 1971 and 2005 have shown a decrease in both the mean annual and winter precipitation and a significant temperature increase (+6.8°C) during December in the same period. As a consequence of the decrease in winter precipitations, which are the main driver for the overall basin water balance (Tekken and Kropp, 2012, 2015), a scenario of increased vulnerability is emerging for the Moulouya River Basin.

A recent map of flood-risk zones has underlined the vulnerability of this particular area due to its coastal geomorphology. Nearly 45% of the coast near the Moulouya River (including tourist development zones) is exposed to a high or very high flood risk (Mouzouri *et al.*, 2013). Similar results are shown by comparing aerial photographs from 1958 to 2014 (Lasgaa *et al.*, 2015), and the authors report 'catastrophic changes', especially at the mouth of the Moulouya, and serious deterioration of the beaches.

Focus: tourism development, vulnerability and participation in the Al Hoceima area.

Parallel to Vision 2010, a tourism development plan for the Al Hoceima area entitled 'Vision 2015' was adopted, targeting 10,000 hotel beds and some 5000 direct jobs. The Souani Mediterranean resort, which is located on the waterfront of Ajdir, near Al Hoceima, is one of these new resorts supported by the Vision 2015 plan (Fig. 24.3). The project is located in an 80-ha forest planted by the Spanish in 1942 to stabilize the sand dunes and as a barrier against the intrusion of sand inland. In 1963, the area became the site of the first Club Med in Morocco, and remained open until 2003 (Fig. 24.4).

The author participated in a 2-year EU-funded project known as 'Destinations' for the implementation of a Tourism Carrying Capacity Assessment (TCCA) in Al Hoceima. At the same time, the region has been the subject of the Moroccan Coastal Area Management Programme (CAMP), which drew its inspiration from the ICZM process (on the ICZM Protocol and Mediterranean practices, see Billé and Rochette, 2015). The Moroccan State Secretariat for Water Resources and the Environment (SEEE) and the UNEP Mediterranean Action Plan (MAP) are the main partners of both projects. ICZM and adaptation to climate change share similar principles and approaches (O'Mahony *et al.*, 2015). The TCCA process is integrated into the ICZM methodology,

Fig. 24.3. Al Hoceima bay from the city. In the foreground Spanish Isla de Mar, Isla de Tierra and Peñón de Alhucemas military fort. In the background is the beach and the forest hosting Souani resort area in Ajdir municipality, May 2014.

and outlines the use of various scenarios for selecting appropriate development options (UNEP, 2009). A Tourism Plan for the project area has been developed, and the Souani project has been chosen as a pilot scheme. In 2009 and 2010, a spontaneous committee, which worked not only parallel to but also as a participant in both projects, challenged the Souani project, underlining the economic, social and environmental impacts and risks related to the project.

This committee secured a number of major achievements, such as halting the project for approximately 18 months and securing a revision of the layout plans. The impacts forecast as a result of the TCCA's activities were adopted as an action agenda by the committee, leading the developers to move construction further inland from the dunes, to modify its implementation plan, the coefficient of land occupation and the number of planned swimming pools, and to guarantee the preservation of an important archaeological site within the resort area. The resort is now operating, and according to the latest version of the project, it will consist of an all-inclusive holiday village with 850 beds, a 200-bed three-star hotel and 276 timeshare properties.

Climate Change: Managing Solutions

Morocco is one of the leading African countries in the area of climate policies. In June 2015, its Intended Nationally Determined Contribution (INDC) was submitted to the UNFCCC and was welcomed by independent observers such as Germanwatch[4] and Climate Action Tracker[5] as a positive advance and an effective commitment. In September 2016, in line with Paris Agreement requirements,

Fig. 24.4. Project hoarding showing the first version of the tourism development plan at Souani resort limit in Ajdir, Al Hoceima, June 2008.

Morocco updated the INDC publishing its Nationally Determined Contribution.[6] As a result of its successful performance and climate diplomacy, Morocco hosted the 22nd UNFCCC Conference of the Parties (COP 22) in Marrakech in November 2016. The 2017 Climate Change Performance Index (CCPI), which is published annually by Germanwatch and Climate Action Network Europe, ranks Morocco's climate performance in eighth place among the 58 countries that are evaluated worldwide, and between the most performing Mediterranean nations (Burck *et al.*, 2017).

Ambitious renewable energy plans are being developed in the country (mainly from solar and wind sources) together with a legal framework supporting renewable sources and restricting support for fossil energy sources. Morocco's key asset, however, is its low level of global emissions (0.2% of global GHG emissions in 2010). The INDC scenario also seemed to be very ambitious: the effective implementation of its conditional targets depends largely on international donors, and doubts arose around the undefined measurement of its business-as-usual (BAU) emission levels. The same strategies emerge from the 2016 Nationally Determined Contribution (Table 24.1). It does, however, place Morocco among the global leaders in the field.

The INDC and the Nationally Determined Contribution recognize coastal tourism as one of the vulnerable sectors requiring adaptation measures, and also sets significant goals for 2030 with regard to these measures, focusing mainly on water (aquifers refill, storage, use rationalization, desalinization and treatment), flooding and erosion risks. Morocco is currently drafting a National Adaptation Plan up to 2030 that aims to include climate change adaptation as an across-the-board issue.

Nevertheless, analyses of three pivotal development strategies, the National Water Strategy, the Green Morocco Plan (agriculture) and Vision 2020 (tourism), have identified a

Table 24.1. Mitigation targets according to Nationally Determined Contribution, 2016.[6]

Unconditional target	A 17% reduction in GHG emissions by 2030 compared to a BAU scenario, with 4% coming from AFOLU actions. Without AFOLU actions, the reduction target is 13%
Conditional target	An additional reduction of 25% achievable under certain conditions, which would bring the total GHG reduction to 42% below BAU emission levels by 2030, including AFOLU actions. Without AFOLU actions, the additional reduction would be 21%, which would bring the conditional reduction target to 34%
Financial needs	Meeting the overall target of 42% requires an investment estimated at US$50 billion between 2010 and 2030. Meeting the conditional component of the target, for which costs are estimated to reach US$24 billion, is conditional upon access to new sources of finance and to additional support, compared to that received over the past years

AFOLU, Agriculture, Forestry, and Other Land Use; BAU, business as usual

number of gaps in long-term impact analysis and regional differentiation. According to some analyses, these three strategies – and especially Vision 2020 – do not explicitly identify the adaptation measures required to address climate change over the long term (Khattabi, 2014). The Vision 2020 Plan only generically recognizes sustainable tourism as the key point of reference for impact assessment and monitoring, and in reality, the call for more effective regulatory standards and financial support for the sustainable development of tourism does not seem to fit in with Morocco's ambition to emerge as a leading country in climate policy implementation.

Conclusions: Future Expectations

The vulnerability of Morocco's coastal systems and low-lying coastal areas is a well-established problem, and tourism development is being called on to face these realities now and, we may presume, more and more in the near future. Morocco has adopted the ICZM approach by signing the ICZM Protocol and has developed an interesting legislative framework. As the Souani resort experience shows, however, the generic call for respect for Moroccan law and international conventions on coastal conservation highlights the internal contradictions of its development strategies. With regard to coastal tourism, the effective implementation of the existing normative tools would represent an effective step forward, and a consistent adoption of ICZM strategy will provide a framework for scientific research at a regional scale, and for policy-making and governance issues that for now are lacking in Morocco (Snoussi *et al.*, 2009; El Mrini *et al.*, 2012). At present, the efforts that are being made appear to be inadequate. One example of this is the World Bank ICZM project (The World Bank, 2012), which was created with US$25.18 million, and has the aim of incorporating the ICZM approach into local development planning and supporting coastal resource co-management along the Mediterranean coast (the pilot actions in Saidia, Al Hoceima, Nador and selected nearby areas). For now, however, according to the project's website, the results have been unsatisfactory. Despite this, in the presence of both ambitious development targets through large coastal projects and the incontestable effects of climate changes in the short and long term, ICZM may be a framework for linking scientific research results with effective planning for adaptation.

Notes

[1] www.tangerfreezone.com
[2] www.agencemarchica.gov.ma

[3] www.nadorwestmed.ma
[4] https://germanwatch.org
[5] http://climateactiontracker.org
[6] www4.unfccc.int/ndcregistry/PublishedDocuments/Morocco%20First/Morocco%20First%20NDC-English.pdf (accessed June 2017)

References

Anfuso, G. and Nachite, D. (2011) Climate change and the Mediterranean southern coasts. In: Jones, A.L. and Phillips, M. (eds) *Disappearing Destinations: Climate Change and Future Challenges for Coastal Tourism*. CAB International, Wallingford, UK, pp. 99–110.

Araque Jiménez, E. and Crespo Guerrero, J.M. (2010) Tourisme, territoire et environnement sur la côte méditerranéenne du Maroc. *Cahiers de la Méditerranée* 81, 331–348.

Barthel, P.A. and Planel, S. (2010) Tanger-Med and Casa-Marina, prestige projects in Morocco: new capitalist frameworks and local context. *Built Environment* 36(2), 48–63.

Berriane, M. (1978) Un type d'espace touristique marocain, le littoral méditerranéen. *Revue de Géographie du Maroc* 2, 5–26.

Berriane, M. (1990) Tourisme national et migration de loisirs au Maroc: etude geographique. Thesis, University of Tours, France.

Billé, R. and Rochette, J. (2015) The Mediterranean ICZM Protocol: paper treaty or wind of change? *Ocean & Coastal Management* 105, 84–89.

Boumeaza, T., Sbai, A., Salmon, M., Benata, M. and Ozer, A. (2010) Impacts écologiques des aménagements touristiques sur le littoral de Saïdia, Maroc oriental. *Méditerranée* 115(2), 94–102.

Bourse, L. (2011) *Summary: the Tetouan Coastal Destination in Morocco. Profile of Sustainability in Some Mediterranean Tourist Destinations, Final version*. Plan Bleu UNEP/MAP Regional Activity Centre, Sophia Antipolis, France.

Brown, S., Kebede, A.S. and Nicholls, R.J. (2011) Sea-Level Rise and Impacts in Africa, 2000 to 2100. Available at: https://www.researchgate.net/publication/307210112_Sea-Level_Rise_and_Impacts_in_Africa_2000_to_2100 (accessed 30 November 2015).

Burck, J., Marten, F. and Bals, C. (2017) *The Climate Change Performance Index: Results 2017*. Germanwatch & Climate Action Network Europe. Available at: www.germanwatch.org/en/ccpi (accessed 12 June 2017).

Dobruszkes, F. and Mondou, V. (2013) Aviation liberalization as a means to promote international tourism: the EU-Morocco Case. *Journal of Air Transport Management* 29, 23–34.

Dobruszkes, F., Mondou, V. and Ghedira, A. (2015) Assessing the impacts of aviation liberalisation on tourism: some methodological considerations derived from the Moroccan and Tunisian cases. *Journal of Transport Geography*, DOI:10.1016/j.jtrangeo.2015.06.022.

Donat, M.G., Peterson, T.C., Brunet, M., King, A.D., Almazroui, M. *et al.* (2014) Changes in extreme temperature and precipitation in the Arab region: long-term trends and variability related to ENSO and NAO. *International Journal of Climatology* 34, 581–592.

Ducruet, C., Mohamed-Chérif, F. and Cherfaoui, N. (2011) Maghreb port cities in transition: the case of Tangier. *Portus Plus* 1, 1–14.

El Badraoui, M.H. and Berdai, M. (2011) *Adaptation of the Water-Energy System to Climate Change: National Study - Morocco*. Plan Bleu, UNEP/MAP Regional Activity Centre, Sophia Antipolis, France, p. 90.

El Mrini, A., Anthony, E., Maanan, M., Taaouati, M. and Nachite, D. (2012) Beach-dune degradation in a Mediterranean context of strong development pressures, and the missing integrated management perspective. *Ocean & Coastal Management* 69, 299–306.

Filahi, S., Tanarhte, M., Mouhir, L., El Morhit, M. and Tramblay, Y. (2015) Trends in indices of daily temperature and precipitations extremes in Morocco. *Theoretical and Applied Climatology*, DOI 10.1007/s00704-015-1472-4.

Gauché, E. (2011) L'exemple du Rif oriental marocain. Le désenclavement, facteur de dynamisation touristique? *Espaces* 293, 38–44.

Hall, C.M., Amelung, B., Cohen, S., Eijgelaar, E., Gössling, S. *et al.* (2015) On climate change skepticism and denial in tourism. *Journal of Sustainable Tourism* 23(1), 4–25.

Hazbun, W. (2010) Modernity on the beach: a postcolonial reading from southern shores. *Tourist Studies* 9(3), 203–222.

Hillali, M. (1994) Le développement du tourisme sur la côte méditerranéenne du Maroc: réalité et perspectives. In: GRG Rif (ed.) *Développement des montagnes rifaines: quelle stratégie?* Publications de la Faculté des Lettres et des Sciences Humaines, Tétouan, Morocco, pp. 329–344.

Hillali, M. (2007) *La politique du tourisme au Maroc: diagnostic, bilan et critique*. L'Harmattan, Paris.

Hinkel, J., Brown, S., Exner, L., Nicholls, R.J., Vafeidis, A.T. *et al.* (2012) Sea-level rise impacts on Africa and the effects of mitigation and adaptation: an application of DIVA. *Regional Environmental Change* 12(1), 207–224.

IPCC (2014) *Climate Change 2014: Impacts, Adaptation, and Vulnerability. Part A: Global and Sectoral Aspects*. Cambridge University Press, Cambridge.

IUCN, Centre for Mediterranean Cooperation (2010) *Integration of Biodiversity in the Decision-making Process: Lessons Learnt from the Moulouya Projects*. Data compiled by Professor Mohammed Melhaoui, University of Oujda, Morocco, p. 65.

Kanai, M. and Kutz, W. (2011) Entrepreneurialism in the globalising city-region of Tangier, Morocco. *Tijdschrift voor Economische en Sociale Geografie* 102(3), 346–360.

Khattabi, M.A. (2014) Vulnérabilités Climatiques et Stratégies de Développement. Institut Royal des Etudes Stratégiques, Rabat, Morocco. Available at: www.ires.ma/wp-content/uploads/2015/11/vulnerabilites_climatiques_et_strategies_de_developpement_0.pdf (accessed 12 June 2017).

Khomsi, K., Mahe, G., Tramblay, Y., Sinan, M. and Snoussi, M. (2015) Trends in rainfall and temperature extremes in Morocco. *Natural Hazards and Earth System Sciences Discussions* 3(2), 1175–1201, DOI:10.5194/nhessd-3-1175-2015.

Khouakhi A., Snoussi M., Niazi S. and Raji O. (2013) Vulnerability assessment of Al Hoceima Bay (Moroccan Mediterranean coast): a coastal management tool to reduce potential impacts of sea-level rise and storm surges. *Journal of Coastal Research* 65, 968–973.

Kingdom of Morocco (2011) Vision 2020, Stratégie de Développement Touristique. Available at: www.orientalinvest.ma/telechargementfichiers/tourisme/Plaquette-2020-FR-bat.pdf (accessed 12 June 2017).

Lasgaa, H., Boumeaza, T. and Sbai, A. (2015) Evolution of the coastline of Saidia - Cap Water (Northeastern Morocco). *Journal of Applied Geology and Geophysics* 3(4), 39–48, DOI: 10.9790/0990-03423948.

Minca, C. and Borghi, R. (2009) Morocco: restaging colonialism for the masses. In: Obrador Pons, P., Crang, M. and Travlou, P. (eds) *Cultures of Mass Tourism: Doing the Mediterranean in the Age of Banal Mobilities*. Ashgate, Aldershot, UK, pp. 21–52.

Mounime, E., Lotfi, M. and Hajbi, A. (2014) Maritime transport and international trade: a study of transhipment opportunities and their impacts on the Mediterranean region growth. *The Business & Management Review* 4(4), 252–257.

Mouzouri, M., Irzi, Z. and Essaddek, A. (2013) Utilisation d'imagerie satellitaire et d'un Modèle Numérique d'Altitude pour la cartographie des zones à risque d'inondation sur le littoral méditerranéen de Saïdia (Nord-est du Maroc). *Revue Française de Photogrammétrie et de Télédétection* 201, 49–63.

O'Mahony, C., Gray, S., Gault, J. and Cummins, V. (2015) ICZM as a framework for climate change adaptation action – Experience from Cork Harbour, Ireland. *Marine Policy*, In Press. DOI:10.1016/j.marpol.2015.10.008.

Pateau, M. (2014) Dynamiques superficielles et gestion des risques de l'arrière-pays du complexe portuaire Tanger-Med (Rif septentrional, Maroc). PhD Thesis, University of Paris Diderot, France.

Raji, O., Niazi, S., Snoussi, M., Dezileau, L. and Khouakhi, A. (2013) Vulnerability assessment of a lagoon to sea level rise and storm events: Nador lagoon (NE Morocco). *Journal of Coastal Research* 65, 802–807.

Re, V., Sacchi, E., Mas-Pla, J., Menció, A. and El Amrani, N. (2014) Identifying the effects of human pressure on groundwater quality to support water management strategies in coastal regions: a multi-tracer and statistical approach (Bou-Areg region, Morocco). *Science of the Total Environment* 500–501, 211–223.

Snoussi, M., Ouchani, T. and Niazi, S. (2008) Vulnerability assessment of the impact of sea-level rise and flooding on the Moroccan coast: the case of the Mediterranean eastern zone. *Estuarine, Coastal and Shelf Science* 77, 206–213.

Snoussi, M., Ouchani, T., Khouakhi, A. and Niang-Diop, I. (2009) Impacts of sea-level rise on the Moroccan coastal zone: quantifying coastal erosion and flooding in the Tangier Bay. *Geomorphology* 107(1–2), 32–40.

Tekken, V. and Kropp, J.P. (2012) Climate-driven or human-induced: indicating severe water scarcity in the Moulouya River basin (Morocco). *Water* 4(4), 959–982.

Tekken, V. and Kropp, J.P. (2015) Sustainable water management – perspectives for tourism development in North-eastern Morocco. *Tourism Management Perspectives* 16, 325–334.

Tekken, V., Costa, L. and Kropp, J.P. (2013) Increasing pressure, declining water and climate change in North-eastern Morocco. *Journal of Coastal Conservation* 17, 379–388.
The World Bank (2012) Integrated Coastal Zone Management Project. Available at: http://projects.worldbank.org/P121271/integrated-coastal-zone-management-mediterranean-coast?lang=en (accessed 12 June 2017).
Tramblay, Y., Badi, W., Driouech, F., Adlouni, S.E., Neppel, L. *et al.* (2012) Climate change impacts on extreme precipitation in Morocco. *Global and Planetary Change* 82–83, 104–114.
UNEP (2009) *Sustainable Coastal Tourism. An Integrated Planning and Management Approach.* UNEP-DTIE and PAP/RAC, Paris, ISBN: 978-92-807-2966-5.
Verdeguer, E. (2005) Turismo en Marruecos: retos y oportunidades. *Economía Exterior* 33, 119–124.

25 Case Study Zanzibar: Climate Change and Tourism in Zanzibar: Interrogating Impacts and Interventions

Thani Said,[1]* Hamzah Muzaini[2] and René van der Duim[1]

[1]*Cultural Geography Group, Wageningen University, the Netherlands;*
[2]*Southeast Asian Studies Department, National University of Singapore*

Introduction

With a population of 1,303,569 steadily growing at a rate of 3% per year (according to the 2012 census), Zanzibar is primarily made up of two major islands, Unguja and Pemba, with over 50 smaller surrounding (habitable and non-habitable) islets. The deep Pemba Channel separates these two islands by about 60 km from the mainland. Zanzibar has a total area of 2654 km². Pemba is small and narrow, characterized by an indented coastal line. Unguja is the largest island, 67 km long and 23 km wide, having an area of 985 km². Zanzibar experiences a humid marine tropical climate. The islands receive heavy rains in March to May and shorter periods of rain mostly from November to December. June to October is normally the dry season. The livelihoods and economy of Zanzibar strongly depend on the coastal and marine ecosystem that provides not only food but also recreational and cultural resources, which are capitalized upon for enhancing tourism development. The marine ecosystem also provides regulating ecosystem services such as the prevention of flood and erosion, thus maintaining the health of the islands' environment. The coastal ecosystem in Zanzibar is crucial for the local economy as well as the livelihoods of the islanders. The economic value provided by the coastal ecosystem services (as a whole) accounts for over 45% of the GDP of Zanzibar (RGoZ, 2014a). Politically, Zanzibar is a semi-autonomous country within the United Republic of Tanzania (URoT), located about 45 km off the mainland coast of Tanzania. Zanzibar has a long political history as it moved from being its own empire (before 1890), a British protectorate in the 1900s, to becoming part of the United Republic of Tanzania (since the mid-1960s). Zanzibar achieved its independence from the British in December 1963, followed by the revolution of January 1964, and then joined with Tanganyika on 26 April 1964 to form the United Republic of Tanzania. Following the union with Tanganyika, the status quo of Zanzibar changed immensely from a sovereign state with full autonomy, to literally a sub-political division of the United Republic of Tanzania (Jumbe, 1994; Ghassany, 2010).

*Corresponding author, e-mail: thani.said@wur.nl

Tourism Development in Zanzibar

After the revolution (from the mid-1960s up to the 1980s) Zanzibar had a centralized form of economy depending chiefly on agriculture and state-controlled commerce and trade sectors, with the majority of its people making their livelihoods from terrestrial and marine ecosystems (i.e. subsistence fishing and agriculture). During this period, Zanzibar was one of the leading global exporters of cloves in the world market. At the beginning of the 1990s, the price of cloves and the production capacity of the spice declined. Even so, the tourism sector was not yet considered as an integral part of the economy of Zanzibar. Although it offered an exotic experience for outsiders, there were few hotels, the main one being a large government-owned hotel with 118 rooms and five meeting venues at Stone Town (now known as the Bwawani Hotel, built in 1970 and recently renovated in 2011).

In fact, it was not until after the economic recession of the 1980s (which resulted in economic reforms to liberalize the local economies allowing for the private sectors to enter the economic arena) that the tourism sector began to take off. This shifting of economic paradigms was facilitated by Structural Adjustment Programs (SAP) introduced in the mid-1980s to ease the economic depression in Zanzibar. The tourism sector is now a top foreign income earning sector in Zanzibar, contributing over 80% of the country's foreign exchange. Despite this success story, the tourism sector has environmental and social impacts due to insufficient enforcement of laws. The key tourism policies and instruments shaping the industry today are many; some are distinctive while others are pertinent to tourism. Some of the most important policies and strategies governing the tourism sector in Zanzibar are: The Zanzibar Tourism Master Plan (2003); Tourism Policy of Zanzibar (2005); The Strategies for Half million Tourists in Zanzibar (2005–2014), Tourism Lab Strategy (see RGoZ, 2014b), Tourism Regulation (2014) and the Tourism Marketing Strategies (2015–2020). Other policies pertinent to tourism, environment and the livelihoods of the communities are: The Zanzibar Environmental Policy (2013), The Zanzibar Strategy for Growth and Reduction of Poverty Strategy (2010–2015), Zanzibar Investment Policy (2004), The National Vision 2020, and the Zanzibar Investment Promotion and Protection Act 2004.

The increasing collaboration and partnership between governmental and private actors has now become evident. For example, the Zanzibar Tourism Cooperation (ZTC), the Zanzibar Association of Tourisms Investors (ZATI) and the Zanzibar Association of Tour Operators (ZATO) have been working hand in hand with the government to help boost and develop the tourism sector. The general aim of this structural shift, among others, was to create a conducive environment for economic growth and diversity, and open the door wider for different sectors' and local communities' engagement in economic processes. Since then, private firms' and individuals' investments in tourism and other sectors have increased. By January 2015, the number of registered tour operators has reached 114, and there are over 349 accommodation establishments. According to the Zanzibar Tourism Commission (ZTC), the number of tourists in recent years has swiftly increased from 42,141 in 1990 to 181,301 in 2013, and rose to 208,651 in 2014. Undoubtedly this can also be attributed to the tourism policies and strategies of mainland Tanzania, since most tourists prefer to combine a visit to Tanzania with a visit to Zanzibar as well.

Methodology

To understand the tourism and climate change policy environment in Zanzibar the first author conducted 16 interviews with key informants from governmental ministries and departments and 34 interviews with non-governmental organizations (NGOs) and community-based institutions (CBOs). Interviews were complemented with a review of policy documents, focusing mainly on tourism, climate change, the environment, and the agricultural and fisheries sectors.

Climate Change and Non-climate Stressors: Assessing the Impacts in Zanzibar

Small Island Developing States (SIDS) are likely to be the most vulnerable to climate change and variability. The sizes of their landmasses, their fragile and ecological landscapes and population density and growth have been among the main factors contributing to their vulnerability to climate change (Nurse *et al.*, 2001; Ebi *et al.*, 2006). The signs and impacts of climate change are now becoming evident everywhere and in almost every economic sector. For example, according to the IPCC (2007), the global sea level is projected to increase by between 0.2 m to over 0.6 m. For islands like Zanzibar, these data are very worrisome as a large proportion of the island (over 328 km^2 or about 19.7%) lies below the 5 m line above sea level (Fig. 25.1). More importantly, most of the tourism activities take place in these areas, especially on the eastern strip of Unguja, thus exposing over 224,000 people (in both tourism and non-tourism sectors) to potential climate change risks.

A recent report (RGoZ, 2014a) on climate change projection shows a significant increase in maximum monthly average temperatures of 1.5–2°C by 2045–2050 (Fig. 25.2). The data also reveal the unpredictability and variability of the rains throughout the islands. In the last two decades, there has been uncertainty about rainfall on the island. During the wet season (March–May) there has been a slight increase of rain and the dry season period has lengthened (September–January). Similarly, the changing of climate, among other factors, has been associated with the recent increase of extreme events like floods and strong winds. The El Niño events have become more frequent and persistent in the last 15 years, tropical cyclones and coastal storms have caused major damage and severe erosion to the marine ecosystems, impacting upon many tourism structures along the eastern side of Unguja (Figs 25.3, Figs 25.4). According to the governmental economic report of 2014/2015, sea-level rise, water intrusion and salinization have been challenging problems along the coastal line in both Unguja and Pemba,

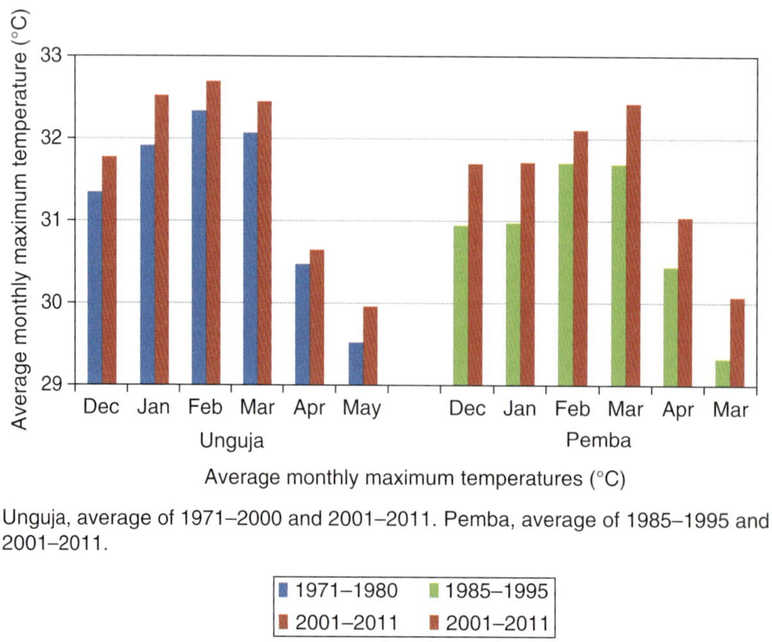

Fig. 25.1. Temperature trends in Zanzibar (Tanzania Meterological Agency; Watkiss *et al.*, 2012).

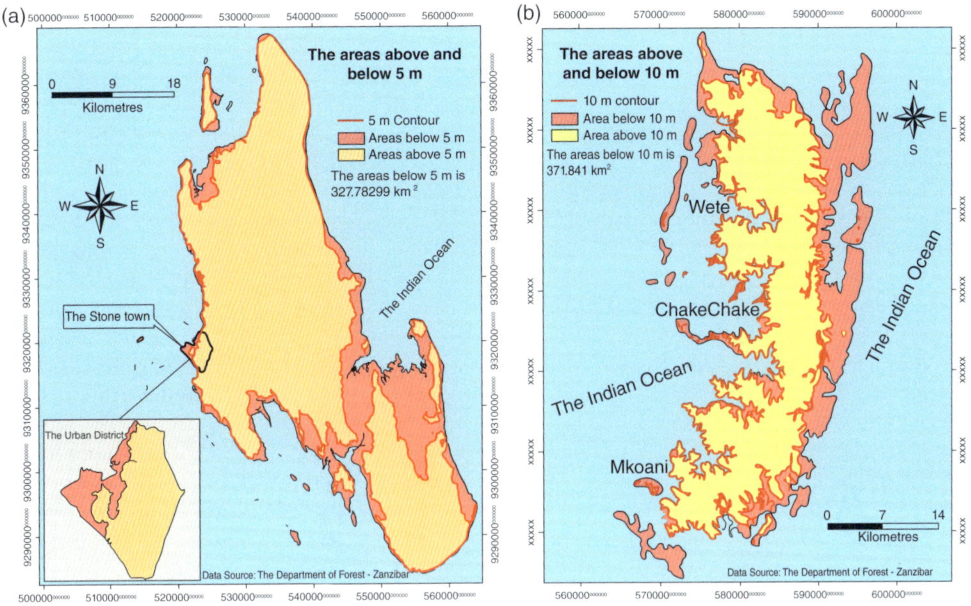

Fig. 25.2. Areas subject to flooding depending on sea-level rise: (a) Unguja, showing the 5 m contour; and (b) Pemba showing the 10 m contour (Watkiss *et al.*, 2012).

Fig. 25.3. Seawater encroachment and massive cliff erosion in Pemba coastal areas at Kichakapumu village, Msuka Mashariki.

where more than 145 localities have been affected by salt-water inundation. The problem of rain variability and rapid weather inconsistencies has also interrupted the farming (of both crops and seaweed) and fishing activities of the local communities in many parts of Zanzibar (RGoZ, 2014a).

Fig. 25.4. Seawater rise and periodical flooding has begun, and potentially continues to threaten the livelihoods and security of the low-lying coastal communities in Zanzibar, such as this area of Tumbe at Bandarini village.

Climate change has not only increased the existing vulnerability of different communities but also generated new vulnerabilities (i.e. inundate the farmland with saltwater, changing of agricultural calendar, rain variability, intensifying water shortage and so forth; Figs 25.5, 25.6). The increase in global greenhouse gases (GHGs) has been indicated as one of the major marine ecological transformational forces of the last millions of years (Hoegh-Guldberg and Bruno, 2010). Marine biodiversity (e.g. corals, fisheries and other marine-based resources), which is one of the main attractions for an increasing number of international tourists in Zanzibar, is also in danger from climate change and its consequent impacts (Füssel, 2009). In Zanzibar, the degradation of the marine system reduces the number of catches, which normally are sold in the tourist hotels and local markets. The recent coral depletion has also reduced tourism attractions such as snorkelling and diving. The IPCC's Fourth Assessment Report (A4R), entitled *Climate Change 2007*, only mentions Zanzibar once and that is in connection with an event of coral bleaching in 1998, which reduced the combined tourist revenue of Mombasa and Zanzibar by an estimated US$12–18 million (see also RGoZ, 2010).

Climate Change Policies

To briefly review the various climate- and tourism-related policy interventions introduced in Zanzibar, we have adopted the Policy Arrangement Approach (PAA) (Wiering and Arts, 2006; Van der Zouwen, 2006; Buizer, 2008; Arts and Buizer, 2009). In this chapter, climate change adaptation is seen as a policy arrangement, which is defined as a temporary stabilization of the organization and substance of a policy domain at a specific level of policy-making (Buizer, 2008; Arts and Buizer, 2009). Under the PAA, the substantial and organizational aspects of a policy domain can be analysed based on four key dimensions: (i) discourses; (ii) the actors; (iii) the operative rules of the game; and (iv) resources. We will highlight some of the climate change adaptation policies in Zanzibar and in doing so also refer to tourism. While tourism in itself is a contributing agent and driver for environment and climate change, it also offers economic and conservation opportunities, which can subsequently build up the adaptive capacity among the community and contribute to the establishment of a resilient economy.

Fig. 25.5. Seawater inundation and intrusion has turned the farming area into a barren land, for example of this area of Msuka Mashariki-Seblini, in Pemba Island.

Fig. 25.6. Seawater inundation in the rice farms is a recently increasing stressor among the subsistence rice farmers in Zanzibar, as at Konde Limani village in Pemba.

Discourse

From the 1990s, there has been a turn-around in environmental discourses. First, due to the adoption of a more neo-liberal economy in the 1980s, the focus has been on economic growth and increasing foreign currency, exemplified by welcoming more foreign investment and tourists under the banner of 'Welcome all, Zanzibar is Kind'.

At that time, there was very little concern for accompanying environmental and socio-cultural impacts. This changed in the late 1990s and early 2000s when the storyline became more focused on conservation and sustainable development, aiming to balance the trade-offs between economic development and environmental and socio-cultural impacts, largely brought about by increasing concerns over global and regional climate change and the need to build resilient economies in Zanzibar. This translated into mainstream climate change adaptation strategies and pro-poor policies. In relation to the domain of tourism policy, Zanzibar has now adopted the so-called 'tourism for all' as an economic paradigm that aims to involve communities more in developments under the arrangement of community–private–government partnerships (see RGoZ, 2014b). In Zanzibar, the idea to mainstream climate change adaptation strategies and create pro-poor policies has been recognized in many economic sectors to ensure the risks from climate change are mitigated and to improve the community resilience to vulnerability, environmental degradation and socio-political disruption.

Actors

As a result of the above-mentioned discourse shift, climate change in Zanzibar is also now considered a crosscutting issue that touches almost every economic sector, including tourism. Therefore, the First Vice President Office (FVPO) is now the key national institution that coordinates climate change alongside other concerns. The FVPO works in coalition with both public and private sectors to adhere to the National Framework for Climate Change Responses, while the Ministry of Finance is tasked with developing the financial mechanism to enhance various projects related to climate change and contributing to the forging of social and economic resilience in both Unguja and Pemba. The Ministry of Agriculture and Natural Resources is responsible for managing and coordinating activities related to the development of agriculture and natural resources on the islands and ensuring the sustainable use of these resources seen as central to the economy.

The Ministry of Agriculture and Natural Resources includes six departments: Department of Planning, Policy and Research; Department of Administration and Human Resources; Department of Agriculture; Department of Forest and Non-renewable Natural Resources (DFNRR); Department of Irrigation; and the Department of Food Security and Nutrition plus two agricultural institutes (i.e. Kizimbani Agricultural Training Institutes and Institute of Agricultural Research and Liaison Office in Pemba). In relation to climate change response, the Ministry works to prevent further water intrusion in the farming areas and carries out research in the areas of green farming, seed improvement and resilient cropping. To respond to the increasing pressure on the fragile coastal biodiversity, the DFNRR undertakes a nationwide mangrove replanting campaign along the coastal areas in collaboration with local communities' associations as a strategy to conserve the environment and maximize the ecosystem services and benefits, as well as enhance nature-based tourism, which is important on the islands of Zanzibar.

The Jonzani Environmental Conservation Association (JECA), Ngezi-Vumawimbi Natural Resources Conservation Organization (NGENARECO) in Pemba, South Environmental and Development Conservation Association (SEDCA) in Unguja, Community Forest Management (CFM) and the Women Empowerment in Zanzibar (WEZA) are some of the examples of other key actors working in collaboration with the government. International organizations such as the Cooperative for Assistance and Relief Everywhere (CARE) address the problems of deforestation, and aim at improving governance, including benefit sharing, and controlling leakages for the purpose of providing the forest-dependent communities with secure property rights, equitable rewards for providing ecosystem services and other livelihood benefits. Figures 25.7 to 25.10 illustrate some examples of these collaborations.

Fig. 25.7. Bee-keeping as a communities income-generating project at Jonzania-Chawaka Bay Conservation Area (JCBA) at Unguja.

Fig. 25.8. Enhancing seaweed harvesting as a source of income among the Zanzibar coastal communities under the support of CARE in Pemba: Makangale, Jiwe Moja village.

Fig. 25.9. Seedlings and agroforestry as a reforestation policy initiative to enhance both conservation and income generation among the rural communities living near the conservation area in Pemba at Makangale-Vumawimbi.

Fig. 25.10. Diversification of livelihood strategies through engagement in vegetable and fruit farming to reduce household poverty and other unpredictable ecological and non-ecological stressors among the rural communities in Zanzibar: Chokaani village, Gando in Pemba.

Within the energy sector, statistics show that the fuel-wood, charcoal and agricultural residues account for 97% of domestic consumption. Over 90% of the local population in Zanzibar rely on the wood fuel for cooking. The erratic power supply is still imported from the mainland. Solar power is used in a very few places, especially in the rural areas. To address the threat on forest resources and use the existing potential for solar, wind and wave power, the Department of Energy is slowly developing alternative sources of energy to reduce excessive use of conventional sources of energy such as wood-fuel, kerosene, charcoal and the mainland electric supply.

Rules

Under the coordination of the FVPO, the various ministries have now set in place a number of policies, which can be clustered into two major groups: (i) policies distinctively structured to address climate change in Zanzibar; and (ii) policies that are not directly structured to address climate change, but have parts/sections which are pertinent to climate change adaptation strategies. The former include: the National Adaptation Program of Actions (NAPA) (2007); the National Environmental and Climate Change Governance Strategy (2012); the Piloting Local Adaptation in Program of Action in Zanzibar (2013) and the Least Development Countries Fund (LDCF) (2013) designed to promote urgent and immediate adaptations in selected vulnerable coastal sites including those in Unguja and Pemba. The Tanzania Climate Change institutional strengthening programmes (from 2007 to 2010), as renewed in 2011 to 2015, were designed to improve Tanzania's access to climate change finance and were used to support climate change resilience among the rural poor, including Zanzibar. Most of the activities translated from these policies centre around capacity building to maximize the benefits from key livelihood-supporting activities like tourism, farming (vegetable) and fishing (aquaculture).

The indistinctive policies include: the Zanzibar Strategy for Growth and Reduction of Poverty Strategy (2007–2010, renewed in 2010–2015), the National Agricultural and Natural Resources Policy (2002), the National Environmental Policy (2012), the Zanzibar Livestock Policy (2009) and the National Development Vision (2020). Most of these environment and sustainable development-related policies resulted from international protocols and declarations and are targeted at supporting a variety of livelihood options, such as tourism, conservation and agriculture. These strategies are also aimed at addressing environmental challenges and to achieve national developmental targets in Zanzibar (RGoZ, 2010). Despite these policies, the constitutional and political landscapes at which the policies operate have constrained Zanzibar from efficiently acting on climate change responses in a timely manner. Currently, this is the United Republic of Tanzania, which represents Zanzibar in different regional and international climate change platforms like in UNFCC, IPCC and in all other international protocols and conferences. This has proven to under-represent Zanzibar as an island, which requires special attention because of its geographic, economic, demographic and ecological factors (Jumbe, 1994; Horton, 2011; Sheikh *et al.*, 2011).

Resources

The main resources needed to be able to implement the policies mentioned above are money, technical expertise and local participation. The power to distribute the financial resources to be invested in climate change adaptation belongs to the Revolutionary Government of Zanzibar (RGoZ) as channelled through the FVPO. The office receives funding from different sources such as the national public finance, international multilateral and bilateral finance. The financial resources are also distributed through a number of intermediaries such as the Ministry of Finance and Planning, and international NGOs. The current governmental climate-change funding is US$500,000 per year. However, the RGoZ has recently developed a new climate-financing basket known as 'Zanzibar Planning and Financing System for Climate Change', which seeks to attract more financial support for climate adaptation policies, and to disburse this into the different economic sectors accordingly. However, to set Zanzibar on a green and resilient economy, the country needs over US$2 million climate financing per year (RGoZ, 2014).

To use properly its human resources the First Vice President Office (FVPO) of the RGoZ is represented by two important committees, the steering and the technical committee. The former advises on matters relating to the Zanzibar policy strategies and action plans with respect to all issues of climate change, and mainly comprises principal secretaries from across government ministries. The technical committee consists of members of different sectors (both public and private, government and non-governmental organizations), and their main task is to implement the directives and plans of action put forward by the steering committee.

Finally, for the effective implementation of climate change policies, the participation of local communities is essential. Despite the fact that communities in Zanzibar have long experienced and felt the pain of the changing of climate, it is normally policy makers, experts, politicians and project managers who have the power to identify, construct and implement major projects on behalf of the communities. Until now, the participation and involvement of local communities in the policy-making process has not been worth mentioning, usually just as invitees to 'rubber-stamp' what has actually already been decided beforehand. Climate change knowledge thus remains the purview of scientists and experts where local communities themselves do not have a voice even as the impacts of climate change on their lives are very real. Based on the interviews and a review of past and existing projects, it seems that much of the money that is accrued for climate change projects has tended to go into the bureaucratic processes with local communities reaping very little. In fact, what has been allocated for climate change planning and what actually goes into policy implementation at the community level remains highly disproportionate.

Conclusion

In the last decades, Zanzibar has been exposed to the increasing threats of climate and non-climate stressors. These climate change impacts have touched upon almost every economic sector, including tourism, which is now a major economic driver in Zanzibar. To address the problems, the RGoZ has developed a large number of policies. The government, NGOs and communities have formed a coalition (though not necessarily very efficient and successful) to engage in adaptation activities across the various economic sectors with the intention of reducing climate stressors and community vulnerability, and building a resilient economy. According to most of the key informants at the governmental level, so far the implementation of these policies has been hampered by insufficient finances and technical expertise. An efficient offset of environmental and climate change effects, and improvement of the livelihoods of the people in Zanzibar, requires a better understanding of the complex economic and political scenarios, proper designing and an efficient implementation of different adaptation strategies from the grassroots to the national level. On top of that, building a strong enabling environment for public and private partnership for addressing the problem of environmental and climate change can be another best option for Zanzibar.

References

Arts, B. and Buizer, M. (2009) Forests, discourses, institutions: a discursive-institutional analysis of global forest governance. *Forest Policy and Economics* 11(5), 340–347.

Buizer, I.M. (2008) *Worlds Apart: Interactions Between Local Initiatives and Established Policy* (No. 22). Alterra Wageningen UR, Wageningen, the Netherlands.

Ebi, K.L., Lewis, N.D. and Corvalan, C. (2006) Climate variability and change and their potential health effects in small island states: information for adaptation planning in the health sector. *Environmental Health Perspectives* 114, 1957–1963.

Füssel, H.M. (2009) An updated assessment of the risks from climate change based on research published since the IPCC Fourth Assessment Report. *Climatic Change* 97(3), 469–482.

Ghassany, H. (2010) *Kwaheri Ukoloni Kwaheri Uhuru! (Goodbye Colonization, Goodbye Freedom!) Zanzibar and the Afrobia Revolution*. Lulu.com, Raleigh, North Carolina.

Hoegh-Guldberg, O. and Bruno, J.F. (2010) The impact of climate change on the world's marine ecosystems. *Science* 328(5985), 1523–1528.

Horton, K. (2011) Climate change governance in Zanzibar – The Need for Autonomy. In: *The First International Symposium on the Impacts, Vulnerability and Adaptation to Climate Change in Small Island Developing States; Implication to Poverty Eradication*, 12–14 December 2011, Zanzibar, Tanzania.

IPCC (Intergovernmental Panel on Climate Change) (2007) *IPCC Working Group II Fourth Assessment Report*. IPCC, Geneva, Switzerland.

Jumbe, A. (1994) *The Partnership: Tanganyika Zanzibar Union - 30 Turbulent Years*. Amana Publishers, Dar es Salaam, Tanzania.

Nurse, L.A., Sem, G., Hay, J.E., Suarez, A.G., Wong, P.P., Briguglio, L. and Ragoonaden, S. (2001) Small island states. In: McCarthy, J.J., Canziani, O.F., Leary, N.A., Dokken, D.J. and White, K.S. (eds) *Climate Change 2001: Impacts, Adaptation, and Vulnerability*. Cambridge University Press, Cambridge, UK, pp. 843–875.

RGoZ – Revolutionary Government of Zanzibar (2007) *The Zanzibar for Growth and Reduction of Poverty 2007–2010*. Revolutionary Government of Zanzibar, Zanzibar, Tanzania.

RGoZ – Revolutionary Government of Zanzibar (2010) *The Zanzibar Strategy for Growth and Reduction of Poverty 2010–2015*. Revolutionary Government of Zanzibar, Zanzibar, Tanzania.

RGoZ – Revolutionary Government of Zanzibar (2014a) *Policy Briefing; Planning and Financing System for Climate Change*. Revolutionary Government of Zanzibar, Zanzibar, Tanzania.

RGoZ – Revolutionary Government of Zanzibar (2014b) *Tourism Lab Program Report; Zanzibar Development Vision 2020; Result for Prosperity/R4P*. Revolutionary Government of Zanzibar, Zanzibar, Tanzania.

Sheikh, M., Raja, Sinani, T., Bakar, A. and Bakari, A.H. (2011) Climate change arrangement in Zanzibar; a paradox equation! In: *The First International Symposium on the Impacts, Vulnerability and Adaptation to Climate Change in Small Island Developing States; Implications to Poverty Eradication*, 12–14 December 2011, Zanzibar, Tanzania, p. 44.

Van der Zouwen, M. (2006) *Nature Policy between Trends and Traditions: Dynamics in Nature Policy Arrangements in the Yorkshire Dales, Doñana, and the Veluwe*. Radboud University Nijmegen, the Netherlands.

Watkiss, P., Pye, S., Hendriksen, G., Maclean, A., Bonjean, M. *et al.* (2012) The Economics of Climate Change in Zanzibar. Study Report for the Revolutionary Government of Zanzibar, Climate Change Committee. Available at: www.economics-of-cc-in-zanzibar.org/reportsandpublications.html (accessed 7 January 2017).

Wiering, M.A. and Arts, B.J.M. (2006) Discursive shifts in Dutch river management: 'deep' institutional change or adaptation strategy? In: *Living Rivers: Trends and Challenges in Science and Management*. Springer, the Netherlands, pp. 327–338.

26 Climate Change and Coastal Tourism – A Global Perspective: Recognizing Problems – Managing Solutions – Future Expectations

A. Jones*

Institute for Tourism Travel and Culture, The University of Malta; and Islands & Small States Institute, The University of Malta

Introduction

The erosion of coastlines and beaches will undoubtedly continue to pose a significant threat to both environment and tourism and, consequently, the tourist economy of many coastal destinations and regions in the future. Conclusions from *Disappearing Destinations* (Jones and Phillips, 2011) highlighted a number of concluding issues, which included:

- Media impact – dissemination responsibilities and negative perceptions;
- Local governance and public policy making – focusing on the need for forward and long-term planning strategies;
- Coordinating the roles and responsibilities between public and private funding streams;
- Making informed choices and decisions between ameliorative protective and/or adaptive measures;
- Compliance with growing legal frameworks, legal litigation threats, insurance hazard and risk adverse strategies; and
- Maintaining the economic, social and environmental well-being of coastal tourism communities.

The new case studies have illustrated that there are some common themes emerging that continue to emphasize some of those earlier reported issues. In this respect there still remains continuing uncertainty regarding climate change and the validity of current predictions. The fundamental question of whether coastal tourism destinations will have viable sustainable futures is a critical one, still facing many destinations today. It is clearly not an easy question to answer but the evidence from not only the case studies contained in this book but also from evolving research elsewhere, shows that there are a number of indicators and lessons from which we can draw some conclusions and assumptions. Critical themes emerging from the cases discussed confirm that there are a number of growing complex relationships that have evolved between different stakeholder interests which will ultimately determine the future sustainability of many coastal tourism destinations. These can be broadly defined in terms of 'Recognizing

*E-mail: andrew.jones@um.edu.mt

Problems', 'Managing Solutions' and 'Future Expectations'.

Recognizing Problems

Evidence from literature and the case studies suggest that there will be an increased incidence of storm surges and a general rise in sea level. As a result, erosion issues will become more prominent and coastal tourism managers and planners will increasingly have to use techniques that are best suited to the locality under specific threat. The cases discussed generally highlight an acceptance of the threats from climate change and present evidence reported from each region of the world. Each of the cases generally illustrate examples of threats which largely relate to sea level rise, incidents of storm surges and storm damage, particularly through flooding and high wind velocities, environmental damage, incidents of severe temperature rises, increased incidents of health-related risk and, in turn, resulting physical, social and economic disruption as a consequence. These are issues illustrated for example in a number of the case studies that highlight that such impacts are not discrete to any particular type of destination, locality or region. Despite such predictions the main concern emerging from the cases is not the general acceptance of potential climate-change threats but the realities and ability to recognize policy frameworks, time frames and the accuracy of current predictions. Who takes ultimate responsibility for the management of mitigation, adaptive and preventative measures are also sentiments widely expressed by a number of the authors of the case studies presented.

The evidence thus far is clear. Figures 26.1 and 26.2 highlight the most common reported incidents of adverse impacts associated with climate change and its immediate effect across the cases. From those discussed there are clearly common occurrences of both physical and socio-economic impacts at local tourism destinations. The most commonly reported incidents tend to relate to environmental damage, beach flooding, erosion, severe weather and the general acceptance and perceptions of issues associated with coastal vulnerability. These are further supplemented by lesser (in number) but equally important and no less relevant incidents associated with concerns for continued tourism growth, industry malaise,

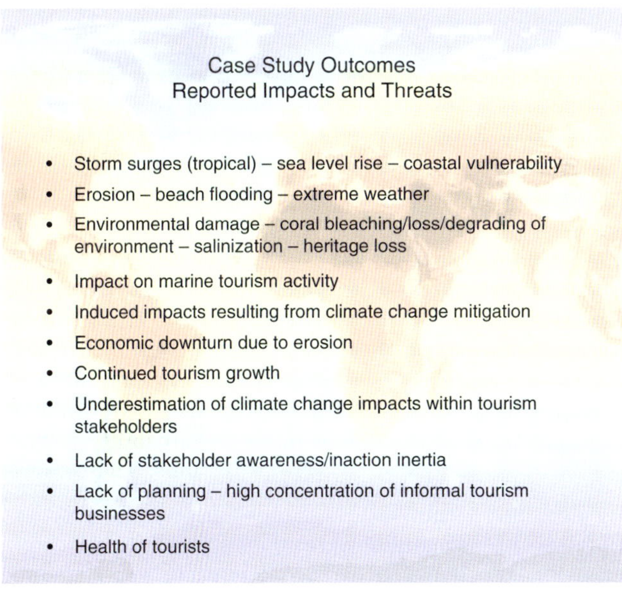

Fig. 26.1. Perceived climate change threats.

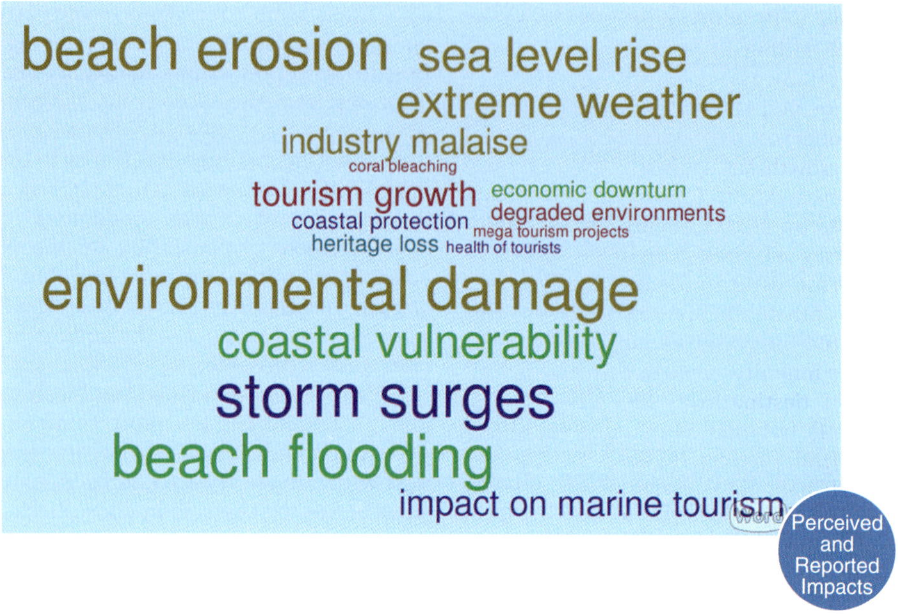

Fig. 26.2. Case studies – cloud diagram of perceived and reported impacts.

impacts to tourism operations – particularly to those relating to marine activities and concerns for the health and well-being of tourists. That said, the notions that continued growth in tourism numbers across the regions and at individual destinations set against increasing threats from climate change do not sit well together and raise fundamental issues associated with future sustainability.

It is difficult to ascertain in quantifiable terms how such incidents can be compared or differentiated between the different cases across the regions or by the type or category of the destination. Each has its unique character and as such its own experiences, needs and priorities. What is clear, however, is that the reported incidents of storm surges, fear of sea level rise and coastal vulnerability are reported across all regions including, for example, Cooper's assessments in Ireland (Chapter 8), Calado's review of the Azores (Chapter 10), Santos Lacueva's study on Mexico (Chapter 14), Bùi's assessments of Hoi An in Vietnam (Chapter 16), Hall's evaluation of Christchurch, New Zealand (Chapter 20) and Perelli's appraisal of Morocco (Chapter 24). When these are linked to broader reported incidents relating to environmental damage these common themes become even more evident and include examples from Malta (Jones, Chapter 11), Iceland (Jenkins, Chapter 12), Turkey (Cenk Demiroglu, Chapter 21), Antarctica (Jabour, Chapter 23) and Zanzibar (Said, Chapter 25).

Curiously a couple of paradoxes would also seem to emerge from these case scenarios. The first paradox focuses on the connection between reported impacts and perceived economic threats. These connections at best would still appear to be tenuous or not fully recognized by or at destinations. When reporting adverse economic impact such predictions, when compared to environmental impacts, fared more positively with only a few cases reporting actual or perceived economic vulnerability. Again such sentiments did not really show regional variation although incidents of economic down-turn were more common from the emerging economies. These included, for example, Sri Lanka (Buultjens, Chapter 17), Vietnam (Bùi, Chapter 16; Truong, Chapter 19) and Mexico (Santos-Lacueva, Chapter 14). Clearly, however, there seems little correlation between perceived impacts and economic well-being with tourism stakeholders, by and large, failing to recognize threats and predicted impacts.

The second illustrates the paradox between concerns expressed for continued tourism growth set against climate change predictions and actual measures being taken to mitigate impacts. Concerns from a number of cases reported growing adverse impact from mitigation measures implemented to offset climate change impact. They also emphasized the resulting negative environmental or amenity damage this has caused at several destinations. Morocco (Perelli, Chapter 24), Ireland (Cooper, Chapter 8) and Vietnam (Bùi, Chapter 16) are cases in point. Calado's (Chapter 10) case of the Azores and perceived conflict relating to the desire to mitigate against carbon emissions through taxation vis-á-vis reliance on air travel for island or remote destinations is another.

In this context there would therefore appear to be a growing disconnect between tourism growth, recognized impacts, management of a coastal destination and environmental and economic well-being. At one level there seems to be a continuing disconnect between perceptions of predicted or actual climate change impacts and recognition of economic vulnerability or the future viability/sustainability of coastal tourism destinations. Denial or malaise is a shared experience in this respect. At the other, there would appear to be a disconnect between the acceptance and implementation of mitigation measures to offset climate impacts against resulting adverse environmental consequences created. In turn, knowledge and expertise would appear to be shared deficits in this respect.

Quintessentially cases show very mixed responses to reported or perceived impacts. A number of impacts are clearly recognized and shared across all cases but the perceptions and experiences between destinations, although sharing some common experiences, largely remain unique to each locality depending on the socio-economic and physical character of the destination. That said, recognizing problems associated with climate change illustrate a multitude of common and shared experiences between the cases and across different regions and climate zones. In this context there seems little differentiation between shared experiences when environmental vulnerability is considered. Overall, destinations located within polar, temperate, subtropical or tropical climate zones share these common concerns and experiences. Differentiation is more evident, however, when more specific impact criteria are considered and this would appear to separate more distinctly destinations located between different climate zones. For example, as would be expected, those destinations located in tropical zones report more on coral bleaching and beach erosion or institutional malaise, whilst destinations in more temperate climes see more shared or common experiences relating to general environmental damage, impacts from large tourism projects and growth threats from increasing tourist numbers. In turn those destinations in more polar regions experience common issues relating to threats to ecosystems and ecological damage. This is not an exact science and in reality the evidence demonstrates that the nature of climate change impacts or threats to a coastal destination would appear to be very much determined by a destination's unique character and setting.

Managing Solutions

Evidence from the cases largely demonstrates that current coastal and destination management practices tend to concentrate on immediate socio-economic and environmental responses, which often give little regard to longer term strategic coastal management concerns linked to long-term strategies that address threats to coastal tourism destinations from climate change. In this respect it would seem that destinations and key tourism stakeholders need to adopt appropriate policy implementation measures that not only aim to integrate more effectively but also aim to 'think' in longer term strategic time-frames.

From the cases discussed there remain critical recognition issues concerning the solutions at hand to mitigate threats from climate change and predicted impacts at destinations. The key responses from the cases illustrate that strategic planning, capacity building and making choices between protective, adaptive and mitigation measures are fundamental to attaining correct solutions. These can be seen in Figs 26.3 and 26.4.

Strategic planning

The issue of strategic planning and associated integrated management concepts were highlighted by several of the cases which put forward notions, on the one hand, of the need to promote conservation and management of the coastal zone and, on the other, of the need to ensure that recreational and tourism assets are protected and strategically managed. In this respect solutions to perceived and reported impacts from the cases show a diverse range of approaches that can be adopted or considered to mitigate against climate threats.

The responses from the cases clearly show that strategic planning, the development of strong planning and design guidelines or national policy guidelines together with, perhaps, the use of CCPI (climate change performance indicators) are common themes expressed across many destinations globally. There is little differentiation, in this respect, between different destinations. The cases illustrate that such concepts are shared in, for example, reviews from Ireland (Cooper, Chapter 8), the Azores (Calado, Chapter 10), Mexico (Santos-Lacueva, Chapter 14), Vietnam (Bùi, Chapter 16), Bangladesh (Hassan, Chapter 18), New Zealand (Hall, Chapter 20), Malta (Jones, Chapter 11), Iceland (Jenkins, Chapter 12) and Morocco (Perelli, Chapter 24).

Clearly the case examples present a set of solutions that suggest that it is becoming more critical that all key stakeholders need to progressively change current patterns of management in order to support future efforts and actions that integrate planning strategies more coherently. The cases make strong recommendations for a more integrated and long-term process of problem recognition, planning, implementation and monitoring. It is probably true to say that only by adopting such approaches the problems of coastal destination management can be fully addressed for the future. The adoption of such an approach will go some way to ensure that coastal environments are conserved, natural and anthropogenic factors and processes are integrated into the decision-making processes, coastal protection measures are fully evaluated and selective strategies for future management of coastal tourism destinations prioritized. This, in turn, will ensure and help secure

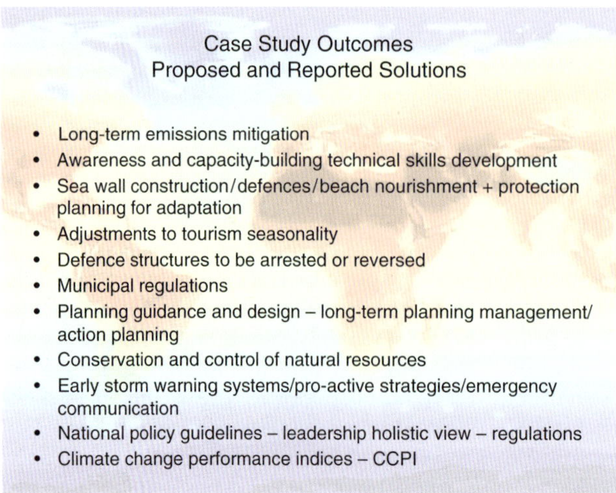

Fig. 26.3. Perceived climate change solutions.

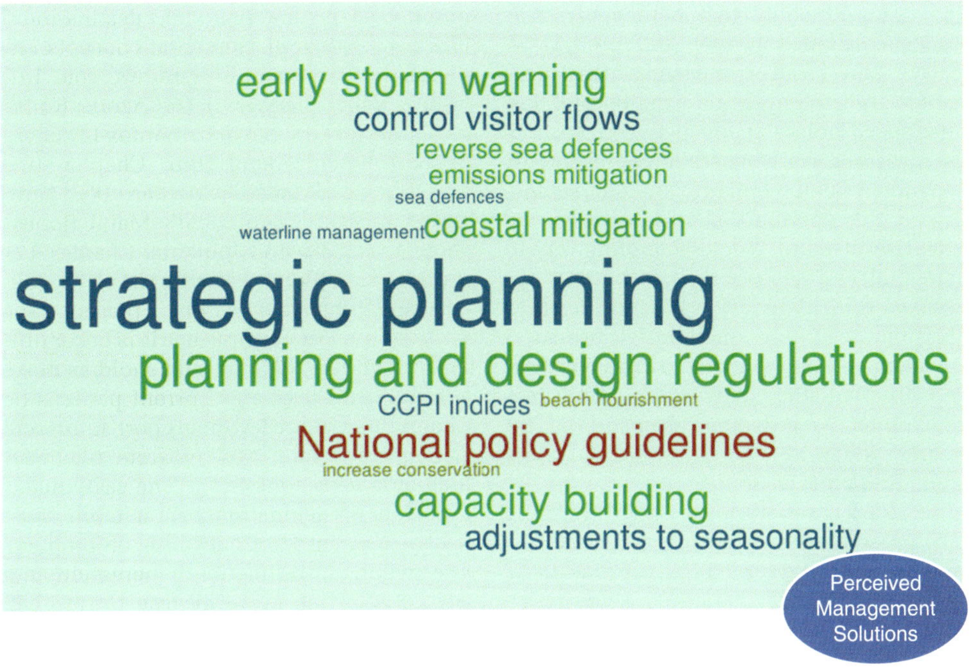

Fig. 26.4. Case studies – cloud diagram of proposed and reported solutions.

the socio-economic and environmental well-being of coastal tourism communities.

Capacity building

Knowledge gaps and the need for capacity building to develop skills to understand and manage consequences from the impacts of climate change were also quite evident from the case responses of Zanzibar (Said, Chapter 25), Barbados (Cumberbatch, Chapter 13) and Malta (Jones, Chapter 11). It can be seen that this lack of recognition spans across key stakeholder groups that include managers and planners at tourism destinations and the managers of tourism facilities and operations on the one hand and the users of tourism facilities – the tourists – on the other. Key to these difficulties can be attributed to: (i) limited professional and technical understanding or indeed full recognition of the problem; (ii) commercial priorities of tourism operations that tend to mask greater recognition of the problem; and (iii) general apathy and policy malaise from key policy makers, managers, consumers and users, which largely tend to ignore the problem or perceive it as a responsibility of others. Despite much global acclaim and reporting on such issues the problem of recognition at local level and empowering local people to act remains at best limited. Building professional capacity and empowerment would therefore appear to be important future tasks and priorities if such problematic issues are to be addressed successfully for destinations.

Protective, adaptive and mitigation measures

Often policy choices for combating the impacts of climate change on coastal tourism destinations will invariably rely upon protective and adaptive measures. These in essence will either focus upon the direct protection of existing coastal environments to address impacts or more broad adaptation mitigation measures to offset impact.

The cases illustrate that both contrasting strategies are widely seen and used as effective ameliorative measures that help reduce climate impact on local coastal destinations. Choices made between each of the measures available are of course largely dependent upon the nature and characteristics of the destination and professional expertise at hand. Socio-economic, environmental and geo-political influences are perhaps critical here in determining how decisions regarding ameliorative measures are taken and implemented. However, there is a differentiation between the more tangible protective measures taken and the less tangible policy-oriented mitigation and adaptation measures desired.

Indeed correlation may exist between how proactive, how well organized and developed a destination is on the one hand and how its success is rated in securing effective and long-term ameliorative measures on the other. In this respect the cases reveal that coastal destinations demonstrate a variety of alternative strategies and examples of feasible and preferred solutions. In this context several themes have emerged from the cases that raise issues and questions concerning the application and implementation of ameliorative measures. They relate in part to, for example, the 'art or science' of making informed choices and the degree to which technical knowledge is available, the availability of accurate data to support and enhance decision making and policy formulation, accepting the consequences and impact of the ameliorative choices made and tensions between long-term and short-term investment choices. The cases demonstrate that measures such as sea defences, beach nourishment and general coastal protective measures are common solutions, for example as seen in the cases from the Azores (Calado, Chapter 10), Vietnam (Truong, Chapter 19) and Israel (Hall, Chapter 22). A paradox in this context might also suggest that there is growing evidence from some cases that physical protective measures already implemented to offset climate change threats are increasingly causing more damage than providing protective advantage. This is reported at a number of destinations such as in Ireland (Cooper, Chapter 8), Israel (Hall, Chapter 22), Bangladesh (Hassan, Chapter 18) and Antarctica (Jabor, Chapter 23). Clearly not all measures taken have been wise or stand up to environmental scrutiny.

When less tangible adaptive and mitigation measures are considered the cases show a very mixed application of choice or preferred solutions available with little pattern emerging between types or locality of a destination. For example, early storm warning systems are advocated by Cenk Demiroglu (Turkey, Chapter 21), Cannas (Sardinia, Chapter 9), Calado (Azores, Chapter 10) and Truong (Vietnam, Chapter19). More broader long-term emissions mitigation is advocated by Jenkins (Iceland, Chapter 12) and Anthonisz (Dubai, Chapter 15). Controlling visitor flows and making adjustments to seasonality are advocated by Hassan (Bangladesh, Chapter 18), Hall (Israel, Chapter 22) and Truong (Vietnam, Chapter 19). The cases illustrate a variety of preferred options and a number of different strategies and measures that can be taken to offset climate impact at destinations. Each of the options 'fit' differently to the different localities and character of destination. Again there are a number of issues that have arisen as a consequence. The key to making choices between alternative ameliorative strategies would appear to be one that specifically relates to knowledge and knowledge gaps as already stated in the need for capacity building. In this context, making choices to ensure the effective amelioration of impacts will increasingly be dependent upon being 'well informed' with current data so that any decisions made on coastal tourism strategies and coastal defences remain appropriate to the sustainable management of the coastal destination. It is a notion that Buultjens explores in the case of Sri Lanka (Chapter 17). Evidence suggests that it is thus important that research and monitoring of natural and man-made processes, including the collection of primary data, is enhanced in order to ensure that accurate decisions are made. This, after all, remains the raw material for understanding the tourism industry and for evaluating protective measures for coastal destinations.

Coastal tourism destinations have evolved from many natural, anthropocentric factors and within this often dynamic perspective it is perhaps logical that management concepts, linked to governance and public policy should aim to be inherently integrated. It is this '*integrated*' management concept that has been explored by several of the cases which put forward notions, on the one hand, of the need to promote conservation and management of the coastal zone and on the other, of the need to ensure that environmental and tourism assets are protected. How this is coordinated and the consequences of the lack of coordination between organizations will bear heavily on achieving ultimate success in combating destination management and tackling climate change predictions. As a consequence it is important to understand not only present-day organization and decision-making processes but also processes linked to the coordination of public and private stakeholder organizations and how decisions concerning the management of destinations from a broad range of socio-economic and environmental stances are achieved. In addition to measures that enhance coordinated effort there is also a need to think with long-term, lateral and strategic vision. Evidence from the cases largely demonstrates that current coastal and destination management practices tend to concentrate on immediate socio-economic and environmental responses that often give little regard to longer term strategic coastal management concerns linked to long-term strategies that will address threats to coastal tourism destinations from climate change. In this respect it is becoming more critical that key stakeholders and appropriate policy implementation measures using institutional, legal, financial and technological frameworks not only aim to integrate more effectively but also aim to 'think' in longer term time-frames.

Meeting Future Expectations

Evidence highlighted from the cases presented demonstrates that coastal tourism destinations initiate a variety of actions that go some way to offset the predicted and actual effects of climate change or to manage expectation of such threats. These actions tend to differ according to the socio-economic, environmental, geographical and political contexts in which each of the destinations are located. In this respect the action taken by local and national organizations vary to quite a degree. Some destinations have a combination of both proactive adaptation and mitigation measures, others concentrate on adaptation, others mitigation and some offer limited responses to all but very basic measures. Each individual destination thus addresses problems associated with its own particular set of circumstances and needs and thus raises quite different sets of issues. However, the cases do illustrate that there are some positive moves in the right direction and a general move to recognize that coastal destinations cannot pander to the status quo. In this respect the cases clearly illustrate a set of emerging expectations. They include lessons that address long-term planning, tailor-made solutions, coordination, integrated coastal zone management (ICZM), capacity building, increased regulation, mitigation innovation, research and information sharing. Indeed most of the cases stress the need to promote such concepts in order to secure the economic, social and environmental well-being of the destinations, in particular in the context of valued coastal environments, and dependent socio-economic systems particularly focused on the tourism industry. Figure 26.5 highlights the key case lessons learned and from which future policies and managed solutions can and may be devised.

All these lessons and expectations would appear to be interrelated. The case studies illustrate that it will become increasingly important to enhance policy- and decision-making processes in order to sustain and protect such destinations from climate change threats. Indeed, ensuring successful actions and outcomes in this context very much belies the future well-being of each destination.

Clearly the socio-economic viability of coastal tourism destinations and climate

Fig. 26.5. Case study outcomes – perceived lessons for the future.

change dynamics are sensitive political issues in the context of local, national and international politics. Whether economic sense will prevail, overriding political sensitivities that include defending local and national interests, remains a difficult issue to resolve. At best most tourism destinations will guard their economic interests and their social and environmental contexts which in turn support their tourist economies. How such destinations will continue to remain sustainable in the future is questionable. What will become increasingly important, however, is the need to balance decisions associated with climate change and the assessed socio-economic impacts these will have on specific tourism destinations. A 'one fits all' policy will clearly not be appropriate. Balanced decision making based upon protection of some areas and retreat in others or seasonal tourist adjustments to tourism flows and the development of new market options will need to be made. The continued development of integrated tourism policies that encourage links to coastal hinterlands might also be an option to help offset vulnerability. Such issues are illustrated in Fig. 26.6.

In summary, the case studies highlight a number of management and policy contexts that still require attention if tourism coastal environments are to be sustained in the longer term. In this respect a number of recurring management and policy themes have emerged. These relate to:

- Market change;
- Tourism planning, management and innovation;
- Infrastructure; and
- Skills, education and capacity.

Market change

The case studies suggest that there will be continuing need to innovate and diversify the tourism 'offer' at a destination. This might, for example, include the development of new segmented or niche markets that offer specialism, short break or inter-regional co-operation to offset over-reliance on existing more traditional tourism markets. It might

Fig. 26.6. Meeting expectations – learning lessons.

also include catering for more discerning consumers looking for enviro-friendly, green alternatives or low impact/low carbon footprint agendas. It will certainly have to respond to risk-averse strategies and create safer environments through adopting new mitigation and adaptation measures. Longer term it might also include spatial redistribution, perhaps retreat, seasonal adjustment and control or limits to tourism growth. Whatever the bespoke outcomes and choices for a destination the long-term strategic coordination of such policies will be fundamental.

Tourism planning, management and innovation

This has been a recurring theme throughout the case studies. The message is one that emphasizes coordination and integration of policy making and management decisions. In this respect the coordination of environmental protection, conservation and tourism development, growth and regulation appears to be imperative. This in turn may necessitate actions that encourage the creation and development of innovative tourism environments that reduce danger, promote safer environments, and encourage risk-averse measures that lessen potential insurance and liability risk warnings. Making informed coordinated choices, for example the inclusion of smart design concepts, low energy, self-sufficiency agendas and ensuring quality environments, should go hand in hand with this approach. Making strategic long-term choices would again appear to be fundamental in this respect.

Infrastructure

The cases clearly demonstrate a very mixed response for the need for mitigation and adaptive measures. What was clear, however, was the recognition of potential climate impacts and the need to manage and offset these threats through adopting mitigation and adaptive measures. A number of solutions were proposed but these generally were differentiated between active and passive adaptive and mitigation investment. Key to this was the need to make informed decisions that prioritized the need to protect vulnerable coastal environments and protect coastal tourism economies. The message from the cases suggests that there is still a need to take decisions based upon informed environmental and economic necessity. In this context ensuring that correct choices are made regarding the building of protective infrastructure is crucial to the sustainability and well-being of a specific destination. Concepts associated with natural environmental solutions, green and enviro-agendas, creating safe environments, mitigation against health risks and making a choice between when to take action and when to not, were perceived as being essential in this respect. The need for further research to make better-informed choices and the need for more strategic responses was also a clear message from the case in this respect.

Skills, education and capacity

The need for better information and to make better-informed choices to tackle threats from climate change were a common theme running through the cases. The role of professional and skills development together taking measures to encourage general capacity building between tourism stakeholders appeared to be a common deficit in this respect. As such the need and desire for 'being richer in skills' was a common sentiment expressed. This was conveyed in a variety of contexts. At one level there appears to be a desire for professional expertise to make informed choices regarding, for example, the implementation of infrastructure, making decisions for tourism planning and development or enhancing environmental management processes. At another level is the desire to encourage professional expertise and skills to manage tourism operations ranging from the improvement of entrepreneurial skills, understanding environmental awareness, increasing technological know-how and extending capacity building activities between tourism stakeholders. All require directed action that needs strategic planning and long-term decision making.

Summary

In summary, the loss or erosion of coastal destinations through predicted climate change impacts will undoubtedly pose a significant threat to both natural ecosystems and tourism infrastructures and consequently the tourist 'communities' at such destinations. Evidence from the cases reviewed illustrates that perceived and actual threats are indeed real, although accurate predictions and current assessments still remain at best ambiguous and at worst speculative. The cases present a variety and varied set of local destination experiences. It would seem fair to say that there remains uncertainty regarding climate change and the validity of current predictions. However, general perceptions demonstrate that there is recognition of increased threats through increased climate anomalies such as storm surges, heat stress and a general rise in sea level. Perceptions also recognize resulting impacts such as ecological change, environmental damage, erosion and socio-economic threats. In this respect, however, mixed responses and knowledge gaps to secure ameliorative actions appear to be a norm. In essence the cases have illustrated a highly complex relationship between climate, environmental and tourism processes that are not always coordinated or entirely in harmony.

It is clear from the case studies that there is still a need to think with long-term and lateral vision. Evidence shows that current coastal destination management practices tend to concentrate on immediate socio-economic and environmental issues, which often give little regard to longer-term strategic destination management concerns addressed. Again, from the evidence, it would appear that strategic longer-term approaches to managing coastal tourism destinations based upon strategic governance, local empowerment and integrating legal and media processes are becoming more critical. Evidence of good practice is demonstrated in many of the cases reviewed; however, the examples are only a small illustration of the growing global threats that climate change can have on such destinations. In this respect the cases remain a small 'patch work' of measures that illustrate indicators for how climate change is perceived, how destinations are managed and policy choices made. As such they cannot provide all the answers but nevertheless go some way to indicate choices and desires for addressing specific threats. In an increasingly uncertain world where climate predictions are questioned and media hyperbole appears to be the dominant conduit for presenting information, such cases are important in providing a rational base from which to assess and analyse current experiences, meet expectations and learn lessons.

Despite remaining uncertainties, what is abundantly clear is that there is a need for all stakeholders operating within coastal tourism destinations to engage and integrate more fully with decision-making and policy processes. Pertinent to this is engagement with appropriate long-term policy implementation measures which connect more closely with governmental frameworks. In essence this means creating or ensuring structures for management and decision making that are fit for purpose. In this respect each case has demonstrated a unique character with bespoke needs to combat climate threats. A 'one fits all' solution to tackle mounting issues and challenges of climate change is probably not appropriate when tackling the complexities within a coastal tourism destination. However, key principles can be considered. At present evidence suggests that organizational structures at coastal tourism destinations remain lacking. In this respect the future success in managing coastal tourism destinations particularly with the onslaught of climate change will increasingly be dependent on effective measures that support a cyclic process of problem recognition, effective planning, skills development, proper implementation, controlled monitoring and research. Only then can appropriate responses to the complex problems associated with climate change, coastal erosion, environmental change and the long-term sustainability and management of coastal tourism destinations be fully addressed.

Reference

Jones, A.L. and Phillips, M. (eds) (2011) Conclusions. In: Jones, A.L. and Phillips, M. (eds) *Disappearing Destinations: Climate Change and the Future Challenges for Coastal Tourism.* CAB International, Wallingford, UK.

Index